Advanced Research on Plant Lipids

Proceedings of the 15th International Symposium
on Plant Lipids

edited by

N. Murata
National Institute for Basic Biology,
Okazaki, Japan

M. Yamada
University of Air,
Chiba, Japan

I. Nishida
The University of Tokyo,
Tokyo, Japan

H. Okuyama
Hokkaido University,
Sapporo, Japan

J. Sekiya
The University of Kyoto,
Kyoto, Japan

and

W. Hajime
The University of Tokyo,
Tokyo, Japan

KLUWER ACADEMIC PUBLISHERS
DORDRECHT / BOSTON / LONDON

A C.I.P. Catalogue record for this book is available from the Library of Congress.

ISBN 1-4020-1105-9

Published by Kluwer Academic Publishers,
P.O. Box 17, 3300 AA Dordrecht, The Netherlands.

Sold and distributed in North, Central and South America
by Kluwer Academic Publishers,
101 Philip Drive, Norwell, MA 02061, U.S.A.

In all other countries, sold and distributed
by Kluwer Academic Publishers,
P.O. Box 322, 3300 AH Dordrecht, The Netherlands.

Printed on acid-free paper

Printed in the Netherlands.

PREFACE

The 15th International Symposium on Plant Lipids was held in Okazaki, Japan, in May 12th to 17th, 2002, at the Okazaki Conference Center. The Symposium was organized by the Japanese Organizing Committee with the cooperation of the Japanese Association of Plant Lipid Researchers.

The International Symposium was successful with 225 participants from 29 countries. We acknowledge a large number of participants from Asian countries, in particular, from China, Korea, Malaysia, Taiwan, Thailand and the Philippines, presumably because this was the first time that the International Symposium on Plant Lipids was held in Asia. We also acknowledge a number of scientists from Canada, France, Germany, UK and USA, where plant lipid research is traditionally very active.

The Symposium provided an opportunity for presentation and discussion of 68 lectures and 93 posters in 11 scientific sessions, which together covered all aspects of plant lipid researches, such as the structure, analysis, biosynthesis, regulation, physiological function, environmental aspects, and the biotechnology of plant lipids. In memory of the founder of this series of symposia, the Terry Galliard Lecture was delivered by Professor Ernst Heinz from Universität Hamburg, Germany. In addition, special lectures were given by two outstanding scientists from animal lipid fields, Professor James Ntambi from University of Wisconsin, USA, and Dr. Masahiro Nishijima from the National Institute for Infectious Diseases, Japan. To our great honor and pleasure, the session of Lipid Biosynthesis was chaired by Dr. Andrew Benson, the Professor Emeritus of the University of California, San Diego, USA, who is, needless to say, the discoverer of phosphatidylglycerol and the Calvin-Benson Cycle. We are proud to announce that the originality and the rigor of the research presented in the Symposium were truly outstanding.

This proceedings book includes almost 100 articles representing the majority of contributed papers in the Symposium. They are rearranged in ten chapters of specific interests. We are sure that this book provides the most exciting and recent knowledge of plant lipid research.

The Editors are most grateful to Ms. Masayo Kawai and Ms. Keiko Oda for their assistance during the organization of the Symposium and the preparation of this book.

June 24, 2002   Editors
Okazaki, Japan

Dedication

This book is dedicated to Professor Jean-Claude Kader, in recognition of his pioneering research into plant lipids field and his leadership in the organization of the International Symposia on Plant Lipids.

Jean-Claude Kader started his doctoral research at the Université Pierre et Marie Curie (Paris VI) under the supervision of Professor Paul Mazliak, who is one of the founders of the series of the International Symposia on Plant Lipids. During his studies of the biochemical properties of oleoyl desaturase, Jean-Claude Kader observed the protein-mediated transfer of phospholipids between organelles in plant cells *in vitro*. After he received his doctoral degree in 1982, he studied protein biochemistry and enzymology as a postdoctoral fellow in the laboratory of Terry Galliard in Norwich (U.K.), to whose memory the Terry Galliard Lecture is dedicated at each Symposium.

After he returned to Paris, Jean-Claude Kader was appointed as a scientist at CNRS (Centre National de la Recherche Scientifique) and initiated his pioneering studies of lipid transfer protein (LTP), which mediates the transfer of membrane lipids among organelles in plant cells. In 1982 he purified LTP, for the first time, from maize seedlings. Subsequently, Mitsuhiro Yamada, at the University of Tokyo, isolated an LTP from castor bean seedlings. Friendly cooperation between these researchers led to successful research on higher-plant LTPs. Jean-Claude Kader succeeded in cloning a cDNA for the LTP of maize in 1988. This was the first time that molecular biology has been introduced into the field of plant lipid research. In 1991, Jean-Claude Kader was appointed to a research director and head of the laboratory at the Université Pierre et Marie Curie (Paris VI). He is now developing a genomic approach for the study of all the genes that encode LTPs and LTP-like proteins in *Arabidopsis thaliana*, with the goal of elucidating the roles *in vivo* of these proteins, which appear to participate in the defense reactions of higher plants and in their acclimation to changes in environmental conditons.

In addition to his outstanding contribution to research on plant lipids, Jean-Claude Kader organized the 11th International Symposium on Plant Lipids in Paris in 1994 and he has been the Editor-in-Chief of the international journal "Plant Physiology and Biochemistry" for 16 years.

Finally, as is evident from his photograph, one of his most important contributions to the field is his smile, which lubricates the development of friendship among researchers just as lipids can serve as lubricants in the plant world.

A Chronology of the International Symposium on Plant Lipids

Sympo- sium	Year	City (country)	Chief Organizer	Terry Galliard Lecturer
1st	1974	Norwich (UK)	Terry Galliard	
2nd	1976	Karlsruhe (Germany)	Hartmut K. Lichtenthaler	
3rd	1978	Göteborg (Sweden)	Conny Liljenberg	
4th	1980	Paris (France)	Paul Mazliak	
5th	1982	Groningen (Holland)	J. F. G. M. Wintermans	
6th	1984	Neuchatel (Switzerland)	Paul-Andre Siegenthaler	
7th	1986	Davis (USA)	Paul K. Stumpf	
8th	1988	Budapest (Hungary)	Peter Biacs	
9th	1990	Wye (UK)	Peter J. Quinn	
10th	1992	Tunis (Tunisia)	Abdelkader Cherif	
11th	1994	Paris (France)	Jean-Claude Kader	Norio Murata
12th	1996	Toronto (Canada)	John P. Williams	John Shanklin
13th	1998	Seville (Spain)	Juan Sanchez	John L. Harwood
14th	2000	Cardiff (UK)	John L. Harwood	John Ohlrogge
15th	2002	Okazaki (Japan)	Norio Murata	Ernst Heinz
16th	2004	Budapest (Hungary)	Peter Biacs	

A List of Volumes of the Proceedings Book

Sympo-sium	Year	Publisher	Editors	Title	Dedication
1st	1974	Academic Press	T. Galliard E. J. Mercer	Recent Advances in the Chemistry and Biochemistry of Plant Lipids	
2nd	1976	Springer-Verlarg	M. Tevini H. K. Lichtenthaler	Lipids and Lipid Polymers in Higher Plants	
3rd	1978	Elsevier /North-Holland Biomedical Press	Lars-Åke Appelqvist Conny Liljenberg	Advances in the Biochemistry and Physiology of Plant Lipids	
4th	1980	Elsevier /North-Holland Biomedical Press	P. Mazliak P. Benveniste C. Costes R. Douce	Biogenesis and Function of Plant Lipids	J. F. G. M. Wintermans
5th	1982	Elsevier Science Publisher	J. F. G. M. Wintermans P. J. C. Kuiper	Biochemistry and Metabolism of Plant Lipids	Paul K. Stumpf
6th	1984	Elsevier Science Publisher	P. A. Siegenthaler W. Eichenberger	Structure, Function and Metabolism of Plant Lipids	Morris Kates
7th	1986	Plenum Press	P. K. Stumpf J. B. Mudd W. D. Ness	The Metabolism, Structure, and Function of Plant Lipids	Andrew Benson
8th	1988	Plenum Press	P. A. Biacs K. Gruiz T. Kremmer	Biological Role of Plant Lipids	Hartmut K. Lichthenthaler
9th	1990	Portland Press London	P. J. Quinn J. L. Harwood	Plant Lipid Biochemistry, Structure and Utilization	J. Brian Mudd
10th	1992	Centre National Pedagogique	A. Cherif D. B. Miled-Daoud A. Smaoui M. Zarrouk	Metabolism, Structure and Utilization of Plant Lipids	Paul Mazliak
11th	1994	Kluwer Academic Publishers	J. -C. Kader P. Mazliak	Plant Lipid Metabolism	Terry Galliard
12th	1996	Kluwer Academic Publishers	J. P. Williams M. U. Khan N. W. Lem	Physiology, Biochemistry and Molecular Biology of Plant Lipids	Grattan Roughan
13th	1998	Secretariado de Publications Universudad de Sevilla	J. Sanchez E. Cerda-Olmedo E. Martinez-Force	Advances in Plant Lipid Research	Ernst Heinz
14th	2000	Portland Press London	J. L. Harwood P. J. Quinn	Recent Advances in the Biochemistry of Plant Lipids	Norio Murata
15th	2002	Kluwar Academic Publishers	N. Murata M. Yamada I. Nishida H. Okuyama J. Sekiya H. Wada	Advanced Researches of Plant Lipids	Jean-Claude Kader

TABLE OF CONTENTS

Chapter 1. The Terry Galliard Lecture

Chapter 2. Lipid Structure and Analysis

Chapter 3. Fatty Acid Biosynthesis

Chapter 4. Fatty Acid Desaturation

Chapter 5. Lipid Biosynthesis

Chapter 6. Waxes, Sphingolipids and Isoprenoids

Chapter 7. Lipase and Lipid Catabolism

Chapter 8. Lipid Trafficking and Signaling

Chapter 9. Lipids and Function

Chapter 10. Biotechnology

Chapter 1:

The Terry Galliard Lecture

STEROL GLUCOSIDES AND CERAMIDE GLUCOSIDES: CLONING OF ENZYMES CONTRIBUTING TO THEIR BIOSYNTHESIS

I. HILLIG, M. LEIPELT, P. SPERLING, P. TERNES , D. WARNECKE,
U. ZÄHRINGER* AND E. HEINZ
*Institut für Allgemeine Botanik, Universität Hamburg, Ohnhorststr. 18,
22609 Hamburg, Germany*
** Forschungszentrum Borstel, Abt. Immunchemie und Biochemische
Mikrobiologie, Parkallee 22, 23845 Borstel, Germany*

1. Introduction

Plant cells contain several groups of glycolipids contributing to the formation and function of plastidial and extraplastidial membranes. The enzymes responsible for the biosynthesis of the three dominating groups of glycosylated diacylglycerols from plastids characterized by monogalactosyl-, digalactosyl- and sulfoquinovosyl headgroups have all been cloned (Dörmann and Benning, 2002) and provide the opportunity to study their functions by using the methods of reverse genetics.

Membranes of the nucleocytoplasmic compartment such as endoplasmic reticulum, Golgi stacks, tonoplast and plasma membrane, but not mitochondria or microbodies (hardly any data are available for the nuclear membrane) contain sterol glucosides and ceramide glucosides as predominating glycolipids. Compared to the simplicity of the three standard glycosyl diacylglycerols, the sterol and ceramide derivatives comprise more complex mixtures. Many of these compounds, from which some representatives are shown in figs. 1 and 2, were isolated and analyzed for the first time in the laboratories of Carter, Fujino and Lester (summarized by Heinz, 1996), but since a functional assignment was not possible, these compounds were set aside as exotic natural products.

For an eventual analysis of possible functions of these extraplastidial glycolipids, we started to clone enzymes contributing to the biosynthesis of sterol and ceramide glucosides. For easier handling of these genes, we also looked for a useful, single-celled model organism, which should produce both sterol glycosides and ceramide glycosides and which, at the same time, should be suitable for genetic manipulation. This screening led to the selection of the yeast *Pichia pastoris* (Sakaki et al., 2001), which in this and other aspects of lipid metabolism is closer to plants than *Saccharomyces cerevisiae:* it

N. Murata et al. (eds.), Advanced Research on Plant Lipids, 3–12.
© 2003 *Kluwer Academic Publishers. Printed in the Netherlands.*

does not only contain these two glycolipids, it also produces polyunsaturated fatty acids and is capable of introducing double bonds into sphingoid long-chain bases.

More and more data from plants and other organisms show that both groups of glycolipids have functions, which link lipid metabolism to completely different fields of cellular activities. In general, one might expect glycolipid involvements in three types of functions. As constituents of membranes, they contribute to the function of the hydrophobic interior as well as of the polar surface of membranes by various interactions with proteins and lipids. A loss of such important building blocks cannot be tolerated, but it is not clear to what extent a glycolipid with a given headgroup can be replaced by another glycolipid with structural differences in the polar or apolar part of the molecule. Glycolipids are also involved as direct substrates in enzymatic reactions (fatty acid modifications, headgroup epimerization, primer elongation, see below) and dependent on the specificity, importance and redundance of these reactions, the loss of such a glycolipid may be lethal. Finally, glycolipids may function in signaling cascades, for example as elicitors and ligands in plant/pathogen interactions or generally as intracellular effectors. With regard to functions as second messengers, particularly the roles of the hydrophobic parts of steryl glycosides and ceramide glycosides, i.e. free sterols, ceramides and long-chain bases, have been investigated in much detail. But hardly any of these investigations have been carried out with the corresponding glycosylated compounds, whereas we will focus on the role of intact glycolipids in plants and microorganisms.

2. Sterol glucosides

Many of the enzymes catalyzing plant sterol biosynthesis have been cloned (Bach and Benveniste, 1997). In several cases, when visible phenotypes were traced back to genes involved in sterol biosynthesis, specific functions could be ascribed to the corresponding enzymes or their products. For example, a block in the C14-reductase reaction in the mainstream biosynthetic sequence affects body organisation in seedling development known as the FACKEL phenotype (Schrick et al., 2000). This block is upstream of a branchpoint, which separates the biosynthesis of the growth-regulating brassinosteroids via campesterol from sito- and stigmasterol synthesis, which do not show these pronounced growth effects. It should be mentioned that the sterols present in free and glycosylated form display very similar profiles and that also campesterol glucoside is a prominent component (Lynch and Steponkus, 1987). At present, no specific signaling functions can be ascribed to any sterol glycoside, but other interesting involvements are being discovered.

The first cDNA encoding a UDP-glucose:sterol ß-glucosyltransferase was cloned (Warnecke et al., 1997) based on the purification of this enzyme from oat microsomal membranes (Warnecke and Heinz, 1994). The N-terminal amino acid of the purified enzyme corresponds to the amino acid 133 of the deduced open reading frame (ORF) comprising a total of 608 amino acids. In Western blot analyses with oat membrane preparations, only the N-terminally truncated protein was detected. The expression of both the full-length cDNA or a fragment corresponding to the isolated protein lacking the first 132 amino acids resulted in the formation of functionally active enzymes in *E. coli.*

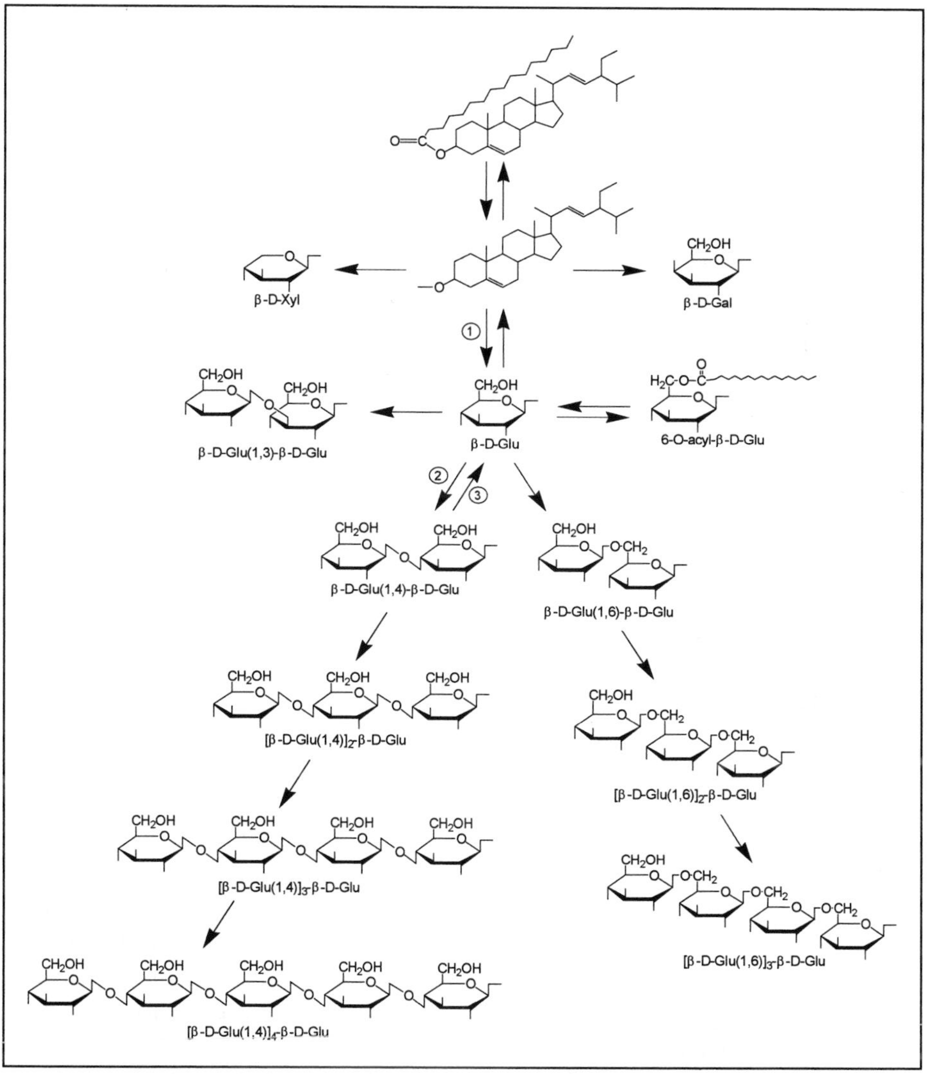

Figure 1. Structural formulae of some sterol derivatives found in plants (from Heinz, 1996). As suggested recently (Peng et al., 2002) and discussed in the text, cellulose biosynthesis is initiated by glucosylation of sterol (reaction 1), processive glucosylation of sterol glucoside (reaction 2) followed by release (3) of sterol glucoside and a "cellodextrin" oligosaccharide to be used as primer for elongation to cellulose.

Similar sterol glucosyltransferases were subsequently cloned from microorganisms such as *Candida albicans, Pichia pastoris, Dictyostelium discoideum* and *Saccharomyces cerevisiae* (Warnecke et al., 1999). The ORFs of these enzymes are roughly twice as large as the plant enzyme, but deletion studies have shown that only a C-terminal fragment of the size of the plant enzyme is required for activity. This part contains the UDP-binding motif, whereas the N-terminal parts are characterized by the presence of GRAM- and pleckstrin sequences believed to be involved in regulatory protein–protein and protein–membrane interactions (Doerks et al., 2000). When expressed in *E. coli* or *P. pastoris,* none of the cloned enzymes showed the processivity characteristic for bacterial UDP-glucose:diacylglycerol glucosyltransferases (Jorasch et al., 2000). Therefore, the synthesis of the higher homologues shown in fig. 1 requires additional enzymes. Similarly, all the expressed enzymes were specific for UDP-glucose, and only low activity was observed with UDP-xylose. Therefore, the enzymes catalyzing the synthesis of sterol mannoside or sterol galactoside still wait for being cloned.

Why do plants, including several eukaryotic algae, invariably synthesize sterol glycosides? For a long time, the only answer was: we do not know. After deletion of the sterol glucosyltransferase genes in *P. pastoris, S. cerevisiae* and *C. albicans*, we could not find any unusual phenotype under normal growth conditions. On the other hand, heat (41°C) and ethanol stress (6 % v/v) increased the level of ergosterol glucoside in *P. pastoris* (Sakaki et al., 2001). Another organism, *Physarum polycephalum*, also responded to a rise in temperature with an immediate increase in sterol glucoside (Murakami-Murofushi et al., 1997), which most likely can only be ascribed to enzyme and not to transcriptional activation. A completely different function of fungal ergosterol glucosides was discovered in the rice pathogen *Magnaporthe grisea*. After insertional mutagenesis and screening for reduced virulence, one of the most severely affected mutants was recognized as being hit in the sterol glucosyltransferase gene (Sweigard et al., 1998). A satisfactory explanation of all these effects is not possible at present.

A most surprising link to another and very important aspect of plant life was elucidated by the recent discovery that the synthesis of cellulose starts with the synthesis of sterol glucoside (reaction 1 in fig. 1, Peng et al., 2002). This glycolipid is used as a substrate by the processive CesA glucosyltransferase (reaction 2) to produce higher homologues of the cellobioside type with ß-1,4-linked glucosyl residues. The resulting oligosaccharide chain is split off by the Kor cellulase (reaction 3) and used as primer for subsequent processive elongation to cellulose by multimeric cellulose synthase complexes operating in the plasma membrane. The released sterol glucoside is recycled by the CesA glucosyltransferase in repetitive rounds for ß-1,4-glucan chain initiation.

With this correlation in mind, the structural work documented in fig. 1 is looked at in a new appreciation. The function of sterol glucoside in cellulose synthesis may also explain the observation that upon protoplast formation of mesophyll cells, nearly all free sterols are converted into sterol glucosides (Kesselmeier et al., 1987) as a response to the loss of the cell wall. It will also be interesting to find out whether sterol glucosides are involved in the biosynthesis of cellulose by *D. discoideum* or even prokaryotes (Read and Bacic, 2002). On the other hand, cell walls of some algae, in EM pictures sometimes with striking similarity to the microfibrillar texture of cellulose layers of higher plants, are made up of mannan and xylan microfibrils (Mackie and Preston, 1974). Does their polymerization also require corresponding sterol glycosides for initiation? *Arabidopsis thaliana* has two genes encoding sterol glucosyltransferases. In view of their important function, it will be interesting to find out whether T-DNA-tagged knock-out mutants do exist and whether it will be possible to define the functions of these redundant activities. Our own experiments on overexpression or inhibition of the sterol glucosyltransferase activity in tobacco plants were not successful (Durst et al., unpublished). In the context of its function as donor of an oligosaccharide chain, it should be mentioned that sterol glucoside has also been suggested to be the immediate glucosyl donor for ceramide glucoside formation. As shown below, our experiments do not support this proposal.

3. Ceramide glucosides

Ceramide glucosides are formed by sequential assembly from a long-chain sphingoid base via glycosylation of the intermediate ceramides (Lynch, 2000). Due to the efforts of several laboratories, particularly of those concentrating on yeast and mammalian systems, most enzymes involved in this assembly have been cloned. In the meantime, also plant sequences are available for several enzymes catalyzing the synthesis and modification of the ceramide backbone: serine palmitoyl transferase (Tamura et al., 2001), fatty acyl amide α-hydroxylase (Mitchell and Martin, 1997), sphingoid C4-hydroxylase (Sperling et al., 2001), sphingoid Δ8-desaturase (Sperling et al., 1998) and a sphingoid Δ4-desaturase homologue (Ternes et al., 2002). The biosynthesis of the very long-chain fatty acids bound in amide linkage in the ceramide backbones is most likely catalyzed by malonyl-CoA-dependent elongases. In yeast, these processive reactions are controlled by two enzymes (Elo2p and Elo3p; Oh et al., 1997). In plants, members of two large and unrelated enzyme groups encoded by KCS and ELO genes (discussed in Zank et al., 2002) could contribute to the biosynthesis of these ceramide-bound long-chain fatty acids.

The cloning of the first UDP-glucose:ceramide ß-glucosyltransferase of plant origin (Leipelt et al., 2001) made use of the sequence of the corresponding mammalian enzyme (Ichikawa et al., 1996), but the sequence identity between the plant and the other proteins, including the mammalian enzyme, is only between 9-11 %. Ceramide glucosyltransferases were also cloned from *P. pastoris*, *M. grisea* and *C. albicans*. When expressed in *S. cerevisiae* or *P. pastoris*, none of these enzymes and in particular not the plant enzyme resulted in the formation of higher homologues. Therefore, the biosynthesis of the higher homologues of ceramide glucoside shown in fig. 2 requires additional enzymes. It is tempting to point out the formal parallelism in the occurrence of higher homologues of sterol and ceramide glycosides, from which only the sterol derivatives have been recognized as playing a vital role in polysaccharide biosynthesis as discussed above. In addition, antibodies against ceramides (Vielhaber et al., 2001) and glucosyl ceramides of different origin have been prepared (Brade et al., 2000; Rodrigues et al., 2000; Toledo et al., 2001), widening the possibilities to study the functions and subcellular locations of enzymes and products.

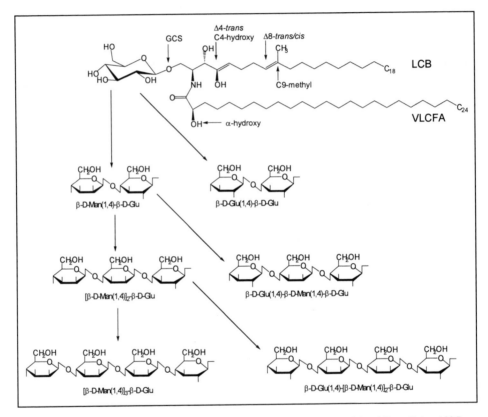

Figure 2. Structural formulae of some ceramide glycosides found in plants and fungi (from Heinz, 1996). In the compound at the top, many of the possible ceramide modifications in the long-chain sphingoid base (LCB) and the very long-chain fatty acid (VLCFA) are indicated. Not included are dihydroxy or 3-*trans*-unsaturated acyl groups. GCS stands for UDP-glucose:ceramide β-glucosyltransferase. Mannosylation (processive ?) seems to be stopped by capping with a terminal glucosyl residue. In contrast to the higher homologues of sterol glucoside, no function can be ascribed to the corresponding ceramide derivatives.

Regarding possible functions, one should differentiate between sphingoid bases in free and phosphorylated form, ceramides and the final ceramide glucosides. As in the case of sterol glycosides, we will concentrate in the present context on functions of the glycolipids in plants and microorganisms, which due to the limited knowledge will result in a rather short discussion. The toxic effects on plants evoked by inhibitors interfering at different steps with the assembly of the ceramide backbone (for example Asai et al., 2000) will not be discussed, because it is not clear whether the effects are due to interference with ceramide/sphingosine signaling or ceramide glucoside functions.

In an attempt to elucidate the roles of ceramide glucoside played in microorganisms, we deleted the ceramide glucosyltransferase genes in *P. pastoris* and *C. albicans* (Leipelt et al., 2001). But as in the case of the sterol glucosides, the resultant knock-out strains did not show any unusual phenotype under the tested growth conditions. In this context, it should be mentioned that the molar proportions of glucosylceramides as well as of ergosterol glucosides in fungal lipid extracts are rather low (0.5-2.0 %; Sakaki et al., 2001). Most sphingoid bases are recruited for the synthesis of inositol phosphorylceramide derivatives ending up in mannosylated form as free lipids or as hydrophobic anchors of membrane proteins (Dickson and Lester, 1999). This partitioning has never been determined in plants, which also contain so-called phytoglycolipids and ceramide-anchored membrane proteins.

The fungal ceramide glucosides are characterized by a specific ceramide backbone. It consists of the long-chain base 4-*trans*-8-*trans*-9-methyl-sphingadienine, which is confined to these glycolipids, and a C16/C18 fatty acid in free or α-hydroxylated form, which in addition may carry a 3-*trans*-double bond. Glycolipids of this structure induce the formation of fruiting bodies when added to mycelia of the mushroom *Schizophyllum commune* (Kawai and Ikeda, 1985) as well as phytoalexin production when sprayed on rice plants (Koga et al., 1998). This effect is particularly interesting, since it implies recognition of the fungal glycolipid by a plant receptor followed by a signaling cascade resulting in activation of nuclear transcription. The only difference between plant and fungal ceramide glucosides is the 9-methyl branching at the *trans*-double bond of the sphingoid base. The mannosylated sphingolipids have different ceramide backbones, which do not contain this methyl-branched sphingoid base and, in addition, carry very long-chain fatty acids. Deciphering the glycosphingolipid crosstalk between plant and fungal pathogen should yield interesting details regarding the function of these compounds in both organisms.

In animal systems, ceramide glucosides are involved in and required for similar cell–cell communications. This is evident from the fact that single cell cultures can be maintained with the ceramide glucosyltransferase gene deleted, whereas the same knock-out results in embryonic lethality very early in embryo development (Yamashita et al., 1999). This is compatible with an exposure of these glycolipids on the cell surface as also found in plants (Cantatore et al., 2000). On the other hand, in specialized cell types (melanoma cells), ceramide glucosides are also required in single cell cultures for intracellular transport of a specific membrane protein (Sprong et al., 2001). Therefore, it will be interesting to find out why plants maintain a certain proportion of these glycolipids in extraplastidial membranes. In *P. pastoris*, we deleted the genes for both the ceramide and the sterol glucosyltransferase, but even this double-knockout mutant did not display any remarkable phenotype. This shows that both glycolipids do not fulfill essential functions under normal growth conditions. Furthermore, we expressed a plant ceramide glucosyltransferase in this double mutant to see whether the plant enzyme requires sterol glucoside or UDP-glucose as a glucosyl donor (Lynch et al., 1997). The expression of the plant enzyme in the double mutant resulted in the accumulation of ceramide glucosides indicating that the plant enzyme is a UDP-glucose:ceramide glucosyltransferase (Hillig et al., in preparation). The same is true for the fungal enzymes, since ceramide glucoside contents were not reduced in mutants with deleted sterol glucosyltransferase genes (Warnecke et al., 1999; Leipelt et al., 2001).

4. References

Asai, T., Stone, J.M., Heard, J.E., Kovtun, Y., Yorgey, P., Sheen, J. and Ausubel, F.M. (2000) Fumonisin B1-induced cell death in *Arabidopsis* protoplasts requires jasmonate-, ethylene-, and salicylate-dependent signaling pathways. Plant Cell 12, 1823-1836.

Bach, T.J. and Benveniste, P. (1997) Cloning of cDNAs or genes encoding enzymes of sterol biosynthesis from plants and other eukaryotes: heterologous expression and complementation analysis of mutations for functional characterization. Prog. Lipid Res. 36, 197-226.

Brade, L., Vielhaber, G., Heinz, E. and Brade, H. (2000) In vitro characterization of anti-glucosylceramide rabbit antisera. Glycobiology 10, 629-636.

Cantatore, J.L., Murphy, S.M. and Lynch, D.V. (2000) Compartmentation and topology of glucosylceramide synthesis. Biochem. Soc. Trans. 28, 748-750.

Dickson, R.C. and Lester, R.L. (1999) Yeast sphingolipids. Biochim. Biophys. Acta 1426, 347-357.

Doerks, T., Strauss, M., Brendel, M. and Bork, P. (2000) GRAM, a novel domain in glucosyltransferases, myotubularins and other putative membrane-associated proteins. Trends Biochem. Sci. 25, 483-485.

Dörmann, P. and Benning, C. (2002) Galactolipids rule in seed plants. Trends Plant Sci. 7, 112-118.

Heinz, E. (1996) Plant glycolipids: structure, isolation and analysis. in Christie, W.W. (ed.), Advances in Lipid Methodology - Three. The Oily Press, Dundee, pp. 211-332.

Ichikawa, S., Sakiyama, H., Suzuki, G., Hidari, K.I. and Hirabayashi, Y. (1996) Expression cloning of a cDNA for human ceramide glucosyltransferase that catalyzes the first glycosylation step of glycosphingolipid synthesis. Proc. Natl. Acad. Sci. U S A 93, 12654.

Jorasch, P., Warnecke, D.C., Lindner, B., Zähringer, U. and Heinz, E. (2000) Novel processive and nonprocessive glycosyltransferases from *Staphylococcus aureus* and *Arabidopsis thaliana* synthesize glycoglycerolipids, glycophospholipids, glycosphingolipids and glycosylsterols. Eur. J. Biochem. 267, 3770-3783.

Kawai, G. and Ikeda, Y. (1985) Structure of biologically active and inactive cerebrosides prepared from *Schizophyllum commune*. J. Lipid Res. 26, 338-343.

Kesselmeier, J., Eichenberger, W. and Urban, B. (1987) Sterols and sterylglycosides of oats (*Avena sativa*). Distribution in the leaf tissue and medium-induced glycosylation of sterols during protoplast isolation. Physiol. Plantarum 70, 610-616.

Koga, J., Yamauchi, T., Shimura, M., Ogawa, N., Oshima, K., Umemura, K., Kikuchi, M. and Ogasawara, N. (1998) Cerebrosides A and C, sphingolipid elicitors of hypersensitive cell death and phytoalexin accumulation in rice plants. J. Biol. Chem. 273, 31985-31991.

Leipelt, M., Warnecke, D., Zähringer, U., Ott, C., Müller, F., Hube, B. and Heinz, E. (2001) Glucosylceramide synthases, a gene family responsible for the biosynthesis of glucosphingolipids in animals, plants, and fungi. J. Biol. Chem. 276, 33621-33629.

Lynch, D.V. and Steponkus, P.L. (1987) Plasma membrane lipid alterations associated with cold acclimation of winter rye seedlings (*Secale cereale* L. cv Puma). Plant Physiol. 83, 761-767.

Lynch, D.V., Criss, A.K., Lehoczky, J.L. and Bui, V.T. (1997) Ceramide glucosylation in bean hypocotyl microsomes: evidence that steryl glucoside serves as glucose donor. Arch. Biochem. Biophys. 340, 311-316.

Lynch, D.V. (2000) Enzymes of sphingolipid metabolism in plants. Methods Enzymol. 311, 130-149.

Mackie, W. and Preston, R.D. (1974) Cell wall and intercellular region polysaccharides. in Stewart, W.D.P. (ed.), Algal Physiology and Biochemistry. Blackwell Sci. Publ., Oxford, pp. 40-85.

Mitchell, A.G. and Martin, C.E. (1997) Fah1p, a *Saccharomyces cerevisiae* cytochrome b5 fusion protein, and its *Arabidopsis thaliana* homolog that lacks the cytochrome b5 domain both function in the alpha-hydroxylation of sphingolipid-associated very long chain fatty acids. J. Biol. Chem. 272, 28281-28288.

Murakami-Murofushi, K., Nishikawa, K., Hirakawa, E. and Murofushi, H. (1997) Heat stress induces a glycosylation of membrane sterol in myxoamoebae of a true slime mold, *Physarum polycephalum*. J. Biol. Chem. 272, 486-489.

Oh, C.S., Toke, D.A., Mandala, S. and Martin, C.E. (1997) ELO2 and ELO3, homologues of the *Saccharomyces cerevisiae* ELO1 gene, function in fatty acid elongation and are required for sphingolipid formation. J. Biol. Chem. 272, 17376-17384.

Peng, L., Kawagoe, Y., Hogan, P. and Delmer, D. (2002) Sitosterol-beta-glucoside as primer for cellulose synthesis in plants. Science 295, 147-150.

Read, S.M. and Bacic, T. (2002) Plant biology. Prime time for cellulose. Science 295, 59-60.

Rodrigues, M.L., Travassos, L.R., Miranda, K.R., Franzen, A.J., Rozental, S., de Souza, W., Alviano, C.S. and Barreto-Bergter, E. (2000) Human antibodies against a purified glucosylceramide from *Cryptococcus neoformans* inhibit cell budding and fungal growth. Infect. Immun. 68, 7049-7060.

Sakaki, T., Zähringer, U., Warnecke, D.C., Fahl, A., Knogge, W. and Heinz, E. (2001) Sterol glycosides and cerebrosides accumulate in *Pichia pastoris*, *Rhynchosporium secalis* and other fungi under normal conditions or under heat shock and ethanol stress. Yeast 18, 679-695.

Schrick, K., Mayer, U., Horrichs, A., Kuhnt, C., Bellini, C., Dangl, J., Schmidt, J. and Jurgens, G. (2000) FACKEL is a sterol C-14 reductase required for organized cell division and expansion in Arabidopsis embryogenesis. Genes Dev. 14, 1471-1484.

Sperling, P., Zähringer, U. and Heinz, E. (1998) A sphingolipid desaturase from higher plants. Identification of a new cytochrome b5 fusion protein. J. Biol. Chem. 273, 28590-28596.

Sperling, P., Ternes, P., Moll, H., Franke, S., Zähringer, U. and Heinz, E. (2001) Functional characterization of sphingolipid C4-hydroxylase genes from *Arabidopsis thaliana*. FEBS Lett. 494, 90-94.

Sprong, H., Degroote, S., Claessens, T., van Drunen, J., Oorschot, V., Westerink, B.H., Hirabayashi, Y., Klumperman, J., van der Sluijs, P. and van Meer, G. (2001) Glycosphingolipids are required for sorting melanosomal proteins in the Golgi complex. J. Cell Biol. 155, 369-380.

Sweigard, J.A., Carroll, A.M., Farrall, L., Chumley, F.G. and Valent, B. (1998) Magnaporthe grisea pathogenicity genes obtained through insertional mutagenesis. Mol. Plant Microbe. Interact. 11, 404-412.

Tamura, K., Mitsuhashi, N., Hara-Nishimura, I. and Imai, H. (2001) Characterization of an *Arabidopsis* cDNA encoding a subunit of serine palmitoyltransferase, the initial enzyme in sphingolipid biosynthesis. Plant Cell Physiol. 42, 1274-1281.

Ternes, P., Franke, S., Zähringer, U., Sperling, P. and Heinz, E. (2002) Identification and characterization of a sphingolipid delta4-desaturase family. J. Biol. Chem. in press.

Toledo, M.S., Suzuki, E., Levery, S.B., Straus, A.H. and Takahashi, H.K. (2001) Characterization of monoclonal antibody MEST-2 specific to glucosylceramide of fungi and plants. Glycobiology 11, 105-112.

Vielhaber, G., Brade, L., Lindner, B., Pfeiffer, S., Wepf, R., Hintze, U., Wittern, K.P. and Brade, H. (2001) Mouse anti-ceramide antiserum: a specific tool for the detection of endogenous ceramide. Glycobiology 11, 451-457.

Warnecke, D., Erdmann, R., Fahl, A., Hube, B., Müller, F., Zank, T., Zähringer, U. and Heinz, E. (1999) Cloning and functional expression of UGT genes encoding sterol glucosyltransferases from *Saccharomyces cerevisiae*, *Candida albicans*, *Pichia pastoris*, and *Dictyostelium discoideum*. J. Biol. Chem. 274, 13048-13059.

Warnecke, D.C. and Heinz, E. (1994) Purification of a membrane-bound UDP-glucose:sterol ß-D-glucosyltransferase based on its solubility in diethyl ether. Plant Physiol. 105, 1067-1073.

Warnecke, D.C., Baltrusch, M., Buck, F., Wolter, F.P. and Heinz, E. (1997) UDP-glucose:sterol glucosyltransferase: cloning and functional expression in *Escherichia coli*. Plant Mol. Biol. 35, 597-603.

Yamashita, T., Wada, R., Sasaki, T.C., Deng, C., Bierfreund, U., Sandhoff, K. and Proia, R.L. (1999) A vital role for glycosphingolipid synthesis during development and differentiation. Proc. Natl. Acad. Sci. U S A 96, 9142-9147.

Zank, T.K., Zähringer, U., Beckmann, C., Pohnert, G., Boland, W., Holtorf, H., Reski, R., Lerchl, J. and Heinz, E. (2002) Cloning and functional characterisation of an enzyme involved in the elongation of Δ6-polyunsaturated fatty acids from the moss *Physcomitrella patens*. Plant J. in press.

Chapter 2:

Lipid Structure and Analysis

SILVER ION LIQUID-CHROMATOGRAPHIC MOBILITY OF PLANT DIACYLGLYCEROLS AS A FUNCTION OF THEIR COMPOSITION AND SPATIAL ARRANGEMENT

A.G. VERESHCHAGIN AND V.P. PCHELKIN
Laboratory of Lipid Metabolism, Institute of Plant Physiology, Russian Academy of Sciences, Moscow 127276, Russia

1. Introduction

It is known that Ag^+ ions reversibly form coordination complexes with the double bonds of neutral lipids of plant origin (Christie, 1994). The complexation results in a considerable and very selective increase in lipid polarity, and for these reasons such complexes are widely used as an important tool in lipid chromatographic analysis. Nevertheless, the mechanism of the interaction between Ag^+ ions and lipid unsaturated centers in liquid-chromatographic (LC) systems is poorly understood at present. To contribute to the elucidation of this problem, we performed separation of *rac*-1,2-unsaturated diacylglycerol (DAG) Ag^+ complexes using both adsorption thin-layer chromatography (AgTLC) and reversed-phase TLC (AgRPTLC).

2. Materials and Methods

For continuous-flow AgTLC, a preparation of DAGs obtained by transesterification of plant oils with glycerol and including palmitic (P), stearic (St), oleic (O), linoleic (L), and linolenic (Le) fatty acid (FA) residues was separated using a silica gel plate impregnated with $AgNO_3$ and a $CHCl_3$/isopropanol (99/1, v/v) mixture as a mobile phase (Pchelkin and Vereshchagin, 1991); DAG zones were detected by 5% phosphomolybdic acid at 120°C, and their mobility (R_1) was determined in relation to that of a standard diacylglycerol, 1,3-LL. For continuous AgRPTLC, the same preparation was separated on the plate impregnated with $AgNO_3$ and 10% tetradecane in C_6H_6 using 5% H_3BO_3 in CH_3OH saturated with $AgNO_3$ and tetradecane as a mobile phase, and DAGs were detected as above; DAG mobility (R_2) was determined in relation to that of 1,3-LeLe (Pchelkin and Vereshchagin, 1992).

15

N. Murata et al. (eds.), Advanced Research on Plant Lipids, 15–18.
© 2003 *Kluwer Academic Publishers. Printed in the Netherlands.*

3. Results and Discussion

Results of AgTLC of DAGs are shown in Table 1. It is seen that, in contrast to a general belief that AgLC discriminates lipids only on the number of double bonds (e), the members of five pairs of DAG species with the same e (*i.e.* with $\Delta e = 0$), *viz.* StL and OO ($e = 2$), PL and OO ($e = 2$), StLe and OL ($e = 3$), PLe and OL ($e = 3$), and OLe and LL ($e = 4$) are separated from each other, and their first members differ from the second ones in a lower R_1, *i.e.* in a higher polarity of the complexes. Thus, on hand are two groups of DAG species: StL, PL, StLe, PLe, and OLe (Group I) and a less polar Group II, OO, OL, and LL. Evidently, the distinctions between the Group I and II DAGs in their polarity are caused by differences in their conformation. The Group I DAG species include FAs with a highly coiled conformation, such as Le and L, and, in each of these species, the FAs are very different in their configuration. In contrast, in Group II, the OO and LL include identical FAs, and OL contains FAs similar to each other in their structure.

TABLE 1. Degree of unsaturation (e) of diacylglycerol molecular species and their relative mobility in the course of AgTLC (R_1) and AgRPTLC (R_2)

DAG species (e)	AgTLC			AgRPTLC		
	Found R_1	R_1 calculated by the equations		Found R_2	R_2 calculated by the equations	
		(1)	(2)		(3)	(4)
1	2	3	4	5	6	7
StO (1)	4.88	-	4.88	0.15	-	0.13
OO (2)	2.50	-	2.42	0.30	-	0.27
StL (2)	1.47	1.47	-	0.30	0.33	-
PL (2)	1.47	1.47	-	0.38	0.33	-
OL (3)	0.94	-	1.20	0.43	-	0.41
LL (4)	0.60	-	0.60	0.51	-	0.54
StLe (3)	0.33	0.44	-	0.51	0.50	-
PLe (3)	0.33	0.44	-	0.60	0.50	-
OLe (4)	0.20	0.13	-	0.69	0.67	-
LLe (5)	0.10	0.04	-	0.83	0.83	-
LeLe (6)	0.03	0.01	-	1	1.00	-
DAG group		I	II		I	II
r^a		0.999	0.997		0.986	0.994

$^a r$ — regression coefficients between the found R_1 or R_2 values and those calculated by the equations (1) and (2) as well as (3) and (4).

From the comparison of e and R_1 in Table 1 with each other it could be expected that the R_1 vs. e relationship has to be an exponential one. For its determination, LLe and LeLe species containing highly coiled Le chains were additionally included in Group I, and StO, which, like OO, is characterized by a conformation close to an extended one, was included in Group II (see Table 1). For Group I, the relationship is expressed as the $R_1 = 4.88 \exp (1.2 - 1.2e)$ (equation 1), and, for Group II, as $R_1 = 4.88 \exp (0.7 - 0.7e)$ (equation 2).

Thus, in both groups, the R_1 vs. e relationship is exponential, but they differ from each other in its quantitative pattern. The results of Table 1 are confirmed by the evidence presented by Nicolova-Damyanova et al. (1995). Our recalculation of their data on the retention of 19 triacylglycerol (TAG) species ($e = 1 - 9$) from vegetable oils on adsorption Ag-HPLC yielded the values of R_{SSO}, i.e. TAG species mobility in relation to that of a standard SSO species (S = St or P), and it turned out that the R_{SSO} vs. e relationship was also exponential: $R_{SSO} = \exp (2.3 - 2.3e)$, $r = 0.997$.

Finally, the real existence of Groups I and II in our DAG mixture is proved by the fact that calculation of the R_1 vs. e relationship according to equations 1 and 2 for total DAG species (Table 1, column 1) yields the $r = 0.970$ and 0.958 values, which are much lower than those for the separate groups (Table 1).

Results of AgRPTLC in Table 1 demonstrate that an increase in e of DAG molecular species from 1 to 6 is accompanied by a 6.7-fold increase in their R_2 values. Thus, on the AgRPTLC, as on RPTLC without Ag^+ (Pchelkin and Vereshchagin, 1981), the R_2 vs. e relationship is close to a direct linear one. Moreover, these results show that, in the DAG pairs with $e = 2, 3, 3$, and 4 (PL and OO; StLe and OL; PLe and OL; OLe and LL, respectively), their first members are more mobile, i.e. more polar, than the second ones.

Thus, as on AgTLC, on hand are two DAG groups, which are characterized, at $\Delta e = 0$, by a higher (PL, StLe, PLe, OLe) and a lower (OO, OL, LL) polarity of their complexes and, in their composition and polarity, are similar, respectively, to the Groups I and II found on AgTLC. Therefore, the R_2 vs. e relationship was found separately for Group I: $R_2 = 0.167e$ (equation 3) and Group II: $R_2 = 0.135e$ (equation 4). It is seen that these groups differ from each other in the angular coefficient of this relationship. Their real existence is also warranted by the fact that calculation of the R_2 vs. e relationship by equations 3 and 4 for total DAGs yields r values (0.983 and 0.966), which are much lower than those for separate groups (Table 1).

Finally, a direct linear pattern of the mobility vs. e relationship on AgRPLC of neutral lipids can be demonstrated from the results of Nicolova-Damyanova et al. (1993). Indeed, our recalculation of their data on the retention of St, O, L, and Le phenethyl and phenacyl esters during AgRPHPLC showed that their relative mobility is equal to $0.15e + 0.15$, $r = 0.996$, and $0.16e + 0.16$, $r = 0.994$, respectively.

It is concluded that the two versions of TLC of DAG complexes are qualitatively different in their mechanisms. Indeed, on AgTLC, the complexes are formed by an indeterminate number of coordination centers of various nature and only at the adsorbent surface; their separation proceeds according to an adsorption mechanism, and R_1 vs. e relationship is an inverse exponential one. Meanwhile, on AgRPTLC, the complexes are formed with double bonds, only in solution, and at the 1: 1 ratio; they are fractionated by lipophilic partition between two liquid phases, and the R_2 vs. e relationship is a direct linear one. Nevertheless, in both cases the R_1 and R_2 values are determined exclusively by polarity of the complexes, which, in turn, depends on both their composition (primarily e) and configuration. In the latter case, regardless of the mode of formation of the complexes, either at the surface or in solution, Group I DAGs characterized by a coiled acyl conformation always greatly exceed in their polarity Group II DAGs with the same e, but with the configuration close to an extended one; in the former group, this excess amounts to $2 - 3$-fold and $30 - 40\%$ for AgTLC and AgRPTLC, respectively. In both methods, Groups I and II differ quantitatively, but not qualitatively, in the pattern of the mobility vs. e relationships of DAG complexes.

This work was supported by the Russian Foundation for Basic Research, project no. 99-04-49208.

References

Christie, W.W. (1994) Silver ion and chiral chromatography in the analysis of triacylglycerols. Progr. Lipid Res. 33, 9-18.

Nicolova-Damyanova, B., Christie, W.W., and Herslöf, B.G. (1993) High-performance liquid chromatography of fatty acid derivatives in the combined silver ion and reversed-phase mode. J. Chromatogr. 653, 15-23.

Nicolova-Damyanova, B., Christie, W.W., and Herslöf, B.G. (1995) Retention properties of triacylglycerols on silver ion high-performance liquid chromatography. J. Chromatogr. A 694, 375—380.

Pchelkin, V.P. and Vereshchagin, A.G. (1981) Reversed-phase thin-layer chromatography of diacylglycerols as their labile dimethylborate esters. J. Chromatogr. 209, 49-60.

Pchelkin, V.P. and Vereshchagin, A.G. (1991) Identification of individual diacylglycerols by adsorption thin-layer chromatography of their coordination complexes. J. Chromatogr. 538, 373-383.

Pchelkin, V.P. and Vereshchagin, A.G. (1992) Reversed-phase thin-layer chromatography of diacylglycerols in the presence of silver ions. J. Chromatogr. 603, 213-222.

BETAINE LIPIDS IN MARINE ALGAE

M. Kato[1], Y. Kobayashi[1], A. Torii [1], M. Yamada[2]
[1]*Graduate School o f Humanities and Sciences,
Ochanomizu University,
2-1-1 Otsuka Bunkyo-ku, Tokyo112-8610, JAPAN*
[2]*The University of the Air,
2-11 Wakaba , Mihama-ku, Chiba-261-8586, JAPAN*

Abstract

Betaine lipids are detected in the limited species of plants including algae. Three types of betaine lipids have been reported up to now. The structure of DGCC(diacylglyceryl-3-*O*-carboxyhydroxymethylcholine) was determined in *Pavlova lutheri* (Haptophyceae). DGCC was found to be one of the common coustituents of Haptophyceae. The aim of the present work was to analyze the fatty acids and polar lipids in CCMP504(Haptophyceae) to clarify the biosynthesis of betaine lipids in Haptophyceae. The principal polar lipid components in CCMP504 were galactolipids (MGDG,DGDG,SQDG) and betaine lipids (DGCC,DGTA). An analysis of the incorporation of ^{14}C-labelled compounds into the algal cells indicated that methionine might be a precursor of polar moieties of DGCC and DGTA. ^{14}C-choline was converted to only DGCC. ^{14}C-DGTS as a intermediate was not detected, suggesting that the biosynthetic pathway of DGTA in CCMP504 was different from that in *Ochromonas danica*. The biosynthesis of DGCC was seemed to be independent of that of DGTA in CCMP504.

Introduction

Betaine lipids wre complex lipids which have a positively charged trimethylammonium group and constitute a group of polar lipids in plants, together with phospholipids and glycolipids. Three types of betaine lipids have been reported in lower plants and algae. The first structure of a betaine lipid, DGTS (1,2-diacylglyceryl-*O*-2'-(hydroxymethyl)-(*N,N,N*-trimethyl)homoserine) was determined by Brown and Elovson(1974). A second betaine lipid, DGTA (1,2-diacylglyceryl-*O*-2'-(hydroxymethyl)-(*N,N,N*-trimethyl)- β -alanine), was

N. Murata et al. (eds.), Advanced Research on Plant Lipids, 19–22.
© 2003 *Kluwer Academic Publishers. Printed in the Netherlands.*

subsequently identified by Vogel et al.(1990) in *Ochromonas danica*. A third betaine lipid, DGCC (1,2-diacylglyceryl-3-*O*-carboxyhydroxymethylcholine) was discovered in *Pavlova lutheri* (Haptophyceae)by our group (Kato et al. 1994). DGCC was found to be one of the common coustituents of Haptophyceae (Kato et al. 1996).

Feeding experiments with methionine labeled at specific carbon atoms in algae and moss consistently suggested that the C_4 backbone and *S*-methyl group of methionine were precursors to the C_4 backbone and *N*-methyl groups of DGTS, respectively . (Sato 1998, Sato and Kato 1988, Sato 1991, Vogel and Eighenberger 1992). The biosynthesis of DGTA from DGTS was supposed to involve a decarboxylation and recarboxylation of polar part and a simultaneous deacylation and reacylation of the glycerol moiety (Vogel and Eighenberger 1992). Recently, the lipid-linked *N*-methylation for the biosynthesis of DGTS in *Rhodobacter sphaeroides* was proposed (Hofmann and Eichenberger 1996) and the related genes have been isolated(Klug and Benning 2001). However, there is no information for the biosynthetic pathway of DGCC up to now. The aim of the present work was to clarify the biosynthetic pathway of a polar part of DGCC in Haptophyceae.

Materials and methods

CCMP504 strain (Haptophyceae) was obtained from Provasoli- Guillard Center for Culture of Marine Phytoplankton(U.S.A.). Culture condition and the analytical methods were referred   to our previous reports(Kato et al. 1995, 1996)

Results and Discussion

Compositio of polar lipids in CCMP504

MGDG(20.2%), DGDG(16.8%) and SQDG(13.6%) were the most abundunt classes of lipids, the levels of DGCC(15.1%) and DGTA(3.2%) being lower than those of the former lipids.   DGCC was almost five times that of DGTA, while no DGTS was detected after TLC. Phospholipids fraction (20.4%) consisted of PG,PE and PA.

Composition of fatty acids in CCMP504

The major fatty acid of the crude lipids were 16:0,16:1, 20:5 and 22:6, and the 35% of the total consisted of polyunsaturated fatty acids (data not shown). The fatty acid 22:6 accounted for more than 30% of the fatty acids in DGTA.   DGCC also containing 22:6, although its major fatty acids were 16:0 and 20:5.   It was notable that the distributions of fatty acids were quite different in these betaine lipids (data not shown).

The biosynthesis of the polar moiety of DGCC

[*Methyl-*[14]C]methionine   was incorporated into DGCC and DGTA with a high degree of specificity of into DGCC (Fig.1 A).   Analysis of the radioactivity in

DGCC molecule obtained after the pulse period showed that more than 90% of the radioactivity was localized in the polar group (data not shown) . No ^{14}C-DGTS was detected during the experiment.　The obtained result demonstrated that methionine was a precursor to DGCC.　[*Methyl-*^{14}C]choline and [1,2-^{14}C]choline were exclusively incorporated into only DGCC(Fig.1B, C) .　[*Methyl-*^{14}C]SAM was also incorporated into DGCC and [*carboxyl-*^{14}C]SAM appeared to be incorporated nonspecifically into many classes of lipids, although the incorporation into DGCC in terms of relative specfic activity was higher than that of other classes of lipids (Fig.2A,B).

The present study clearly demonstrated that methyl groups of SAM which was synthesized from methionine were incorporated into DGCC.　However, the biosynthesis of trimethylammonium group of DGCC was seemed to be derived from choline..Choline was reported to be synthesized from ethanolamine and SAM

Figure 1　Changes in the distribution of radioactivity from [*methyl-*^{14}C]methionine(A), [*methyl-*^{14}C]choline chloride(B) and [1,2-^{14}C]choline chloride in various lipid classes.　A concentrated suspension of cells(10 ml, 3-7 x 10^7 cells/ml) was incubated with 370kBq of the each radiorabelled compound.　After 2 hours, cells were harvested and then resuspended in fresh medium without radiorabelled compounds for 3 hours.

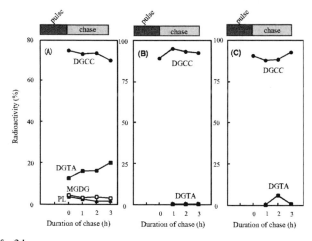

Figure 2　Changes in the distribution of radioactivity from [*methyl-*^{14}C]SAM(A) and [*carboxyl-*^{14}C]SAM (B) in various lipid classes.　A concentrated suspension of cells(10 ml, 3-4 x 10^7 cells/ml) was incubated with 185kBq of [*methyl-*^{14}C]SAM or 74kBq of [*carboxyl-*^{14}C]SAM.　After 2 hours, cells were harvested and then resuspended in fresh medium without radiorabelled compounds for 24 hours.

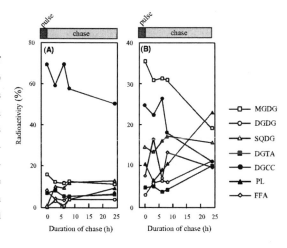

22

(Weretilnyk et al. 1995). Moreover, the biosynthesis of DGTA was thought to be independent of that of DGCC. To elucidate the biosynthetic pathway of DGCC in detail, the tracer experiment with stable isotopes and the analysis by NMR are now in progress.

References

Brown, A.E. and Elovson, J. (1974) Isolation and characterization of a novel lipid,1(3),2-diacylglyceryl-(3)-O-4-(N,N,N-trimethyl)homoserine,from *Ochromonas danica. Biochemistry* **13**, 3476-3482.

Hofmann, N. and Eichenberger, W. (1996) Biosynthesis of diacylglyceryl–N,N,N-trimethyl homoserine in *Rhodobacter sphaeroides* and evidence for lipid-linked N methylation. *J. Bacteriol.* **178**, 6140-6144.

Kato, M., Sakai, M., Adachi, K., Ikemoto, H. and Sano, H. (1996) Distribution of betaine lipids in marine algae. *Phytochemistry* **42**, 1341-1345.

Kato, M., Hajiro-Nakanishi, K., Sano, H. and Miyachi, S. (1995) Polyunsaturated fatty acid and betaine lipids from *Pavlova lutheri. Plant Cell Physiol.* **36**, 1607-1611.

Kato, M., Adachi, K., Hajiro-Nakanishi, K., Ishigaki, E., Sano, H. and Miyachi, S. (1994) A betaine lipid from *Pavlova lutheri. Phytochemistry* **37**, 279-280.

Klug, R.M. and Benning, C. (2001) Two enzymes of diacylglyceryl-O-4'-($N,N, N,$-trimethyl)-homoserine biosynthesis are encoded by *btaA* and *btaB* in the purple bacterium *Rhodobacter sphaeroides. Proc. Natl. Acad. Sci. USA* **98**, 5910-5915.

Sato, N. (1988) Dual role of methionine in the biosynthesis of diacylglyceryltrimethyl- homoserine in *Chlamydomonas reinhardtii. Plant Physiol.* **86**, 931-934.

Sato, N. (1991) Lipids in *Cryptomonas* CR-1. II. Biosynthesis of betaine lipids and galactolipids. *Plant Cell Physiol.* **32**, 845-851.

Sato, N. and Kato, K. (1988) Analysis and biosynthesis of diacylglyceryl-N,N,N-tri-methylhomoserine in the cells of *Marchantia* in suspension culture. *Plant Sci.* **55**, 21-25.

Vogel, G. and Eichenberger, W. (1992) Betaine lipids in lower plants. Biosynthesis of DGTS and DGTA in *Ochromonas danica* (Chrysophyceae) and the possible role of DGTS in lipid metabolism. *Plant Cell Physiol.* **33**, 427-436.

Vogel, G., Woznica, M., Gfeller, H., Muller, C., Stampfli, A.A., Jenny, T.A. and Eichenberger, W. (1990) 1(3),2-Diacylglyceryl-3(1)-O-2-(Hydroxymethyl)(N,N,N-trimethyl)- β- alanine (DGTA) :A novel betaine lipid from *Ochromonas danica*(Chrysophyceae). *Chem. Phys. Lipids* **52**, 99-109.

Weretilnyk. E.A., Smith, D.D., Wilch, G.A. and Summers, P.S. (1995) Enzymes of choline synthesis in spinach. Response of phosphobase N-methyltransferase activities to light and salinity. *Plant Physiol.* **109**, 1085-1091.

STUDIES ON THE LIPID MOLECULAR SPECIES AND MINOR COMPONENTS OF SOME PHILIPPINE SEED AND NUT OILS

LAURA J.PHAM, EMMANUEL REVELLAME
and PRECY M. RASCO

*Oils and Fats Laboratory,National Institute of Molecular
Biology and Biotechnology (BIOTECH)at U.P.Los Baños,
College,Laguna,Philippines*

Introduction

Several underexploited or unexplored plants and trees are provided us by the tropical world and studies have demonstrated that Southeast Asia avails to us a large variety of oils and fats which if developed commercially can provide enormous potential for the vegetable seed oils and fats industry worldwide

With the advances in the chemistry and biotechnology of fats and oils, it is now possible to prepare tailored fats or specialty oils which could represent a revolutionary possibility for upgrading Philippine seed and nut oils. We have a vast potential of seed oil resources which could be excellent sources of oils that have commercial value due to their glyceride and fatty acid profile and minor components in the oil which are now the focus of interest because of their nutritional and pharmaceutical applications.

In view of the importance of lipid molecular species and minor components in food and pharmaceuticals, good and viable sources of these components are being continuously explored. This research work looked into selected Philippine seed and nut oil components such as t he triglyceride species, regiospecific distribution of their fatty acids in the glycerol backbone for their possible modification through enzymatic interconversion into high value products and the characterization of the minor components such as the carotenoids,sterols and tocopherols for a wider application of the oils.

Materials and methods

Materials

Coconut (Cocos nucifera Linn) Pilinut (*Canarium Ovatum Engl.*), Lumbang (*Aleurites moluccana*),), Bunga de Tsina (*Veitchia mernilli*),MacArthur Palm (*Phytosperma macarthuri*), Cashew (*Anacardium occidentale L.*), and Talisay (*Terminalia catappa L.*) were obtained from the Horticulture Dept. of the University of the Philippines at Los Baños,Laguna Philippines.

N. Murata et al. (eds.), Advanced Research on Plant Lipids, 23–26.
© 2003 *Kluwer Academic Publishers. Printed in the Netherlands.*

Methods

Triglyceride molecular species analysis

Triglycerides were fractionated in terms of PN by HPLC equipped with a JASCO intelligent pump The column used was a reversed phase μ –Bondapak C_{18} column (4 mm i.d. x 30 cm, 4 um). The mobile phase was hexane:isopropyl alcohol:acetic acid (15:15:1 v/v/v) and flowrate was isocratically controlled at 1.0ml/min. Peaks were detected with a JASCO RI detector. The triglyceride specie was determined in terms of partition number.

Regiospecific analysis of fatty acids(Pham and Kwon, 1992)

After hydrolyzing the triglycerides with pancreatic lipase, which has a 1,3-position specificity, the resulting sn-2 MG were separated from FFA by TLC on silica gel inpregnated with boric acid and bands were scraped off separately. The fatty acid composition of sn-2-MG was analyzed by Gas Chromatography after BF_3 methylation.

Analysis of Tocopherols, Carotenoids, and Sterols (Mordret, 1996)

Silica gel plates were spotted with 20-50 mg of standards and samples and developed using a solution of hexane-ethyl acetate (80:20) and sprayed with 0.01% Rhodamine-6G solution in ethanol. The spots were visualized under UV light and quantified using the QuantiScan software. Spots were compared with authentic standards, scraped and extracted from the silica gel with 10 ml of diethyl ether-methanol (9:1) and injected to the HPLC for the different components.Tocopherol analysis by HPLC was carried out using a Finepak SIL –5 column with n-hexane-isopropyl alcohol-acetic acid (100: 0.5: 0.5) as the mobile phase at a flow rate of 1.0 ml/min and a UV detector set at 295 nm. Analysis of carotenoids was performed using a C_{18} column,eluting solvent was acetonitrile:dichlorimethane:methanol(70:20:10,v/v/v)and UV detector at 450 nm.

Results and discussion

Oil Seeds and Nuts

The seed and nut oils chosen for this research work were selected for their availability,abundance and importance. The information on the lipid molecular species and minor components of the oils from the selected seed oils and nuts are very important for their restructuring into high value products. Coconut(*Cocos nucifera Linn*) for example is an important agro-industrial product in the Philippines. Because of stiff competition from other oils,there is a vigorous effort to develop our oils for the nutraceutical industry and for specialty fats and oils that would meet the demand of consumers for products with desired characteristics.

Fatty acid and Triglyceride Components

Fractionation of the glyceride species by thin layer chromatography showed that majority of the lipid molecular species is that of the triglycerides. The partial glycerides (diglycerides and monoglycerides) are low in comparison and some are present in not more than trace amounts. These results are in agreement with findings of other researchers which reported 99.8% (Bezard,1971)and 90%(Banzon and Velasco,1982) triglycerides of the total oil glycerides. The fatty acid composition for the triglyceride of each oil is shown in Table 1. Fatty acid analysis of coconut triglycerides shows that it has a low degree of unsaturation .Most of the triglycerides have oleic acid ($C_{18:1}$) as the major fatty acid followed by linoleic acid ($C_{18:2}$) and a few contain linolenic acid($C_{18:3}$).

Table 1. Fatty acid Profile of the different Seed and nut oil Triglycerides

Fatty Acid	Coconut	Cashew	Pili Nut	(% Composition) Bunga de Tsina	Lumbang	MacArthur Palm	Talisay
C_6	0.52						
C_8	8.21						
C_{10}	6.45						
C_{12}	47.35						
C_{14}	19.57	0.28	0.10	0.19		0.11	
C_{16}	8.98	11.38	34.58	37.00	10.02	40.59	34.30
$C_{16:1}$	0.53				0.45		
C_{18}	3.21	8.98	9.97	5.83	5.77	3.15	4.53
$C_{18:1}$	4.82	59.20	44.06	28.98	44.11	12.69	32.39
$C_{18:2}$	0.79	18.28	10.31	24.02	29.17	38.05	28.33
$C_{18:3}$		0.10	0.59	2.98	13.67	2.70	
C_{20}		0.56	0.23	0.69	0.24	0.38	0.46

A detailed analysis of the triglyceride fraction of each oil was made by HPLC in terms of Partition Number (PN) Table 2 shows nine PN components in coconut oil triglycerides, the smallest being PN 28 and the highest is PN 44. The predominant component is PN36 followed by PN 38 and PN 34. Pilinut triglycerides has five components with PN 48 dominating followed by PN 46 and PN 50.Cashew on the other hand has four components with PN 48 dominating and followed by PN46. Results also show that most of the seed oils have triglyceride species of intermediate molecular weight. More triglyceride species have high molecular weight as compared to thelow molecular weight triglyceride species. This profile makes the oil samples potential materials for high value products.

Table 2.Triglyceride profile in terms of Partition Number (PN) by HPLC

PN	Coconut	Composition of Cashew	Pilinut	Fractions(%) Bunga de Tsina	Talisay
28	0.39	-	-	-	-
30	2.67	-	-	-	-
32	11.82	-	-	-	-
34	17.25	-	-	-	-
36	22.11	-	-	-	-
38	19.08	-	-	-	-
40	14.26	-	-	-	1.85
42	8.01	1.70	-	14.78	15.39
44	4.37	8.70	3.29	4.56	25.79
46	-	26.37	15.14	32.78	15.08
48	-	47.51	60.22	41.56	36.64
50	-	15.45	21.39	6.31	3.93

Regiospecific Analysis

Pancreactic lipase which is specific for the primary ester bonds was used for the hydrolysis of the triglycerides. With coconut, the major fatty acid which is lauric is preferentially attached to t he second position. Interestingly with cashew all fatty acids have equal distributions between the sn-2 and sn1,3 positions. Pilinut on the other hand has the same trend for the major fatty acids such as oleic and palmitic.Stearic is however preferentially attached to the second position while the polyunsaturated fatty acids are esterified to the primary positions.This is quite a deviation from Brockerhoff's general rule that saturated acids tend to occupy the 1 and 3 postions. The information in this distribution pattern should be useful as their position could be highly important aspects in the biosynthetic process and in the tailoring of fats and oils.

Unsaponifiables(Minor Components)

The unsaponifiable matter which contains the minor components is the constituent that has very low solubility in water after basic hydrolysis. Percent unsaponifiable matter(UM) and the fractions of the UM determined by TLC is shown in Table 3. The sterol fraction is the highest minor component present in the unsaponifiable matter .

Table 3. Unsaponifiable Matter (UM) and minor component Fractions in the UM

Seed/Nut oil	UM(%)	% of Total UM		
		Carotenoid	Tocopherol	Sterol
Pilinut	0.552	5.02	6.52	66.71
Cashew	0.546	7.02	41.66	51.32
Bunga de Tsina	3.11	45.74	13.58	40.68
Lumbang	0.542	9.61	49.15	41.24
Talisay	0.292	-	33.46	66.54
MacArthur Palm	2.21	40.93	-	59.09

The presence of these minor components in the different oil samples is very interesteresting as they have been recognized as important components in the preparation of specialty fats and oils.

REFERENCES

Banzon,J.A. and J.R.Velasco.(1982).Coconut Production and Utilization. Philippine Coconut and Development Foundation ,Inc.(PCRDFP),Pasig Metro Manila,Philippines.p.247.

Bezard,J.(1971). Triglyceride Composition of Coconut Oil.JAOCS.48(3):134-139

Brokerhoff,H.1971.Stereospecific analysis of Triglycerides.Lipids.12:942-956.

Mordret,E.(1996)Analysis of the Minor Components:in A.Karleskind(ed),Oils and Fats Manual.Lavoisier Publishehing,Paris. Pp.1183-1191.

Pham,L.J. and D.Y.Kwon.(1992).Regiospecific Analysis of Triacylglycerols of Some Seed Oils by Pancreatic Lipase.Foods and Biotechnology.Vol.1.No.2.pp75-78.

ANTIOXIDANT ACTIVITY OF SOME NATURAL EXTRACTS IN CORN OIL

F. ANWAR[1], M. I. BHANGER[2] AND S. YASMEEN[1]

[1.] *PCSIR Labs Complex, Karachi-75280, Pakistan*
[2.] *Center of Excellence in Analytical Chemistry, University of Sindh, Jamshoro, Pakistan*

Abstract:

Antioxidant activity of seven plant extracts was evaluated using accelerated aging of refined, bleached and deodorized corn oil at 65 °C. Methanolic extracts of coffee beans, rice bran, guava leaf, roasted wheat germ, rosemary (*Rosemarinus officinalis*), Peppermint (Mentha piperita), common basil (*Ocimum basilicum)* were added at concentration of 0.15% (wt/wt). The thermoxidation process was followed by measurement of peroxide value (P.V), primary (conjugated diene hydroperoxides) and secondary (volatile hexanal, pentanal, ethylenic diketones) oxidation products, antioxidant activity index (AI) and free fatty acids (FFA).

Rosemary (AI; 3.50), coffee beans (AI; 3.32) , rice bran (AI;2.98) were most effective in retarding PV, secondary oxidation products and showed highest antioxidant potential as indicated by AI. Roasted wheat germ and *Ocimum basilicum* inhibited generation of secondary products more than increase in PV. The investigated plant extracts showed good antioxidant activity in relation to corn oil and their overall efficacy was; rosemary, coffee beans, rice bran, roasted wheat germ, *Ocimum basilicum*, guava leaf, *Mentha piperita* in a decreasing order.

Introduction:

In recent years there is an increasing evidence that changing one's diet to increase the intake of food 'relatively high' in selected natural antioxidants, such as plant polyphenols, vitamin C and flavonoids reduce the incidence of degenerative diseases e.g., cancer, CVD and aging (Wilson 1999 ; Mery, 1999).

Thus, there is keen interest among the food scientists to identify antioxidants that are safe and natural origin and in this area a diverse group of plant compounds called natural antioxidants is the focus of intense research. A significant number of

N. Murata et al. (eds.), Advanced Research on Plant Lipids, 27–30.

plants have been evaluated for their antioxidant activities using different assays (Bandomene et al ., 2000 ; Lious et al., 1999).

The present work reports the antioxidant activity of seven natural extracts, derived from different indigenous plant sources in corn oil at accelerated aging conditions.

Materials and Methods:

Coffee beans (*Coffea arabica*), rosemery (*Rosemarinus officinalis*, peppermint (*Mentha piperita*), common basil (*Ocimum basilicum*), Rice bran (*Oryza sativa*), guava leaves (*Psidium gugava*), wheat germs (*Triticum durum)* and refined, bleached and deodorized corn oil were got from local botanical and industrial sources. All other reagents and chemical used were of E. Merck.

Extraction of Antioxidant Materials:
All the plant material were dried at room temperature and ground to pass a 1 mm sieve. Extraction was carried by soaking the materials with distilled methanol for six days at ambient conditions (4 changes for each material). The combined filtrates were concentrated to dryness after rotary evaporation at 45 C. The crude, viscous extracts were stored under nitrogen -18 untill tested.

Antioxidant Assay of the Extracts and Measurement of Oxidative Deterioration:
Crude extracts were added separately to refined, bleached and deodorized corn at level of 0.15% (wt/wt). The mixture were stirred for 30 minutes at 50 °C. Control samples were also prepared under the same conditions. The samples were stored at controlled temperature of 65 °C in an oven for 15 days. The oxidative deterioration level was followed by the measurement of various oxidation parameters after every three days.

Determination of peroxide value (P.V), free fatty acid (FFA) and specific extinctions at 232 and 268 nm was made according to the IUPAC standard methods (Paquot, 1979). A prgrammed Metrohm Rancimat Model 679 was used for the determination of induction periods of control and stabilized (treated with plant extracts) corn oil at the initial stage. Thus, antioxidant activity index (AI) of the plant extracts were calculated from the induction periods (Laubi et al., 1988).

Results and Discussion:

Antioxidant activity index (AI) of different oil treatments (control and stabilized oil samples) determined at the initial stage, as plotted versus storage period, is shown in Fig 1. It is evident that addition of methanolic extracts of plant materials quite significantly improved the oxidative stability of corn oil. Rosemary, coffee beans and rice bran extracts prolonged the induction periods of corn oil appreciably, where as, guava leaf and peppermint extracts seemed less effective.

Fig. 2 shows the relative increase in PV of oil treatments during storage. A typical pattern in the rise of PV was observed for all the treatments. The control (without addition of any plant extract) had the higher PV of all the treatments, indicating the highest intensity of oxidation. Among the different extracts, rosemary and peppermint extracts were found to be the most and least effective respectively in retarding PV of corn oil. A very slow rise in PV of the stabilized corn oil as compared with control reflected a good antioxidant activity of the extracts.

Fig 3 shows the relative increase in free fatty acids of the oils formed as a result of hydrolysis and oxidation. Generally a slow followed by a gradual, rapid increase in the formation of FFA was observed in all the treatments. Rosemary, rice bran and coffee beans extracts were more effective in retarding FFA as compared to others. However there was lack of significant variation and correlation in the values of this parameter among some of the treatments. There are reports in the literature, which revealed that regression analysis of the experimental data did not show a general correlation between FFA and oxidative stability of the oils (Frega, et al., 1999).

Averaged conjugated dienes and trienes values plotted against period of storage have been shown in Fig 4 and 5 respectively. A typical pattern was also seen in the rise of primary and secondary oxidation products (conjugated dienes, and trienes) for all the treatments. The control had the greatest rise in conjuable oxidation products as compared with those of stabilized ones. The lowest slop was observed for the rosemary and coffee beans extracts, showing their highest antioxidant activity.

It was concluded from the results of our study that the rate of oxidation of corn oil stabilized with plant extracts was significantly retarded during accelerated aging. This was attributed due to antioxidant activity of the extracts. Extraction and identification of natural antioxidant compounds from plant species is recommended for further studies.

Reference

Bandoniene, D., Pukalskas, A., Venskutomis, P. R., Gruzdiene, D. (2000) Preliminary screening of anioxidants of some plants extracts in rapeseed oil, *Food Res. Int.* 23 (9), 785-791.

Frega, N., Mozzon, M., Lercker, (1999) Effects of free fatty acids on oxidative stability of vegetable oils, J. Am. Oil Chem., Soc., 76, 325-329.

Laubi, M.W., Bruttle, P.A., and Schalach, E. (1988) A modern method for the determining oxidative stability of fats and oils, Int. food Marketing& Technology, 1, 16-18.

Lious, B-K., Chen, H-Y., Yen, G-C. (1999) Antioxidant activity of methanolic extracts from various traditionally edible plants, *Zhongguo Nongye Huaxue Huizhi*, 37 (1), 105-116.

Mery, A.S. (1999) Antioxidant activity of phenolic compounds in grape juice & purine juice on human low density lipoproteins *Fruit Process* 9 (11) 426-430.

Paquot, C. (1979) IUPAC Standards Methods for the Analysis of Oils, Fats and Derivatives, 6th Edition, Pergamon Press, Oxford.

Wilson, T. (1999) Whole foods, antioxidants and health, *Anitooxid. Hum. Health Dis.* 141-150.

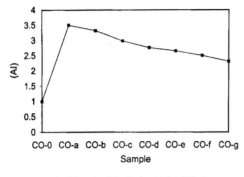

Fig. 1. Antioxidant Activity Index (AI) of Plant
Extracts in Corn Oil at the Initial Stage.

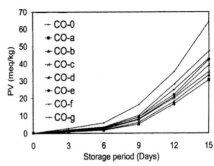

Fig. 2. Relative Increase in Peroxide Value
(PV) of Corn Oil During Storage.

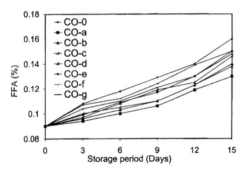

Fig. 3. Relative Increase in Free Fatty Acids
(FFA) of Corn Oil During Storage.

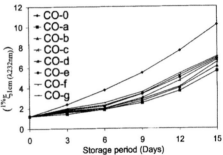

Fig. 4. Relative Increase in Specific Extinction
($^{1\%}\xi_{1cm(\lambda232)}$) of Corn Oil During Storage.

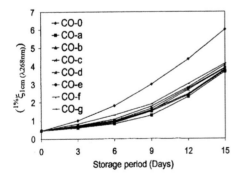

Fig. 5. Relative Increase in Specific Extinction
($^{1\%}\xi_{1cm(\lambda268)}$) of Corn Oil During Storage.

Lengends Key :

CO-0 Corn oil with out any extract (control).
CO-a Corn oil stabilzed with rosemary extract.
CO-b Corn oil stabilzed with coffee beans
         extract.
CO-c Corn oil stabilzed with rice barn extract.
CO-d Corn oil stabilzed with roasted wheat
         germ extract.
CO-e corn oil stabilized with *Ocimum basilicum*
         Extract.
CO-f corn oil stabilzed with guava leaf extract.
CO-g Corn oil stabilzed with *Mentha piperita*
         extract.

STUDY OF THE ASYMMETRIC DISTRIBUTION OF SATURATED FATTY ACIDS IN SUNFLOWER OIL TRIACYLGLYCEROLS.

NOEMÍ RUIZ-LÓPEZ, ENRIQUE MARTÍNEZ-FORCE, RAFAEL GARCÉS
Instituto de la Grasa. Consejo Superior de Investigaciones Científicas.
Avda. Padre García Tejero, 4. 41012 Seville. (SPAIN)

1. Introduction

Triacylglycerols (TAG) are the main components of vegetable oils, accounting for almost 100% of the weight. Each oil has a characteristic pattern of TAG and the chemical, physical and nutritional properties of a particular oil are determined mainly by the abundance of its different TAG molecular species. Plant triacylglycerols are produced by the classic Kennedy pathway which involves, initially, two acylations of glycerol 3-phosphate in *sn*-1 and *sn*-2 positions with acyl-CoA esters to produce phosphatidate by the enzymes, glycerol 3-phosphate acyltransferase (GPAT) and lysophosphatidate acyltransferase (LPAAT), respectively. The phosphatidate is then hydrolysed to diacylglycerol by the enzyme phosphatidate phosphohydrolase and, subsequently, the diacylglycerol can be further acylated by acyl-CoA to yield triacylglycerol (a reaction catalysed by diacylglycerol acyltransferase, DAGAT). The last enzyme being unique to triacylglycerol biosynthesis. Those acyltransferases regulate fatty acid stereochemical distribution. Therefore the acyltransferases specificity and the acyl-CoA pool control the proportion of the different TAG species formed. In order to study variability in the system of acyl-transferases in sunflower seeds a new extraction method has been developed. Using this method we were able to carry out triacylglycerols variability studies on sunflower seeds with high saturated fatty acid content (stearic >12%).

2. Material and methods

Sunflower (*Helianthus annuus* L.) seeds from normal, high oleic and high stearic acid content (Osorio et al., 1995) were used in this work. Seeds were harvested at 35 days

N. Murata et al. (eds.), Advanced Research on Plant Lipids, 31–34.
© 2003 *Kluwer Academic Publishers. Printed in the Netherlands.*

after flowering (DAF) and processed immediately. For the half-seed analysis a piece of the end of the seed (achene without hull and seed coat) corresponding with a third of the total seed was sliced. Oil was extracted from the half-seed with heptane and analysed by GLC (Fernandez-Moya et al., 2000). TAG were transmethilated following the method of Garcés and Mancha (1993) and analysed by GLC.

3. Results and discussion.

Stereospecific analysis of sunflower oil TAG had shown that saturated fatty acids (Sat) were located mainly in positions *sn*-1 and *sn*-3 of the glycerol molecule. In high saturated fatty acid sunflower mutants saturated fatty acids are mainly found in positions *sn*-1 and *sn*-3 of TAG, they were found in *sn*-2 in very low amount, being the main fatty acids in this position the oleic and linoleic acids (Álvarez-Ortega et al., 1997). However the acyl groups are not distributed according with the 1,3-random, 2-random theory (Vander Wal, 1960). The saturated fatty acids, palmitic and stearic, showed preference for position *sn*-3 over the *sn*-1 (Reske and Siebrecht, 1997).

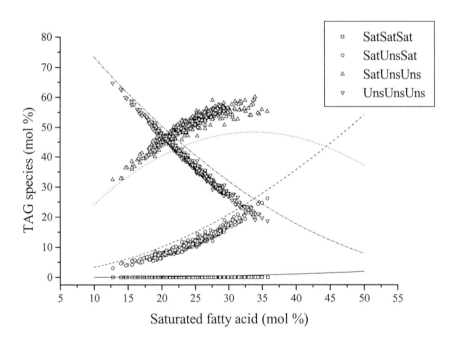

Figure 1. Theoretical and observed distribution in seeds segregating for high oleic and high stearic characters.

In high stearic sunflower mutants TAG species containing two molecules of linoleic acid and one saturated fatty acid are more abundant that the expected by the 1,3-random 2-random theory (Fernández-Moya et al., 2000). The increment of SatUnsUns and the reduction of UnsUnsUns TAG species were directly correlated with the total stearic acid content in the oil. The expected values for sunflower TAG species groups (SatSatSat, SatUnsSat, SatUnsUns and UnsUnsUns) for different saturated fatty acid content, based on the observed composition in position sn-2 and total fatty acid content applying the 1,3-random 2-random theory are represented in Figure 1 as well as the TAG compositions of sunflower oils from more than 500 analyses of seeds differing in the stearic acid content and in the oleic/linoleic ratio. The TAG composition has been determined by GLC using the above method and the data of these TAG species grouped by level of unsaturation are shown in Figure 1.

As Figure 1 shows, sunflower saturated fatty acids follow an asymmetric distribution in TAG, being always the observed values for SatUnsUns higher than the expected and the values for SatUnsSat and UnsUnsUns lower than the anticipated by a non-specific distribution in positions sn-1 and sn-3.

Using the total saturated fatty acid content, the saturated fatty acid content in position sn-2 and the TAG species profile a saturated fatty acid distribution coefficient, α, could be calculated. A symmetric distribution in positions sn-1 and sn-3 would give us a value of 0.5 (50% of saturated fatty acids in each position).

As shown in Table 1, for different stearic and saturated fatty acids contents the distribution coefficient of the saturated fatty acids (α) between positions sn-1 and sn-3 is always between 0.2 and 0.4 in normal linoleic lines and between 0.15 and 0.3 in high oleic lines.

Table 1. Stearic, total saturated fatty acid content, the different TAG groups, and the value of the distribution coefficient α in several normal and mutant lines of sunflower are shown.

Line	Type	18:0	Sat	SatUnsSat	SatUnsUns	UnsUnsUns	α
RHA-274	Normal	5,0	11,7	2,9	29,5	67,7	0,31
CAS-3	HSHL	25,6	37,0	28,6	48,9	21,6	0,33
CAS-29	HSHL	33,2	42,0	28,9	55,8	14,7	0,37
CAS-4	MSHL	12,9	20,3	8,3	44,5	47,3	0,29
G-8	HO	4,9	9,4	1,1	22,7	76,2	0,19
CAS-15	HSHO	23,4	33,5	19,3	61,9	18,8	0,23
	MSHO	19,1	29,8	16,8	55,6	27,5	0,27

4. Conclusion

The fatty acid distribution in TAG of sunflower oils are not distributed according to the 1,3-random 2-random theory. These must be due to a differential specificity in the acyl transferases of the Kennedy pathway over the saturated fatty acids. This irregular distribution is more obvious in the high saturated lines.

5. References

Álvarez-Ortega. R., Cantisán, S., Martínez-Force, E. and Garcés, R. (1997) Characterization of polar and non-polar seed lipid classes from highly saturated fatty acid sunflower mutants. Lipids 32, 833-837.

Fernández-Moya, V., Martínez-Force, E. and Garcés, R. J. (2000) Identification of triacylglycerol species from high-saturated sunflower (*Helianthus annuus*) mutants. Agric. Food Chem. 48, 764-769.

Garcés, R. and Mancha, M. (1993) One-step lipid extraction and fatty acid methyl esters preparation from fresh plant tissues. Anal. Biochem. 211, 139-143.

Osorio, J., Fernández-Martínez, J., Mancha, M. and Garcés, R. (1995) Mutant sunflowers with high concentration of saturated fatty acids in the oil. Crop Sci. 35, 739-742.

Reske, J., Siebrecht, J. and Hazebroed, J. (1997) Triacylglycerol composition and structure in genetically modified sunflower and soybean oils. J. Am. Oil Chem. Soc. 74, 989-998.

Vander Wal, R.J. (1960) Calculation of the distribution of the saturated and unsaturated acyl groups in fats, from pancreatic lipase hydrolysis data. J. Am. Oil Chem. Soc. 37, 18-20.

OKRA(*Hibiscus esculentus L.*) SEED OIL: CHARACTERIZATION AND POTENTIAL USE FOR HIGH VALUE PRODUCTS

PATRISHA J. PHAM*, MILAGROS M. PERALTA* and LAURA J. PHAM**
* *Institute of Chemistry, College of Arts and Sciences (CAS), University of the Philippines Los Baños, Laguna*
** *National Institute of Molecular Biology and Biotechnology at University of the Philippines Los Baños, Laguna*

Introduction

In the Philippines, as in other tropical countries, the main sources of edible and industrial oils are coconut, oil palm and soybean. Oils of commercial importance from these species have received appreciable research attention because of their nutritional, medical and health benefits. However, increase in the human population, shortage of fertile land and restrictions in the importation of food have contributed to the high prices of these traditional sources of edible industrial oils in tropical developing countries. Thus, the focus of interest has been geared toward finding alternatives to prevent hunger and malnutrition. One approach that can be used is the exploitation of less familiar plant sources, but lack of data on their chemical composition limits the prospects for their utilization. Here in the Philippines, there is a growing interest in okra cultivation. About 3,000 hectares have been planted to okra and three companies are in the business of exporting this crop (BAS, 1998) [1]. Hence, local sources are abundant and within easy reach. Since there is no available data on the composition of local okra seed oil, it is the aim of this research work to look into the lipid molecular species of our local okra seed oil for its development into products for the food, pharmaceutical, and cosmetics industry.

Materials and Methods

Materials

Mature okra (*Hibiscus esculentus L.*) seeds of the smooth green or farmers' variety were obtained from the Vegetable Department of the Institute of Plant Breeding at U.P. Los Baños. Organic solvents were obtained from JT Baker, Aldrich Co. and Merck. Triglyceride standard, fatty acid methyl ester standards, visualizing agents, carotenoid & tocopherol standards were purchased from Supelco, Inc. & Sigma Chemical Co.

Preparation of Sample and Determination of Physico-Chemical Characteristics of the Oil

Extraction of oil from okra seeds was adapted from a general method suggested by AOAC (1977) for oilseeds. The Physico-chemical Characteristics of the oil were determined by the methods listed by AOAC (1977) [2] for seed oils.

N. Murata et al. (eds.), Advanced Research on Plant Lipids, 35–38.
© 2003 *Kluwer Academic Publishers. Printed in the Netherlands.*

Determination of Mono-, Di- and Triglycerides

The method used for the determination of glycerides was adapted from the standard method as cited by Pham *et al*, (1996) [3].

Analysis of Triglyceride Components and Fatty Acids

The triglyceride fraction was analyzed for its different components, which were fractionated in terms of increasing partition number (PN) by HPLC *(Kwon, 1996)* [4]. The fatty acid methyl esters (FAME) were prepared using the boron trifluoride-methyl esterification method (Association of Official Analytical Chemists, 1995) [5].

Determination of Carotenoids and Tocopherols in the Unsaponifiable Fraction

The carotenoid and tocopherol components of the unsaponifiable fraction were analyzed using tlc on silica gel after extraction of the unsaponifiable matter *(Diethyl Ether Method, AOAC, 1977)* [6]. Individual tocopherols were identified using a JASCO sil-5 column. The mobile phase was a 1.5 % mixture of isopropyl in hexane. UV detection was carried out at a wavelength of 295 nm. Individual carotenoid components on the other hand, were analyzed using a column packed with Zorbax ODS (25 x 0.46 cm). The mobile phase consisted of acetonitrile: dichloromethane: methanol (70:20:10 v/v/v) with a flow rate of 1 mL/min and detection at 450 nm. *(Neils and Leenheer, 1983)* [7].

Results and Discussion

Physico-Chemical Characteristics

The oil from okra seeds is a light, yellowish oil resembling that of many conventional seed oils such as soybean oil and cottonseed oil. Table 1 shows the main chemical and physical characteristics of okra seed oil as determined from the experiment. From the physico-chemical data, okra oil can be characterized as a source of acylglycerols of long chain, unsaturated fatty acids. This characteristic is common to seed oils in mostly temperate countries.

Table 1. Physical and chemical properties of okra seed oil.

Property	Average values $\pm$ d
Specific Gravity (25°C)	0.9495 $\pm$ 0.0005
Saponification Value (mg KOH/g sample)	170.82 $\pm$ 1.55
Estimated Molecular Weight	985 $\pm$ 9
Iodine Value (mg/100 g sample)	120.512 $\pm$ 2.152
Peroxide Value (meq/kg sample)	6.709 $\pm$ 1.010
% Unsaponifiable Matter	0.8184 $\pm$ 0.0793

Glyceride (Tri-, Di- and Monoglyceride Species) and Fatty Acid Profile

The glyceride profile of okra oil is shown in Table 2.Triglycerides constitute the major glyceride components (75%) of the oil.Small proportions of diglycerides and monoglycerides are present at 10% and 7.5 %, respectively. The partial glycerides are minor components in comparison with the triglycerides. A detailed analysis of triglyceride fraction based on the fatty acid composition is shown in Table 3. About 1/3 of the triglycerides is represented by fractions with PN 46. Considering the fact that oleic, linoleic, and palmitic constitute the major fatty acids (from the Fatty acid Analysis), triglyceride molecular species with PN 46 and 48 constitute almost 68% of the triglyceride component of okra oil.

Table 2. Results of the analysis of okra oil by TLC on silica gel G using hexane: diethyl ether: formic acid (80:20:2 v/v/v).

Glycerides	Average Wt.% $\pm$ d	R f
Triglycerides	75% $\pm$ 5%	0.59
Diglycerides	10% $\pm$ 0%	0.41
Monoglycerides	7.5% $\pm$ 2.5%	0.26
Fatty acids	7.5% $\pm$ 2.5%	0.23

Table 3. Possible triglyceride molecular species of okra oil

Retention Times, mins	PN	Triglycerides	Actual Wt%**	Expected Wt %***
4.650-5.687	36-40	LnLnO, LnLL, LnLnP, PoLnL,MPoLn,MLLn	9.335	0.2072
6.521-7.554	42	LLL, PLLn, MPLn, MMM, PoPoPo	12.214	15.575
7.829	44	PLL, MPoP, MPL, PPoLn, LnLnEr, LnLnAr	13.337	18.789
9.225-9.550	46	OOL,PPL,BeLnLn,ArLnL ArLnM,ArLnPo, ErLnL, ErLnM, ErLnPo	34.329	33.115
10.971-11.425	48	OOO,PPP, SSLn,BeLnL, BeLnPo, BeLnM	29.652	34.449

The fatty acid methyl esters were identified by comparing the retention times with those of a reference standard mixture of known fatty acid composition. The fatty acid methyl ester compositions with their corresponding weight percentages are shown in Table 4. The fatty acid profile shows that okra oil is an unsaturated oil. The ratio of saturated to unsaturated fatty acids is 1:2.

Table 4. Fatty acid composition of okra seed oil by gas liquid chromatography.

Fatty Acid	Wt%
Myristic Acid (C14:0)	0.2973
Palmitic Acid (C16:0)	32.23
Palmitoleic Acid (C16: 1)	0.3089
Stearic Acid (C18:0)	3.943
Oleic Acid (C18: 1)	30.13
Linoleic Acid (C18: 2)	30.05
Linolenic Acid (C18: 3)	0.3814
Arachidic Acid (C20:0)	0.3619
Behenic Acid (C22:0)	0.8145
Erucic Acid (C22: 1)	0.2754
Unidentified	1.213
Saturated	37.64
Unsaturated	61.14
TOTAL	98.78

Composition of the Unsaponifiable Fraction

The thin layer chromatogram (tlc) of the unsaponifiable fraction of the oil confirmed the presence of carotenoids and tocopherols having the same Rf values for the carotenes as well as α, β,γ, & δ tocopherols. The tocopherol content appears to be relatively high and is comparable or even higher with conventional oils reported in literature. This content of tocopherols could give the oil its oxidative stability. HPLC chromatograms revealed that among the four tocopherols present in the unsaponifiable fraction, the most abundant is γ - tocopherol followed by α-tocopherol. δ tocopherol exists in significant amounts while β tocopherol is present in trace levels. The HPLC chromatogram of the separated carotenoids revealed the presence of xanthophylls (lutein), carotenes (α, and β), and two unknown peaks.

References

1. Bureau of Agricultural Statistics, 1998
2. Pacquot, C. (1977) Standard Methods for the Analysis of Oils, Fats and Derivatives.6th edition.p.32-50.
3. Pham, L.J., Casa , E.P. , and Gregorio, M.A. (1996) Pili nut (*Canarium Ovatum* Engl.) oil triglycerides & its Potential for modification in specialty fats and oils through Biotechnology. The Philippine Agriculturist. 79:3 & 4,137-143.
4. Kwon, D.Y., Song H.N. and Yoon, S.H. (1996) Synthesis of Medium chain glycerides by lysase in Organic Solvent, JOACS.73, 1521 -2525.
5. Association of Analytical Chemists (1980) Official Methods of Analysis of the Association of Analytical Chemists. 13th edition, p. 218.
6. Pacquot, C. (1977) Standard Methods for the Analysis of Oils, Fats and Derivatives.6th edition.p.118.
7. Neils, H.J. and De LeenHeer, A.P. (1983) Isocratic Nonaqueous Reversed-Phase Liquid Chromatography of Carotenoids.American Chemical Society.55: 2, 270-275.

Chapter 3:

Fatty Acid Biosynthesis

CELLULAR AND DEVELOPMENTAL IMPLICATIONS FOR CARBON SUPPLY TO FATTY ACID SYNTHESIS IN SEEDS

S. RAWSTHORNE, L.M. HILL, M.J. HILLS, S.E. KUBIS, P.C. NIELD, M.J. PIKE, J.-C. PORTAIS[1], S. TROUFFLARD[2]
[1] *John Innes Centre, Colney Lane, Norwich, NR4 7UH, UK*
[2] *Universite Picardie Jules Verne, F-80039 Amiens cedex, France*

Introduction

The supply of carbon to oil in the developing oilseed embryo implicates compartmentation of metabolism at the whole organ, tissue, cellular and subcellular levels. We have begun to address this by studying the metabolism of sucrose in whole seeds of oilseed rape (*Brassica napus* L.). Sugars are carbon sources for growth of the embryo within the seed and also signal molecules that control the expression of genes. Understanding the sugar environment of the developing embryo is therefore of dual importance.

Following the uptake of sugars by the oilseed embryo they are metabolised in the cotyledons to a range of metabolites that can be imported into the plastid and used for fatty acid synthesis and other plastidial pathways (Kang and Rawsthorne, 1996; Eastmond and Rawsthorne, 2000). We have previously shown the importance of pyruvate and glucose 6-phosphate (Glc6P) as carbon sources for plastidial fatty acid synthesis (Kang and Rawsthorne, 1996; Eastmond and Rawsthorne, 2000). The relative utilisation of these metabolites by plastids changes during the development of the embryo (Eastmond and Rawsthorne, 2000). These changes in utilisation are determined largely by changes in the activities of transporters, located on the plastid inner envelope, that are specific for these metabolites (Eastmond and Rawsthorne, 2000). Recent genomics-based approaches have revealed that the phosphoenolpyruvate (PEP) / inorganic phosphate transporter (PPT) may also be important for fatty acid synthesis in the developing Arabidopsis embryo based on the expression of ESTs (expressed sequence tags) (White et al., 2001). The activity and expression of the PPT has been measured in the developing oilseed rape embryo and compared to that of other transporters. We have also begun to address the importance of the transporters in vivo

41

N. Murata et al. (eds.), Advanced Research on Plant Lipids, 41–44.
© 2003 *Kluwer Academic Publishers. Printed in the Netherlands.*

by manipulating the activity of the Glc6P transporter (GPT) through over- and antisense-expression.

While pyruvate has been shown to be important as a carbon substrate for fatty acid synthesis in oilseed rape embryos, the metabolite that gives the highest rate of fatty acid synthesis with isolated plastids can vary with the species and the tissue from which the plastids have been isolated (Rawsthorne, 2001). For example, we have shown that plastids isolated from endosperm of castor seeds utilise malate preferentially (Eastmond et al., 1997). To investigate differences between species further we have measured how different metabolites are used by plastids isolated from developing embryos of soybean and flax.

Materials and Methods

Plant material, reagents and radiochemicals, determination of metabolite uptake and incorporation of carbon from metabolites by isolated plastids were essentially as described by Eastmond and Rawsthorne, (2000) with minor modifications for incubation conditions (see Fox et al., 2001). Radiolabelled substrates were obtained from NEN Life Science Products, (Houndslow, UK).

Results and Discussion

Sugar metabolism in the endosperm of developing oilseed rape embryos.
To determine how the sugar content of the uncellularized endosperm changed during seed development seeds were punctured with a fine needle, the exuding liquid collected and the content of sucrose, glucose and fructose was determined by enzymatic analysis. This revealed a transition from a predominantly hexose content (>95%) to one of sucrose (~90%). The activity of invertase in the testa (including cellularized endosperm) and in the uncellularized endosperm declined during seed development, correlating with the decrease in hexose and increase in sucrose content (data not shown). However, because the activity of the uncellularized endosperm (on a seed basis) is more than twice that of the testa plus cellularized endosperm it is unclear to what extent the activities of invertase in these two tissues contribute to the change in endosperm sugar composition. Moreover, the uptake of the different sugars by the embryo is also likely to contribute to the overall changes in sugars that we have measured. These data suggest that a simple model for determination of the sugar environment of the embryo are more complex than proposed for the developing *Vicia* embryo in which hexoses are proposed to promote cell division while sucrose promotes storage product synthesis (Wobus and Weber, 1999).

Uptake and utilization of metabolites for fatty acid synthesis by isolated plastids

The uptake of PEP by plastids isolated from embryos doubles during development to 1.1 $\mu mol.unit^{-1}$ $GAPDH.h^{-1}$ as embryos progressed into oil synthesis. Carbon from PEP was incorporated into fatty acids by the plastids at a rate that was of similar magnitude to that of pyruvate (Eastmond and Rawsthorne, 2000). PEP therefore represents an additional source of carbon for plastidial fatty acid synthesis in the oilseed rape embryo.

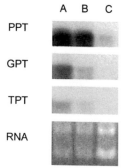

Figure 1. Developmental changes in abundance of the PEP (TPT), Glc6P (GPT), and triose phosphate transporter (TPT) mRNAs in oilseed rape embryos. Stages of development progress from (A) pre–oil to (C) late–oil synthesis.

We studied the expression of the PPT mRNA in the developing embryo through oil synthesis and compared this to the expression patterns of the GPT and triose phosphate transporter mRNAs (Fig. 1). Both PPT and GPT expression decreased considerably during oil synthesis, in contrast to their activities which respectively increased or only decreased by ~30%. These data show that expression of the transporter mRNAs cannot be used to predict changes in the activities of the proteins.

The incorporation of a Glc6P, malate, PEP, pyruvate, and acetate into fatty acids was measured using plastids isolated from developing embryos of soybean and flax. For these substrates, the rate of incorporation of carbon from pyruvate into fatty acids was substantially greater than that from any other substrate (Pike et al.; Troufflard et al., this volume). However, while the in vitro rate from pyruvate represented 46% of that required in vivo for soybean, it was much less for flax. The reason for the discrepancy with flax is not understood at present as other data have revealed that these plastids are intact and utilise Glc6P efficiently for starch synthesis. Collectively, these data imply that pyruvate may be of wider importance as a carbon source for fatty acid synthesis in the developing embryo than solely for oilseed rape (Eastmond and Rawsthorne, 2000).

Manipulation of the activity of GPT in vivo

In order to understand the role of the plastid envelope transporters in the developing embryo in vivo we have expressed the GPT in the antisense orientation driven by the oleosin promoter (Nield et al., this volume). Lines containing multiple copies of the antisense construct had severe phenotypes in the seed, including strongly wrinkled mature seed and an absence of developed cotyledons. However, lines with a single antisense construct had less wrinkled seed and a range of lipid content in mature seed of between 85% and 21% of that in control lines. Collectively, these phenotypic changes show how important the GPT may be in vivo. However, we now need to study the extent to which changes in the GPT activity in the plastids in these transgenic lines determine plastidial metabolic activities, the partitioning of Glc6P between them, and the rates of storage product synthesis.

References

Eastmond P.J. and Rawsthorne S. (2000) Plant Physiol **122**, 767-774.

Eastmond P.J., Dennis D.T. and Rawsthorne S. (1997) Plant Physiol **122**, 767-774.

Fox S.R., Hill LM., Rawsthorne S. and Hills M.J. (2000) Biochem J **352**, 525-532.

Kang F. and Rawsthorne S. (1996) Planta **189**, 516-521.

Rawsthorne S. (2001) Progress in Lipid Research **41**, 182-196.

White J.A., Todd J., Newman T., Focks, N., Girke T. et al., (2001) Plant Physiol. **124**, 1582-1594.

Wobus U. and Weber H. (1999) Biol Chem **380**, 937-944.

WHAT IS NEW ABOUT GLYCEROL METABOLISM IN PLANTS?

MARTINE MIQUEL

Seed Biology Laboratory, INRA Versailles-Grignon
Route de St Cyr
78026 Versailles cedex, France

1. Introduction

Glycerol-3-phosphate (G3P) is an essential component of all glycerolipids. Despite this importance, little is known about its synthesis and supply to the different sites of lipid synthesis in plants. G3P is present in (at least) the plastids and cytoplasm of cells. It has been demonstrated that it can be synthesized via two metabolic pathways (Frentzen, 1993). It can be formed from dihydroxyacetone phosphate (DHAP) by the action of a NAD^+-G3P oxidoreductase in both the chloroplast and cytoplasm of leaves (Santora et al., 1979; Gee et al., 1988a, b, c and 1989; Kirsch et al., 1992), and in immature seeds of *Brassica campestris* (Sharma et al., 2001). G3P can also be generated from DHAP by the glycerokinase-mediated, glycerol-phosphorylation pathway in the cytoplasm (Gosh and Sastry, 1988). Except in developing groundnut seeds (Gosh and Sastry, 1988), both pathways can be found, therefore raising the question of the respective role and importance of the different pathways and their regulation.

The *Arabidopsis* mutants designated *gly1* exhibit a reduced carbon flux through the prokaryotic pathway that is compensated for by an increased carbon flux through the eukaryotic pathway (Miquel et al., 1998). Biochemical approaches reveal that the *gly1* phenotype cannot be explained by a deficiency in the enzymes of the prokaryotic pathway. The chemical complementation of the mutant phenotype by exogenous glycerol

N. Murata et al. (eds.), Advanced Research on Plant Lipids, 45–47.
© 2003 *Kluwer Academic Publishers. Printed in the Netherlands.*

treatment of *gly1* plants suggests a lesion affecting the G3P supply within the chloroplast (Miquel et al., 1998).

Because the *gly1* mutants may help in determining the pathway for chloroplast G3P and as an alternative to their biochemical study we mapped and cloned the *GLY1* gene.

2. Results

2.1. *Mapping and cloning of GLY1*

A strategy based on the polymorphism existing between *Arabidopsis* ecotypes (Columbia for *gly1* mutants and Landsberg *erecta*) was used to map the *GLY1* locus. We first mapped *GLY1* to the bottom of chromosome 2 down to a region spanning 3.7 Mb in which two putative glycerol-3-phosphate dehydrogenase (GPDH) genes could be found. We sequenced both regions from Columbia wild type and two alleles of *gly1*, *gly1-1* and *gly1-2*. For the GPDH located on the T7D17 BAC a point mutation in *gly1-1* suppresses a splicing site whereas in *gly1-2* another point mutation changes a very conserved glutamic acid residue to a lysine residue. No mutation was found for any allele for the second GPDH on chromosome 2.

The full-length cDNA is 1.55 kb and encodes a polypeptide of 420 amino acids. The N-terminal end is enriched in serine residues which is indicative of a transit peptide.

2.2. *Multiple GPDH isoforms in the Arabidopsis genome*

During this work, the complete sequence of *Arabidopsis* genome was released revealing four open reading frames encoding putative GPDH, two on chromosome 2, one on chromosome 3 and one on chromosome 5. The gene encoding the isoform on chromosome 5 was recently cloned (Wei et al., 2001). The protein is a plastidic GPDH which existence could explain the residual amount of 16:3 fatty acid present in the *gly1-1* mutant (Miquel et al., 1998).

Biochemical data have shown that both a cytosolic and a plastidial isoforms of GPDH are present in plant cells (Kirsch et al., 1992) and could sustain glycerolipid biosynthesis through both prokaryotic and eukaryotic pathways. It is therefore intriguing that at least four putative GPDH would exist in *Arabidopsis*. More work is thus needed to understand the respective roles of these isoforms. Concerning the GLY1 protein current work is under progress to precise its localization, to confirm its function and analyze its expression.

References

Frentzen, M. (1993) Acyltransferases and triacylglycerols in T.S. Moore (ed.), Lipid Metabolism in Plants. CRC Press, Boca Raton, Florida, pp. 195-231.

Gee, R.W., Byerrum, R.U., Gerber, D.W. and Tolbert, N.E. (1988a) Dihydroxyacetone phosphate reductase in plants. Plant Physiol. 86, 98-103.

Gee, R.W., Byerrum, R.U., Gerber, D.W. and Tolbert, N.E. (1988b) Differential inhibition and activation of two leaf dihydroxyacetone phosphate reductases. Plant Physiol. 87, 379-383.

Gee, R., Byerrum, R.U., Gerber, D. and Tolbert, N.E. (1989) Changes in the activity of the chloroplastic and cytosolic forms of dihydroxyacetone phosphate reductase during maturation of leaves. Plant Physiol. 89, 305-308.

Gee, R., Goyal, A., Gerber, D.W., Byerrum, R.U. and Tolbert, N.E. (1988c) Isolation of dihydroxyacetone phosphate reductase from *Dunaliella* chloroplasts and comparison with isozymes from spinach leaves. Plant Physiol. 88, 896-903.

Ghosh, S. and Sastry, P.S. (1988) Triacylglycerol synthesis in developing seeds of groundnut (*Arachis hypogaea*): pathway and properties of enzymes of *sn*-glycerol 3-phosphate formation. Arch. Biochem. Biophys. 262, 508-516.

Kirsch, T., Gerber, D.W., Byerrum, R.U. and Tolbert, N.E. (1992) Plant dihydroxyacetone phosphate reductases. Plant Physiol. 100, 352-359.

Miquel, M., Cassagne, C. and Browse, J. (1998) A new class of Arabidopsis mutants with reduced 16:3 fatty acid levels. Plant Physiol. 117, 923-930.

Santora, G., Gee, R. and Tolbert, N.E. (1979) Isolation of *sn*-glycerol 3-phosphate:NAD oxidoreductase from spinach leaves. Arch. Biochem. Biophys. 196, 403-411.

Sharma, N., Phutela, A., Malhotra, S.P. and Singh, R. (2001) Purification and characterization of dihydroxyacetone phosphate reductase from immature seeds of *Brassica campestris* L. Plant Sci. 160, 603-610.

Wei, Y., Perappuram, C., Datla, R., Selvaraj, G. and Zou, J. (2001) Molecular and biochemical characterizations of a plastidic glycerol-3-phosphate dehydrogenase from arabidospis. Plant Physiol. Biochem. 39, 841-848.

GENES AND PATHWAYS INVOLVED IN BIOSYNTHESIS OF EICOSAPENTAENOIC AND DOCOSAHEXAENOIC ACIDS IN BACTERIA

MICHIRU OOTAKI[1], NAOKI MORITA[2], TAKANORI NISHIDA[1], MIKA TANAKA[1], AKIRA HASE[3], YUTAKA YANO[4], AKIKO YAMADA[5], REIKO YU[5], KAZUO WATANABE[5], and HIDETOSHI OKUYAMA[1*]

[1] *Hokkaido Univ., Sapporo 060-0810, Japan;* [2] *Natl. Inst. Adv. Ind. Sci. Technol. (AIST), Sapporo 062-8517, Japan;* [3]*Hokkaido Univ. Education, Hakodate 040-8567, Japan;* [4]*Natl. Res. Inst. Fish. Sci., Yokohama 236-8648, Japan; and* [5]*Sagami Chem. Res. Center, Ayase 252-1193, Japan*

Corresponding author, e-mail address: hoku@ees.hokudai.ac.jp

1. Introduction

Some eubacteria produce long chain polyunsaturated fatty acids (PUFAs) such as eicosapentaenoic acid (EPA) and/or docosahexaenoic acid (DHA) by the mechanism differing from the combination of elongation and oxygen-dependent desaturation of fatty acids. Metz et al. (2001) proposed a possible EPA biosynthetic pathway of bacteria, where EPA can be produced by polyketide synthase (PKS)-like enzyme encoded by the EPA gene cluster. Accordingly, a *trans* double bond at Δ2 of acyl intermediates is isomerized to *cis* configuration either at Δ3 or at Δ2, when it is not reduced to the C-C bond.

The homologous gene clusters involved in biosynthesis of EPA and DHA have been cloned from *Shewanella* sp. SCRC-2738 (Yazawa, 1996) and *Moritella marina* MP-1 (Tanaka et al., 1999), respectively. The EPA gene cluster conferred an ability to *Escherichia coli* to synthesize EPA, however, no DHA was synthesized in *E. coli* transformed with the DHA gene cluster.

To aim at production of DHA in *E. coli* we tried to coexpress the DHA and EPA gene clusters in *E. coli*. Trace levels (less than 0.5% of the total) of PUFAs in the EPA-producing bacterium *Shewanella* sp. SCRC-2738 were analyzed to obtain information on the biosynthetic mechanism of EPA and DHA.

49

N. Murata et al. (eds.), Advanced Research on Plant Lipids, 49–52.
© 2003 *Kluwer Academic Publishers. Printed in the Netherlands.*

2. Materials and Methods

2.1. *Bacterial strains and growth conditions*

EPA-producing *Shewanella* sp. IK-1 (Yano, unpublished) and *Shewanella* sp. SCRC-2738 (Yazawa, 1988), and DHA-producing *Moritella marina* MP-1 (ATCC 15381; Urakawa et al., 1998) were grown in LB medium containing 3% NaCl at 15 and 10°C, respectively, for 36 h with shaking. *E. coli* DH5α and its transformants were cultivated in LB medium at 15°C for 96 h with shaking.

2.2. *Expression of EPA and DHA gene clusters in E. coli*

A cosmid clone responsible for EPA production was obtained from the *Shewanella* sp. IK-1 library by the procedure of Tanaka et al. (1999), and then the EPA cluster of *Shewanella* sp. IK-1 was cloned into plasmid vector, pBluescript II SK+ (Morita et al., unpublished). The DHA cluster of *M. marina* MP-1 and the EPA cluster of *Shewanella* sp. IK-1 were introduced into *E. coli* DH5α cell by electroporation.

2.3. *Analysis of minor PUFAs of Shewanella sp. SCRC-2738*

To determine the double bond positions of minor PUFAs, fatty acid methyl esters prepared from *Shewanella* sp. SCRC-2738 were analyzed as their pyrrolidide derivatives by gas-liquid chromagraphy-mass spectrometry analysis (Andersson and Holman, 1974).

2.4. *Fatty acid abbreviations*

Fatty acids are abbreviated like 12:3(3,6,9), where the numbers before and after the colon indicate the length of carbon chain and the number of double bonds in the chain, respectively, and the distance of the *cis* double bond from the carboxylic end of the fatty acid is indicated in the parenthesis.

3. Results and Discussion

3.1. *Expression of EPA and DHA gene clusters in E. coli*

The EPA gene cluster was cloned from *Shewanella* sp. IK-1 (Morita et al., unpublished). Its structure consisting of five essential open reading frames (ORFs) was almost the same as the EPA gene cluster from *Shewanella* sp. SCRC-2737 (Yazawa, 1996).

E. coli transformed with the EPA gene cluster from *Shewanella* sp. IK-1 and the DHA gene cluster from *M. marina* MP-1 produced EPA and a very low level of DHA (data not shown). The production of EPA can be eventuated to the EPA gene cluster itself, while the production of DHA would be due to coordination of the EPA and DHA gene clusters. Since the DHA gene cluster lacks the ORF corresponding to a gene encoding phosphopantetheine transferase (see Tanaka et al., 1999), this enzyme of both the PKS-like enzyme systems may be compatible. This can be supported by the finding that acyl carrier protein domains deduced from the EPA and DHA gene clusters show significantly high homology in their amino acid sequences (Tanaka et al., 1999; Metz et al., 2001).

3.2. Identification of minor PUFAs of Shewanella sp. SCRC-2738

Watanabe et al. (1999) previously identified 18:4(6,9,12,15) and 20:4(8,11,14,17) in *Shewanella* sp. SCRC-2738. In this study other PUFAs including dienes, trienes, tetraenes, and pentaenes except for EPA, were found. Almost all these minor PUFA components were also found in *E. coli* transformed with the EPA gene cluster but not in its wild type cells (data not shown), suggesting that these PUFAs are products of the EPA gene cluster. In addition to n-3 PUFAs, n-6 PUFAs such as 18:3(6,9,12), n-7 PUFAs such as 20:3(7,10,13), and propylene-interrupted 18:3(6,11,14) and 20:3(8,13,16) were identified.

3.3. Possible biosynthetic pathways of EPA and DHA

The EPA biosynthetic pathway proposed by Metz et al. (2001) is basically acceptable. However, a *cis* double bond may be directly introduced at Δ2 and Δ3, because the PKS-like enzymes responsible for EPA and DHA biosynthesis have two bifunctional dehydrase domains (Tanaka et al., 1999; Metz et al., 2001). Here we propose pathways of EPA and DHA biosynthesis in bacteria (Fig. 1). The main route of this pathway is

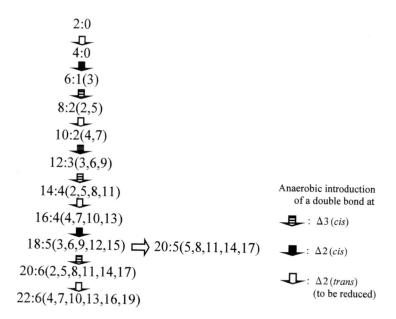

Figure 1. Possible pathways of EPA and DHA biosynthesis in bacteria

52

leading to the synthesis of DHA by the combination of chain elongation and anaerobic introduction of a *cis* double bond at Δ2 or Δ3. EPA can be formed by one-round elongation of 18:5(3,6,9,12,15) (Fig. 1.). The synthesis of unusual PUFAs detected in this study would be due to ambiguous specificity of the PKS-like enzymes toward acyl intermediates. To elucidate the biosynthetic mechanism of PUFAs by PKS-like enzymes future work should be focused on isolation and identification of short-chain acyl intermediates having a *cis* double bond at Δ2 or Δ3.

4. References

Andersson, B.A. and Holman R.T. (1974) Mass spectrometric determination of the double bonds in polyunsaturated fatty acid pyrrolidides. Lipids 10, 215-219.

Metz, J.G. Roessler, P., Facciotti, D., Levering C., Dittrich, F., Lassner, M., Valentine, R., Lardizabal, K., Domergue, F., Yamada, A., Yazawa, K., Knauf, V. and Browse, J. (2001) Production of polyunsaturated fatty acids by polyketide synthases in both prokaryotes and eukaryotes. Science 293, 290-293.

Tanaka, M., Ueno, A., Kawasaki, K., Yumoto, I., Ohgiya, S., Hoshino, T., Ishizaki, K., Okuyama, H. and Morita, N. (1999) Isolation of clustered genes that are notably homologous to the eicosapentaenoic acid biosynthesis gene cluster from the docosahexaenoic acid-producing bacterium *Vibrio marinus* strain MP-1. Biotechnol. Lett. 21, 939-945.

Urakawa, H., Kita-Tsukamoto, K., Steven, S.E., Ohwada, K. and Colwell, R.R. (1998) A proposal to transfer *Vibrio marinus* (Russell 1891) to a new genus *Moritella* gen. nov. as *Moritella marina* comb. nov. FEMS Microbiol. Lett. 165, 373-378.

Watanabe, K., Ishikawa, C., Yazawa, K., Kondo, K. and Kawaguchi, K. (1996) Fatty acid and lipid composition of an eicosapentaenoic acid-producing marine bacterium. J. Mar. Biotechnol. 4, 104-112.

Yazawa, K. (1996) Production of eicosapentaenoic acid from marine bacteria. Lipids 31, S-297-S300.

Yazawa, K. Araki, K., Watanabe, K., Ishikawa, C., Inoue, A., Kondo, K., Watabe, S. and Hashimoto, K. (1988) Eicosapentaenoic acid productivity of the bacteria isolated from fish intestines. Nippon Suisan Gakkaishi 54, 1835-1838.

Brassica KCS IS CAPABLE OF ELONGATING POLYUNSATURATED FATTY ACIDS

JIXIANG HAN[1], FRANK P. WOLTER[2], JAN G. JAWORSKI[1] and
MARGRIT FRENTZEN[3]

[1]*Donald Danforth Plant Science Center, St. Louis, MO 63132, U.S.A.*
[2]*GVS mbH, 53115 Bonn, Germany*
[3]*RWTH Aachen, 52056 Aachen, Germany*

1. Introduction

Polyunsaturated fatty acids (PUFAs) with more than 18 carbons in length are essential constituents functioning in membrane fluidity, signal transduction, and eicosanoid production. The biosynthesis of PUFAs requires an alternating series of desaturation and elongation reactions.

The elongation reactions are catalyzed by microsomal elongase systems which are generally considered to consist of four distinct enzymatic activities, namely ß-keto-acyl-CoA synthase (KCS), ß-ketoacyl-CoA reductase, ß-hydroxyacyl-CoA dehydratase and *trans*-2-enoyl-CoA reductase. In the first step KCS catalyzes the condensation of malonyl-CoA with a long chain acyl group. This condensation reaction is not only the rate limiting step but also the substrate specific step of the elongase.

Desaturases responsible for PUFA formation have been extensively studied in recent years and many desaturase genes have been cloned. More recently, several genes encoding PUFA-elongating activities have been identified (Beaudoin *et al.*, 2000; Parker-Barnes *et al.*, 2000; Leonard *et al.*, 2000; Zank *et al.*, 2000; Qi *et al.*, 2002). These genes are related to the *ELO* gene family of *Saccharomyces cerevisiae* but show no similarity to the *FAE1* genes.

We have demonstrated that the *Brassica napus FAE1.1* gene encodes a KCS which is able to elongate both monounsaturated and saturated fatty acids up to C26 when expressed in yeast (Han *et al.*, 2001). Here, we report a new function of the *B. napus* KCS involved in PUFA elongation. In fact, the *B. napus* KCS can use a range of PUFAs as substrates including 18:2 $\Delta^{9,12}$, 18:3 $\Delta^{9,12,15}$, 20:3 $\Delta^{8,11,14}$, and 20:4 $\Delta^{5,8,11,14}$. This finding revealed that the *B. napus* KCS has a quite broad specificity for PUFA elongation.

N. Murata et al. (eds.), Advanced Research on Plant Lipids, 53–56.
© 2003 *Kluwer Academic Publishers. Printed in the Netherlands.*

2. Materials and Methods

The coding region of the *B. napus FAE1.1* gene was cloned into the yeast expression vector pYES2 (Invitrogen) under the inducible GAL1 promoter. The resulting construct pYKCS and the empty vector pYES2 were transformed into *Saccharomyces cerevisiae* strain INVSC1 (Invitrogen) by a lithium acetate-based method. The yeast cells were grown in complete minimal drop-out uracil medium, containing 2% (w/v) raffinose as a carbon source at 28°C. After induction with 2% (w/v) galactose, the cells were further cultivated for 48 h with 0.3 mM of individual fatty acid substrates in the selective medium.

For the determination of the cell fatty acid composition, fatty acid methyl esters (FAMEs) were made with methanolic H_2SO_4. FAMEs were analyzed by GC-MS (ThermoFinnigan) at an ionization voltage of 70 eV with a scan range of 50-500 Da. The RtX-5 MS capillary column (Restek Corp.) was operated with helium carrier gas at a constant flow rate of 1.0 ml/min and an injection mode of splitless with surge. The oven temperature was increased from 100 to 280 °C at a rate of 16 °C/min, and held for additional 10 min. The GC-MS data were acquired and processed by an on-line PC with Xcalibur data system (ThermoFinnigan).

3. Results and Discussion

Our previous data revealed that the *B. napus* KCS is functionally expressed in yeast cells and catalyses the elongation of the endogenous saturated and monounsaturated acyl groups of the host (Han *et al.*, 2001). This finding encouraged us to further examine the substrate specificity of the *B. napus* KCS specially with regard to PUFA elongation. To this end, PUFA substrates were individually added to the transgenic yeast cells expressing pYKCS and the substrate elongation was measured by cellular fatty acid analysis. These experiments revealed that the *B. napus* KCS could indeed elongate various PUFAs. Fig. 1 for instance shows the fatty acid composition of yeast cells cultivated in the presence of 20:3 $\Delta^{8,11,14}$. Unlike the control cells harboring only the empty vector pYES2 (Fig. 1A), the yeast cells expressing the *B. napus* KCS produced very long chain fatty acids of up to 26 carbons in which 22:3 $\Delta^{10,13,16}$ was identified as main product (Fig. 1B). These data clearly show that the *B. napus* KCS expressed in yeast not only catalyses the elongation of the endogenous saturated and monounsaturated fatty acids but also the elongation of the exogenously added 20:3 $\Delta^{8,11,14}$.

As summarized in Fig. 2, the *B. napus* KCS was found to utilize not only C_{20} but also C_{18} PUFAs as substrates and can elongate not only n-6 but also n-3 PUFAs. The various elongation product profiles indicate that the *B. napus* KCS has distinctly higher activities with C_{18} and C_{20} PUFAs than with C_{22} and C_{24} PUFAs (Fig. 2). In that way the enzyme shows a chain length specificity with PUFAs similar to that observed with monounsaturated acyl groups but different to that observed with saturated acyl groups (Han *et al.*, 2001). As depicted in Fig. 2, the *B. napus* KCS was slightly more active with 20:3n-6 and 18:2n-6 than with 20:4n-6 and 18:3n-3 while it hardly used

(A)

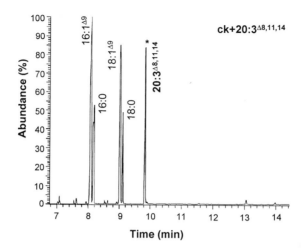

(B)

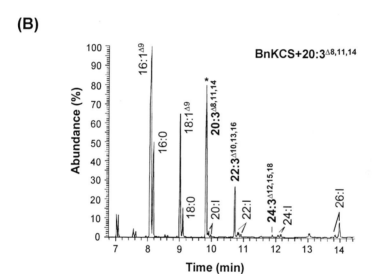

Figure 1. Identification of elongated PUFA products by GC-MS analysis. Yeast cells containing either the empty vector pYES2 (A) or the construct pYKCS (B) were cultivated 48 h in the presence of 0.3 mM 20:3 $\Delta^{8,11,14}$ (marked *) and the cellular fatty acids were then analyzed. The elongation products were identified by the mass spectra of standards.

56

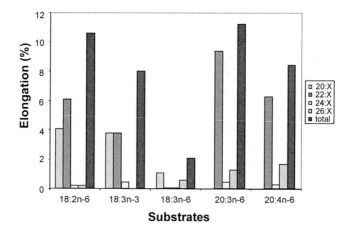

Fig. 2. PUFA products formed from the given substrates by the *B. napus* KCS expressed in yeast. Percent elongation was calculated as [product/(substrate + product)] × 100. X indicates number of double bonds corresponding to the respective substrates added.

18:3n-6 as substrate although this fatty acid showed the highest uptake (63%) into the yeast cells among the five different fatty acids tested. These data suggest that the enzyme elongates PUFAs with the first double bond at Δ8 or 9 at higher rates than PUFAs with the first double bond at Δ5 or 6.

In summary, the data presented here show that the *B. napus* KCS can elongate not only saturated and monounsaturated fatty acids but also PUFAs. Since this KCS displays no obvious sequence similarity to the other PUFA elongating enzymes, mentioned above, there seems to exist distinct mechanism involved in PUFA elongation.

4. References

Beaudoin, F., Michaelson, L.V., Hey, S.J., Lewis, M.J., Shewry, P.R., Sayanova, O.and Napier, J.A. (2000) Heterologous reconstitution in yeast of the polyunsaturated fatty acid biosynthetic pathway. Proc. Natl. Acad. Sci. USA 97, 6421-6426

Han, J., Lühs, W., Sonntag, K., Zähringer, U., Borchardt, D.S., Wolter, F.P., Heinz, E. and Frentzen, M. (2001) Functional characterization of β-ketoacyl-CoA synthase genes from *Brassica napus* L. Plant Mol. Biol. 46, 229-239

Leonard, A.E., Bobik, E.G., Dorado, J., Kroeger, P.E., Chuang, L.-T., Thurmond, J.M., Parker-Barnes, J.M., Das, T., Huang, Y.-S., and Mukerji, P. (2000) Cloning of a human cDNA encoding a novel enzyme involved in the elongation of long-chain polyunsaturated fatty acids. Biochem. J. 350, 765-770

Parker-Barnes, J.M., Das, T., Bobik, E., Leonard, A.E., Thurmond, J.M., Chuang, L.-T., Huang, Y.-S. and Mukerji, P. (2000) Identification and characterization of an enzyme involved in the elongation of n-6 and n-3 polyunsaturated fatty acids. Proc. Natl. Acad. Sci. USA 97, 8284-8289

Qi, B., Beaudoin, F., Fraser, T., Stobart, A.K., Napier, J.A. and Lazarus, C.M. (2002) Identification of a cDNA encoding a novel C18-Δ^9 polyunsaturated fatty acid-specific elongating activity from the docosahexaenoic acid (DHA)-producing microalga, *Isochrysis galbana*. FEBS Letters 510, 159-165

Zank, T.K., Zähringer, U., Lerchl, L. and Heinz, E. (2000) Cloning and functional expression of the first plant fatty acid elongase specific for Δ6-polyunsaturated fatty acids. Bioc. Soc. Trans. 28, 654-658

RE-CONSTITUTION OF THE GLUCOSE 6-PHOSPHATE TRANSLOCATOR IN LIPOSOMES: INHIBITION OF GLUCOSE 6-PHOSPHATE AND INORGANIC PHOSPHATE EXCHANGE BY OLEOYL-COENZYME A

Metabolic regulation of plastidial substrate translocators

SIMON FOX
Brookhaven National Laboratory, NY, USA

ULF-INGO FLÜGGE
Botanical Institute, University of Cologne, Germany

STEPHEN RAWSTHORNE AND MATTHEW HILLS
John Innes Centre, Norwich, UK

Introduction

One of the prime requirements for fatty acid synthesis in plastids are carbon skeletons for the generation of acetyl-CoA, the substrate of acetyl-CoA carboxylase, the enzyme that initiates the pathway by supplying fatty acid synthase with malonyl-CoA. Previous experiments by Johnson *et al.* (2000) demonstrated how *in vitro* the rate long-chain fatty acid synthesis—palmitic and oleic acids—in oilseed rape plastids was reduced by 75% in the presence of ATP and the cofactor CoASH; both the latter are required by long-chain fatty acid CoA ligase for synthesis of long-chain acyl-CoA thioesters (lcACoAs). This inhibitory effect was alleviated by ACBP (acyl-CoA binding protein), a molecule that binds lcACoAs including oleoyl-CoA (18:1-CoA). Later experiments (Fox *et al.*, 2000) showed that addition of 18:1-CoA at sub-micro-molar concentrations (0.1-0.3 μM) significantly inhibited [1-^{14}C]Glc6P (glucose 6-phosphate) uptake into developing oilseed rape embryo plastids. This effect was dependent on the acyl chain length of the lcACoA (no effect <C_{12}, lauryl-CoA) and was reversed by addition of either ACBP or BSA at equimolar concentration. Overall, the conclusion from both experiments was that lcACoAs, intermediates in the synthesis of lipids (the Kennedy Pathway enzymes) at the endoplasmic reticulum where they are substrates for acyltransferase enzymes, were themselves inhibitory to the protein responsible for the transport of Glc6P into the stroma, i.e. the Glc6P translocator. The latter resides on the

N. Murata et al. (eds.), Advanced Research on Plant Lipids, 57–60.

plastid envelope of heterologous tissues and functions through a strict 1:1 exchange of Glc6P mainly with PO_4 or triose phosphates (Kammerer *et al.*, 1998). Here we report results of experiments to further test this idea; that the effect of 18:1-CoA in inhibiting fatty acid synthesis, witnessed in earlier experiments, is due to a direct effect of the acyl-thioester on the Glc6P translocator. For such an analysis we employed purified Glc6P protein reconstituted into artificial membranes, namely, liposomes (spheres composed of phospholipid bilayers).

Methods

Reconstitution of the Glc6P translocator in liposomes

The His-tagged Glc6P protein from maize endosperm, expressed in *Schizosaccharomyces pombe*, was extracted from the yeast cells using NTA columns by standard techniques. Liposomes were created using soybean phospholipids (Sigma type IV), refined by an acetone/chloroform/acetone washing regime and throughout, phosphate free buffers were utilized. The purified phospholipids, once mixed with the membrane-His tag Glc6P translocator were sonicated (30 pulses, 20% duty cycle, Branson 250 Sonifier) with one of the counter exchange substrates (i.e. either Glc6P or PO_4) before purifying the resultant liposomes on PD10 columns removing unbound Glc6P protein and excess substrate that had not been incorporated into the lumen of the liposomes.

Assay of G-6P translocator activity

The Glc6P translocator is a phosphate (PO_4) antiporter, so activity of the protein in liposomes was measured using either radiolabeled [1-^{14}C]Glc6P or $^{32}PO_4$ in exchange for the appropriate unlabeled substrate. Incubations of 1-2 min duration, in a total volume of 250 µl, were terminated by addition of a pyridoxal 5-phosphate solution also containing DIDS and imesalyl. The separation of the bathing medium from the liposomes, enabling determination of the uptake of radiolabel was conveniently achieved by passing the reactions through Dowex anion exchange columns maintained in Pasteur pipettes.

As described the liposomes were incubated ± 18:1-CoA and the effect of BSA, a lcACoA and fatty acid binding protein, tested. The negative control, effectively the background substraction, consisted of the liposomes mixed with the stop solution of pyridoxal phosphate.

Results

1.1 Effect of 18:1-CoA on exchange of PO_4 and Glc6P

The uptake of $^{32}PO_4$ in exchange for unlabeled Glc6P (25 mM) in liposomes (all containing the purified G-6P translocator; approx. 5 µg per incubation) was inhibited by up to 40% on addition of 18:1-CoA to a final concentration of 35 µM (Figure 1).

Uptake of [1-^{14}C]Glc6P in exchange for unlabeled PO$_4$ was inhibited by 72% in the presence of 35 μM 18:1-CoA (Figure 1). The K$_{iapp}$ for this inhibition was calculated from further experiments to be 15-20 μM 18:1-CoA.

Analysis of the uptake of of [1-^{14}C]Glc6P in the presence of different concentrations of PO$_4$ and 18:1-CoA demonstrated that the inhibition was non-competitive and the apparent K$_i$ of approximately 15 μM (Figure 2).

1.2 Effect of BSA on the inhibition of the Glc6P translocator by 18:1-CoA
In experiments in which the uptake of ^{32}PO$_4$ was measured in exchange for Glc6P and in the presence of 18:1-CoA (70 μM) the addition of BSA (bovine serum albumin) to final concentrations of 25, 50 and 150 μM completely alleviated the recorded inhibition caused by 18:1-CoA (30%) in the absence of BSA.

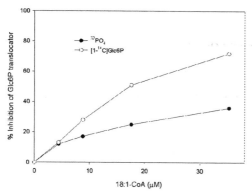

Figure 1. The effect of 18:1-CoA (oleoyl-CoA) on the uptake of either ^{32}PO$_4$ or [1-^{14}C]Glc6P uptake into liposomes containing purified glucose 6-phosphate translocator.

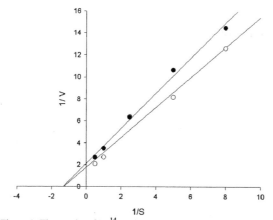

Figure 2. The uptake of [1-^{14}C]Glc6P by glucose 6-phosphate translocator re-constituted into liposomes; measurements were made in the presence of 20 μM 18:1-CoA(●) and in the absence of 18:1-CoA (o).

Conclusions

Data presented on the kinetics of either Glc6P or PO_4 exchange by purified Glc6P translocator proteins demonstrate the latter is inhibited directly and non-competitively by 18:1-CoA; as discussed, this supports conclusions reached by previous work (Johnson *et al.*, 2000; Fox *et al.*, 2000). The K_i obtained here is much higher than that found with purified plastids. However, the liposome experiments contain very large amounts of lipid that will bind the lcACoAs. Therefore the actual concentration of free lcACOAs in the liposome assays may be much lower. The details of the interaction of lcACoA and Glc6P translocator remain to be elucidated, however it is possible that inhibition provides a means of coordinating lipid synthesis at the endoplasmic reticulum with one of the main supply routes for importing carbon into the stroma, the plastidial Glc6P translocator. Moreover, the latter may share many structural features, based on the published amino acid sequence of Glc6P translocators, with other PO_4 antiporters, i.e. the triose phosphate translocator and the phosphoenolpyruvate translocator. The latter are also inhibited by lcACoAs in a chain-length dependent manner (Fox *et al.*, unpublished results), as is the adenine nucleotide transporter from pea root plastids (Fox *et al.*, 2001). Therefore the actions of unbound lcACoAs are likely to affect a number of transport events that are shared by related proteins residing on the plastid envelope. It should be noted that the glucose and pyruvate transporters are completely unaffected by the presence of long chain acyl-CoAs (Fox *et al.*, 2000). The potential concentration of lcACoA available for inhibition presumably is dependent on the ratio of ACBP to lcACoA. The latter have been reported to be close to equilibrium (Fox *et al.*, 2000), so it is envisaged that events affecting the flux of carbon to fatty acids could lead to rapid imbalances in the pool of ACBP and lcACoA thereby releasing lcACoAs to directly affect plastidial membrane transport.

References

Fox, S.R., Hill, L.M., Rawsthorne, S. and Hills, M.J. (2000) Inhibition of the Glucose-6-Phosphate Transporter in Oilseed Rape (Brassica napus L.) Plastids by Acyl-CoA Thioesters Reduces Fatty Acid Synthesis. Biochem. J. 352, 525-532.

Fox, S.R., Rawsthorne, S. and Hills, M.J. (2001) Fatty Acid Synthesis in Pea Root Plastids is Inhibited by the Action of Long-Chain Acyl-CoAs on Metabolite Transporters. Plant Physiol. 126, 1259-1264.

Johnson, P.E., Fox, S.R., Hills, M.J. and Rawsthorne, S. (2000) Inhibition by Long-Chain Acyl-CoAs of Glucose 6-Phosphate Metabolism in Plastids Isolated from Developing Embryos of Oilseed Rape (Brassica napus L.). Biochem. J. 348, 145-150.

Kammerer, B., Fischer, K., Hilpert, B., Schubert, S., Gutensohn, M., Weber, A. and Flügge, U.I. (1998) Molecular characterization of a carbon transporter in plastids from heterotrophic tissues: the glucose 6-phosphate/phosphate antiporter. Plant Cell 10, 105-117.

STRUCTURAL AND FUNCTIONAL ANALYSIS OF THE *FAS1* GENE ENCODING FATTY ACID SYNTHASE β-SUBUNIT IN METHYLOTROPHIC YEAST *HANSENULA POLYMORPHA*

Y. KANEKO[1], K. RUEKSOMTAWIN[2], T. MAEKAWA[1] and S. HARASHIMA[1]

[1]*Department of Biotechnology, Graduate School of Engineering, Osaka University,* [2]*Pilot Plant Development and Training Institute, King Mongkut's University of Technology Thonburi*
[1]*Yamadaoka 2-1, Suita, Osaka 565-0871, Japan,* [2]*Thung Kharu, Bangkok 10140, Thailand*

1. Introduction

The biosynthesis of saturated fatty acids is catalyzed by acetyl-CoA carboxylase (ACC) and fatty acid synthetase (FAS). In yeast and lower fungi, the FAS is a heteromultimeric complex of two multifunctional proteins ($\alpha_6\beta_6$) which are encoded by the two unlinked genes, *FAS1* (subunit β) and *FAS2* (subunit α) (Schweizer et al., 1978). The α-subunit contains three catalytic activities, β-ketoacyl synthase, β-ketoacyl reductase and acyl carrier protein, while the β subunit contains five activities, acetyl transferase (AT), enoyl reductase (ER), dehydrase (DH) and malonyl/palmitoyl transferase (MT/PT) (Stoops et al., 1978). The *FAS* genes have been cloned from various organisms. It was shown that the sequential order of the catalytic domains is similar only among related species such as yeast and lower fungi (Kajiwara et al., 2001; Kottig et al., 1991; Weisner et al., 1988). The methylotrophic yeast, *Hansenula polymorpha*, has a great biotechnological potential and is widely used for studying various metabolic pathways such as peroxisome biogenesis and methanol metabolism, and for producing several useful bioproducts (Gellissen, 2002). Recent year, it also has become favorite host to produce recombinant proteins (Faber et al., 1995; Agaphanov et al., 1995). Moreover, *H. polymorpha*, unlike *Saccharomyces cerevisiae*, is able to synthesize polyunsaturated fatty acids (UFAs) in addition to mono-UFA (Anamnart et al., 1998). Wishing to use *H. polymorpha* as an eukaryotic model to study genetic control of saturated and unsaturated fatty acid biosynthesis, we decided to clone *H. polymorpha FAS1* gene. In this study, we report the isolation, characterization and functional analysis of *H. polymorpha FAS1* gene.

N. Murata et al. (eds.), Advanced Research on Plant Lipids, 61–64.

2.   Results and Discussion

2.1.   *Cloning of the* FAS1 *gene of* H. polymorpha

Southern hybridization analysis of genomic DNA of *H. polymorpha leu1-1* (Anamnart et al., 1998) using the *S. cerevisiae FAS1* gene (*S-FAS1*) as a probe suggested that a 9.6-kb *Bam*HI-*Kpn*I fragment contains the *FAS1* gene. Then, a DNA fraction containing this fragment was recovered and used to make a small gene library with pBluescriptII KS+. On the other hand, we obtained a partial *FAS1* gene fragment (0.4 kb) from *H. polymorpha leu1-1* genomic DNA by PCR with mixed primers (FAS1-1 and FAS1-2, Figure1). Among approximately 200 *E. coli* transformants of the gene library, one clone showed a positive signal in colony hybridization with the PCR amplified *FAS1* fragment. A plasmid in this clone was designated pFA1. The nucleotide sequence of the insert in pFA1 indicated that the 5' portion of the *FAS1* gene was not contained in this fragment. From Southern analysis data, we also found two *Xba*I fragments (6.7 and 4.8 kb) which showed homology to the *S-FAS1* probe.   Comparison of these data and the restriction map of pFA1 revealed that 1.8-kb *Bam*HI-*Xba*I region of pFA1 overlapped with the 4.8-kb *Xba*I fragment and pFA1 contained entire 6.7-kb *Xba*I fragment (Figure 1). Thus, the 4.8-kb *Xba*I fragment was cloned into pBluescriptII KS+ by screening approximately 300 *E. coli* transformants using the 1.8-kb *Bam*HI-*Xba*I fragment of pFA1 as a probe. One positive clone was isolated and its plasmid was designated pFA2. The nucleotide sequence from one end of the *Xba*I insert in pFA2 was matched to the sequence from pFA1. It was concluded that the entire *FAS1* ORF of *H. polymorpha* was located on the joint fragment of two *Xba*I fragments.   Restriction map of *H. polymorpha FAS1* gene is shown in Figure 1.

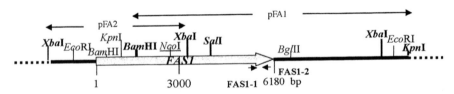

Figure 1. Restriction map of the *FAS1* gene.
Arrows of FAS1-1 and FAS1-2 indicate mixed oligonulceotide primers designed based upon the homology of amino acid sequence from the conserved regions in the β subunit of FAS among different organisms (Kottig et al., 1991; Zhao and Cihlar, 1994; Niwa et al., 1998).

2.2.   *The sequence of* H. polymorpha FAS1 *gene*

The combined nucleotide sequence from two cloned fragments in pFA1 and pFA2 was deposited to DNA Data Bank of Japan (DDBJ) with accession number AB032743. Sequence analysis revealed an intron-free open reading frame of 6,180 bp encoding a protein of 2,060 amino acids with molecular weight of 228,366. The two serine residues (positions 276 and 1817) which are essential for the acetyl transferase, malonyl/palmitoyl transferases and the histidine residue at 1358 of dehydrase active center (Schweizer et al., 1978) were found. The consensus sequence of the NADPH binding site and the putative FMN binding site in enoyl reductase (Stoops et al., 1978) were also found to be located from amino acid residues 770 to 786, and from 672 to 694, respectively. The order of five catalytic domains was identical with those of other yeast species. *H. polymorpha FAS1* gene exhibits high identities 65 % at the amino acid level to the FAS β subunit of *Candida albicans* (Zhao et al., 1994).

2.3. *Transcription of* H. polymorpha FAS1 *gene*

In *S. cerevisiae*, the transcription of *FAS* genes seemed to be irrelevant to the availability of free fatty acids in the medium (Schuller et al., 1992). However, Chilara (1992) showed that *FAS* genes are repressed in *S. cerevisiae* grown for several generations in synthetic medium containing fatty acids. To investigate the regulation of the *H. polymorpha FAS1* expression, we performed Northern analysis with the cloned *FAS1* gene as a probe. Transcription of *FAS1* was examined for cells grown in rich medium supplemented with or without 2 mM myristic acid (C14:0). The result showed that the *FAS1* gene was expressed in the wild-type strain *leu1-1* at low level and its transcript was 6.2 kb in size. Densitometric analysis of the hybridization signals between two conditions revealed that the transcriptional level of *FAS1* did not change significantly, suggesting that transcription of *H. polymorpha FAS1* gene is not repressed by C14:0.

2.4. *Construction of* fas1 *disruptant and its growth phenotype*

In order to obtain a *fas1* gene disruptant of *H. polymorpha*, diploid strain H25 (*leu1-1 ade11-1 /leu1-1 ura3-1*) was transformed with a 6.7-kb *Xba*I fragment of pFA107 containing N-terminal half of the *FAS1* gene disrupted by inserting with *S. cerevisiae LEU2* gene (*S-LEU2*). If replacement of the wild-type *FAS1* locus with the *Xba*I fragment (*fas1::Sc-LEU2*) occurs, it is expected that two Leu$^+$ segregants carrying the disrupted *fas1* locus (*fas1::Sc-LEU2*) of the tetrad do not grow on the nutrient medium supplemented with no fatty acid (Fas$^-$ phenotype). Out of 28 Leu$^+$ transformants tested, only one (designated KYC632) showed 2Fas$^+$Leu$^-$:2Fas$^-$Leu$^+$ segregation in the tetrads, indicating that KYC632 is a *fas1*-disrupted heterozygous diploid. Southern blot analysis also confirmed existence of the disrupted *FAS1* locus in KYC632 and its tetrad. Growth ability of the *fas1* disruptant was tested on rich medium supplemented with various kinds of saturated fatty acid (Table 1). While the *fas1* disruptant grew on rich medium supplemented with 1 mM of C14:0 or palmitic acid (C16:0) as described in *S. cerevisiae fas1* mutant (Schweizer et al., 1978), the growth of the disruptant on lauric acid (C12:0) and stearic acid (C18:0) was not observed even at higher concentration (3 mM). This result suggests that a fatty acid elongation system from C14:0 to C18:0 actually exists in this yeast and C16:0 is indispensable for growth.

FAS1 gene and *fas1* disruptant that we have cloned and created in this study should be useful for various future studies such as those for regulatory mechanism of fatty acid biosynthesis, fatty acid desaturase and fatty acyl chain elongation.

TABLE 1.   Growth characterisitics of the *fas1* disruptant

Strain	Genotype	Growth phenotype on rich medium supplemented with:				
		none	C12-3mM	C14-1 mM	C16-1mM	C18-3mM
KYC632	*FAS1/FAS1*	+	+	+	+	+
632-11A	*FAS1*	+	+	+	+	+
632-11C	*fas1::Sc-LEU2*	-	-	+	+	-

3.   References

Agaphanov, M.O., Beburov, M.Y., Ter-Avanesyan, M.D. and Smirnov, V.N. (1995) A disruption-replacement approach for the targeted integration of foreign genes in *Hansenula polymorpha*. Yeast 11, 1241-1247.

Anamnart, S., Tolstorukov, I., Kaneko, Y. and Harashima, S. (1998) Fatty acid desaturation in methylotrophic yeast *Hansenula polymorpha* strain CBS 1976 and unsaturated fatty acid auxotrophic mutants. J. Ferment. Bioeng. 85, 476-482.

Chirala, S.S. (1992) Coordinated regulation and inositol-mediated and fatty acid-mediated repression of fatty acid synthase genes in *Saccharomyces cerevisiae*. Proc. Natl. Acad. Sci. USA 89, 10232-10236.

Faber, K.N., Harder, W., Ab, G. and Veehuis, M. (1995) Review: Methylotrophic yeasts as factories for the production of foreign proteins. Yeast 11, 1331-1344.

Gellissen, G. (2002) *Hansenula polymorpha*, biology and applications. Wiley-VCH, Weinheim.

Kajiwara, S., Oura, T. and Shishido, K. (2001) Cloning of a fatty acid synthase component *FAS1* gene from *Saccharomyces kluyveri* and its functional complementation of *S. cerevisiae fas1* mutant. Yeast 18, 1339-1345.

Kottig, H., Rottner, G., Beck, K.F., Schweizer, M. and Schweizer, E. (1991) The pentafunctional *FAS1* genes of *Saccharomyces cerevisiae* and *Yarrowia lipolytica* are co-linear and considerably longer than previous estimated. Mol. Gen. Genet. 226, 310-314.

Niwa, H., Katayama, E., Yanagida, M. and Morikawa, K. (1998) Cloning of the fatty acid synthetase β subunit from fission yeast, coexpression with the α subunit, and purification of the intact multifunctional enzyme complex. Protein Expression and Purification. 13, 403-413.

Schuller, H.J., Fortsch, B., Rautenstrauss, B., Wolf, D.H. and Schweizer, E. (1992) Differntial proteolytic sensitivity of yeast fatty acid synthetase subunit α and β contributing to a balanced ratio of both fatty acid synthetase components. Eur. J. Biochem. 203, 607-614.

Schweizer, E., Werkeneister, K. and Jain, M.K. (1978) Fatty acid biosynthesis in yeast. Mol. Cell. Biochem. 21, 95-106.

Stoops, J.K., Awad, E.S., Arslanian, M.J., Gunsberg, S., Wakil, S.J. and Oliver, R.M. (1978) Studies on the yeast fatty acid synthetase subunit composition and structural organization of a large multifunctional enzyme complex. J. Biol. Chem. 253, 4464-4475.

Weisner, P., Beck, J., Beck, K.F., Ripka, S., Muller, G., Luche, S. and Schweizer, E. (1988) Isolation and sequence analysis of the fatty acid synthetase *FAS2* gene from *Penicillium patulum*. Eur. J. Biochem. 177, 69-79.

Zhao, X.J. and Cihlar, R.L. (1994) Isolation and sequence of the *Candida albicans FAS1* gene. Gene 147, 119-124.

CHARACTERIZING KAS ENZYMES USING THE *E. COLI* DOUBLE MUTANT STRAIN CY244

ANNE VINTHER RASMUSSEN[1], RIE YASUNO[2] AND PENNY VON WETTSTEIN-KNOWLES[1]

[1] *Department of Genetics, Institute of Molecular Biology, University of Copenhagen, Oester Farimagsgade 2A, DK-1353 Copenhagen K, Denmark*

[2] *Department of Life sciences, 301A 15 Bldg, Graduate School of Arts and Sciences, Tokyo University, Kamaba3-8-1, Meguro-ku, Tokyo 153-8902, Japan*

1. Introduction

In *Escerichia coli* as in plants, fatty acid biosynthesis is catalyzed by discrete enzymes of the FAS II system. The Claisen condensation of C_2 units is carried out by β-ketoacyl ACP synthase (KAS) I and KAS II. These catalyze all elongation steps with the exception of the initial condensation of two C_2 units catalyzed by KAS III. *Ec*KAS I, however, is specific for elongation of $C_{10:1}$ making it essential for unsaturated fatty acid biosynthesis whereas *Ec*KAS II is specific for elongation of $C_{16:1}$ to $C_{18:1}$. In plants, KAS I catalyzes the elongation steps from C_4 to C_{16} while KAS II is responsible for elongation of C_{16} to C_{18}. In contrast to the bacterial FAS, double bonds are only introduced after elongation in plants. *Ec*KAS I and II differ in their sensitivities to cerulenin, an inhibitor of condensing enzymes, *Ec*KAS I being very sensitive while *Ec*KAS II is intermediately sensitive; the same pattern is observed for barley KAS I and II. Fatty acid biosynthesis in plants has also been revealed to take place in mitochondria and a KAS enzyme from *Arabidopsis thaliana* has been cloned and localized to the mitochondria (Yasuno, R *et al.*, *in preparation*). Both *E. coli* KAS'es have been expressed in soluble active form and their biochemical as well as basic structural characteristics have been revealed (Huang, W *et al.*, 1998, Olsen, JG *et al.*, 1999). With the exception of KAS II from *Synechocystis sp.* (Moche, M *et al.*, 2001) all other KAS I & II orthologs have proved recalcitrant to expression in soluble active form when overexpressed in *E. coli*. In this study, we have systematically explored a range of expression systems to try and overcome this problem for different KAS enzymes. By use of the *E. coli* temperature sensitive strain CY244 defective in fatty acid biosynthesis at 42°C we have characterized these enzymes by their ability to complement CY244 for growth at 42° *in vivo* and to restore elongation activity in CY244 soluble protein extracts at 42°C *in vitro*.

N. Murata et al. (eds.), Advanced Research on Plant Lipids, 65–68.

TABLE 1. KAS enzymes investigated in this study

Enzyme	Gene	Description	% Identity to EcKAS II
EcKAS I	fabB	E. coli KAS I	34.8 %
EcKAS II	fabF	E. coli KAS II	100 %
EcFabB/F	fabB/F	Hybrid E. coli KAS whose N-terminal is encoded by fabB and C-terminal by fabF.	-
EcFabF/B	fabF/B	Hybrid E. coli KAS whose N-terminal is encoded by fabF and C-terminal by fabB.	-
HvKAS I	HvKAS I	Hordeum vulgare KAS I, mature	43.8 %
HvKAS II	HvKAS II	H. vulgare KAS II, mature	44.5 %
AtmtKAS	AtmtKAS	Arabidopsis thaliana mitochondrial KAS, mature	46.9 %

2. Production of recombinant KAS enzymes

All the enzymes in Table 1 utilize a cysteine and two histidines for catalysis (Olsen, JG et al., 2001) as well as showing a high degree of identity and similarity. In addition to the E. coli and plant KAS enzymes, we constructed hybrid E. coli KAS enzymes splicing the N-terminal coding region of one gene to the C-terminal coding region of the other and vice versa, the splice site being between codons 242 and 243 using the number system of EcKAS I. The splice site represents the junction between the two structurally similar halves of the polypeptide (Olsen, JG et al., 2001). The resulting hypothetical enzyme will have the active site cysteine at position 163 encoded by the 5' fab segment while the 3' fab segment encodes the two histidines at positions 298 and 333 as well as the lysine at position 328 whose importance for the catalytic activity has also been revealed (McGuire, KA et al., 2001).

Except for the E. coli enzymes, all the others in Table 1 were detected solely as inclusion bodies when expression was carried out in E. coli M15 at 25°C with the pQE30 vector from Qiagen using conditions described previously (McGuire, KA et al., 2001). Reducing the induction temperature to 15°C and using E. coli C41(DE3) (Miroux, B et al., 1996) as the expression host resulted in some soluble protein, but no more than a maximum of 5%. To try and overcome this problem the KAS ORFs were cloned into the pMALcH vector (Pryor, KD et al., 1997) from which the recombinant protein of interest is expressed under control of the lac promoter as C-terminal fusions to maltose binding protein (MBP) with a 6xHis affinity tag between MBP and the fusion partner. When expression was carried out using E. coli C41(DE3) as the expression host, the fusion proteins were soluble to different degrees, but biochemically inactive even when the induction temperature was reduced to 15°C. HvKAS II was also subcloned to pPICZ from Invitrogen and expressed in the methylotrophic yeast Pichia pastoris following the recommandations supplied in the Invitrogen manual. KAS protein was recovered exclusively in the insoluble fraction as opposed to EcKAS II that was soluble when expressed in this system.

3. Complementation assays

The genes encoding the KAS enzymes listed in Table 1 were subcloned into pKK233-2 which confers resistance to ampicillin (Amp) and transformed into E. coli CY244.

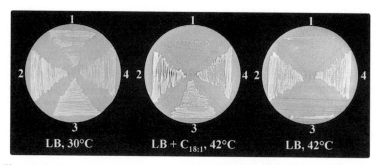

Figure 1. *In vivo* complementation of *E. coli* CY244 on LB + Amp ± C18:1 by *Hv*KAS II.
1 = pKK233-2, 2 = *Ec*KAS I, 3 = *Ec*KAS II, 4 = *Hv*KAS II.

(Ulrich, AK *et al.*, 1983). This strain is *fabF⁻* (Ser220Asn, Gly262Met), *fabB^{ts}* (Ala329Val) making it suitable for complementation studies with recombinant KAS enzymes.

3.1. In vivo *complementation of* E. coli *CY244*

The recombinant KAS enzymes were assayed for their ability to restore growth of CY244 on LB + Amp and LB + Amp supplemented with 0.1mg/mL $C_{18:1}$ at 42°C. *Ec*KAS I activity will restore growth under both conditions while *Ec*KAS II activity will restore growth only when supplemented with $C_{18:1}$ since this strain is impaired in unsaturated fatty acid biosynthesis. *At*mtKAS complements *Ec*KAS II by supplying the missing C8 and C14 chains required for lipoic acid and lipid A (Yasuno, R *et al.*, *in preparation*). Surprisingly, *Ec*KAS I was complemented by *Hv*KAS II (Figure 1) and *Ec*FabF/B (not shown). These results suggest that *Hv*KAS II has some hitherto unexpected attributes with respect to substrate specificity. From the recently revealed structure of *Ec*KAS I complexed with fatty acids (Olsen, JG *et al.*, 2001) it is clear that the residues surrounding most of the fatty acid binding pocket are encoded by the N-terminal subunit. Only at the base of the pocket do any residues from the C-terminal unit come close. Unfortunately structural studies have this far failed to reveal the basis for substrate specificity differences between *Ec*KAS I and II (Olsen, JG *et al.*, 2001). *Ec*FabB/F and *Hv*KAS I complemented neither of the *E. coli* KAS enzymes *in vivo*.

3.2 *Elongation assays:* In vitro *complementation of* E. coli *CY244*

The KAS enzymes were tested for their ability to restore elongation activity to CY244 soluble protein extracts at 42°C, as described previously (McGuire, KA *et al.*, 2001). Briefly, to soluble protein extracts made from CY244 transformed with pKK233-2 containing the appropriate KAS gene and grown at 30°C was added a mix including ^{14}C-malonyl-CoA and acetyl-CoA. After 20 minutes at 42°C acyl-ACPs were recovered by acid precipitation, resolved using urea-PAGE, transferred to a PVDF membrane and visualized by PhophorImaging. All recombinant KAS'es were able to complement the CY244 extract in vitro at 42°C as illustrated in Figure 2. Compare the 0 lanes to the V (vector only) lane. To confirm that the observed activities were attributable to the recombinant KAS their sensitivity to cerulenin, an antibiotic that

68

binds covalently to the active site cysteine in KAS, was investigated. As expected, *Ec*KAS I was very sensitive to cerulenin whereas *Ec*KAS II required 10–20 fold higher concentrations of the antibiotic for complete inhibition (Figure 2). Also as expected, an analogous difference was observed for *Hv*KAS I and *Hv*KAS II (not shown). The hybrid KAS'es, interestingly, both showed intermediate sensitivity as illustrated for *Ec*FabF/B, although the complement of acyl-ACPs is different from that characterizing either *Ec*KAS I or *Ec*KAS II. Namely, the hybrid has increased amounts of C_{12}-ACP *vs* C_{16}-, C_{18}- and longer acyl-ACPs.

While the present assay can only test for *Ec*KAS I and II activities vital for *E. coli*, addition of suitable substrates and /or expression constructs will make it a valuable tool for studying a diverse range of KAS enzymes.

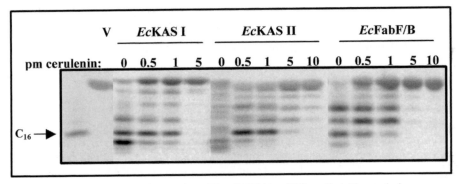

Figure 2. *In vitro* complementation of *E. coli* CY244 at 42°C as affected by cerulenin plus *Ec*KAS I, *Ec*KAS II and *Ec*FabF/B. V = pKK233-2.

References

W. Huang, J. Jia, P. Edwards, K. Dehesh, G. Schneider, and Y. Lindqvist. Crystal structure of β-ketoacyl-acyl carrier protein synthase II from *E. coli* reveals the molecular architecture of condensing enzymes. *EMBO J.* 17:1183-1191, 1998.

K. A. McGuire, M. Siggaard-Andersen, M. G. Bangera, J. G. Olsen and P. Wettstein-Knowles. β-Ketoacyl-(acyl carrier protein) synthase I of *Escherichia coli*: Aspects of the condensation mechanism revealed by analyses of mutations in the active site pocket. *Biochemistry* 40:9836-9845, 2001.

B. Miroux and J. E. Walker. Over-production of proteins in *Escherichia coli*: mutant hosts that allow synthesis of some membrane proteins and globular proteins at high levels. *J.Mol.Biol.* 260:289-298, 1996.

M. Moche, K. Dehesh, P. Edwards and Y. Lindqvist. The crystal structure of β-ketoacyl-acyl carrier protein synthase II from *Synechocystis sp.* at 1.54 A resolution and its relationship to other condensing enzymes. *J.Mol.Biol.* 305:491-503, 2001.

J. G. Olsen, A. Kadziola, P. Wettstein-Knowles, M. Siggaard-Andersen, Y. Lindquist and S. Larsen. The X-ray crystal structure of β-ketoacyl [acyl carrier protein] synthase I. *FEBS Lett.* 460:46-52, 1999.

J. G. Olsen, A. Kadziola, P. Wettstein-Knowles, M. Siggaard-Andersen and S. Larsen. Structures of β-ketoacyl-acyl carrier protein synthase I complexed with fatty acids elucidate its catalytic machinery. *Structure* 9:233-243, 2001.

K. D. Pryor and B. Leiting. High-level expression of soluble protein in *Escherichia coli* using a His6-tag and maltose-binding-protein double-affinity fusion system. *Protein Expr.Purif.* 10:309-319, 1997.

A. K. Ulrich, D. de Mendoza, J. L. Garwin and J. E. Cronan, Jr. Genetic and biochemical analyses of *Escherichia coli* mutants altered in the temperature-dependent regulation of membrane lipid composition. *J.Bacteriol.* 154:221-230, 1983.

R. Yasuno, P. von Wettstein-Knowles and H. Wada. Mitochondrial fatty acid synthesis: Identification of an *Arabidopsis* cDNA for mitochondrial β-ketoacyl-acyl carrier protein synthase. In preparation.

CARBON METABOLISM IN DEVELOPING SOYBEAN EMBRYOS

M.J. PIKE, C.J. EVERETT, S. RAWSTHORNE
Department of Metabolic Biology
John Innes Centre, Colney Lane, Norwich, NR4 7UH, UK

Introduction

Soybean (*Glycine max*) is an agronomically important leguminous crop that is rich in seed protein (40%) and oil (20%). Breeding efforts to increase soybean oil from a 20% level have been unsuccessful as oil content and seed yield have a negative relationship (Burton, 1985). Using a biochemical approach, some understanding of the pathways of carbon flux to fatty acids has been made using isolated plastids from oilseed rape (Kang and Rawsthorne, 1996; Eastmond and Rawsthorne, 2000). Eastmond and Rawsthorne (2000) have shown that in oilseed rape embryos the accumulation of storage products (protein, lipid and starch) changes during development and carbon flux from different metabolites into different pathways in the plastids changes accordingly. Using similar techniques as those used for oilseed rape (Kang and Rawsthorne 1996; Eastmond and Rawsthorne, 2000). This article presents results on the carbon substrates that are important both directly and indirectly for fatty acid synthesis in soybean embryo plastids.

Materials and Methods

Plant material, reagents and radiochemicals
Soybean plants (*Glycine max* [L.] cv A3244) were grown in a glasshouse environment with 22°C night/26°C day temperature, 16 hour photo-period with supplementary lighting provided between October and March. Plants were tagged for flowering time and pods were harvested from day 15 – 60 at three day intervals for protein, lipid and starch analysis. At each harvest, embryos were weighed and stored at –20°C until being analysed for storage products. Embryos were harvested between 25 and 35 DAF (100 – 200 mg embryos fresh weight) for plastid experiments. Sources of chemicals and reagents were as described in Eastmond and Rawsthorne, (2000). Radiolabelled substrates were obtained from NEN Life Science Products, (Houndslow, UK).

N. Murata et al. (eds.), Advanced Research on Plant Lipids, 69–71.
© 2003 *Kluwer Academic Publishers. Printed in the Netherlands.*

Plastid preparation and incubation of plastids with radiolabelled substrates
Preparation of soybean plastids was as described in Fox et al, (2000). For soybean, measurement of marker enzyme recovery and latency was as described in Kang and Rawsthorne, (2000) using NADP-GAPDH for recovery experiments and alkaline pyrophosphatase for latency (Tetlow et al, 1993). Plastid feeding experiments and measurement of incorporation of carbon into fatty acids and the release of CO_2 via the oxidative pentose phosphate pathway according to Eastmond and Rawsthorne, (2000).

Results and Discussion

Substrate utilization for fatty acid synthesis by isolated plastids
The storage product profile of developing soybean embryos was used to define an appropriate time for preparation of plastids particularly in relation to the maximal rate of oil accumulation. Plastids were prepared using the method of Fox et al (2000) using embryos of 100 – 200 mg weight. At this stage plastids are actively producing fatty acids but this is prior to the main period of starch deposition. The presence of starch was found to be detrimental to plastid preparation. To determine the substrates important for fatty acid synthesis, plastids were incubated with ^{14}C-labelled Glc-6-P, malate, pyruvate, phosphoenoylpyruvate (PEP) or acetate and the rate of incorporation determined. Of the substrates supplied, carbon from pyruvate was found to be most readily incorporated into fatty acids accounting for up to 46% of the rate of fatty acid synthesis required in vivo. Similarly, pyruvate has been shown to be an important carbon source for fatty acids in plastids isolated from oilseed rape (Eastmond and Rawsthorne, 2000) and flax embryos (Troufflard et al., this volume). Pyruvate incorporation was shown to be concentration dependent and saturated at 2 mM providing indirect evidence that a pyruvate transporter is present in soybean plastids. Other carbon sources were unable to support substantial rates of fatty acid synthesis except for acetate, which accounted for 13% of calculated in vivo rates (Table 1).

TABLE 1. Incorporation of ^{14}C-substrates into fatty acids in soybean embryo plastids
Values are the mean ± SE of measurements made on three separate plastid preparations
or the mean of two similar experiments where the SE is not given.

Substrate	nmol acetate.h^{-1}.embryo	% of in vivo rate
Glc-6-P	2.7 ± 0.8	3
Pyruvate	58.9 ± 16.7	46
PEP	6.9	6
Malate	3.4	3
Acetate	19.5 ± 7.4	15

Flux of Gl-6-P through the oxidative pentose phosphate pathway for fatty acid synthesis
In oilseed rape plastids the flux of Glc-6-P through the oxidative pentose phosphate (OPP) pathway provides reducing power for fatty acid synthesis in the form of NADPH (Kang and Rawsthorne, 1996). Similarly, for soybean embryo plastids flux of Glc-6-P through the OPP pathway could be measured by the release of $^{14}CO_2$ from [1-^{14}]Glc-6-P. When pyruvate was added to the Glc-6-P containing incubation media as a carbon source for fatty acid synthesis the release of $^{14}CO_2$ was stimulated by up to 48%. These experiments provide direct evidence that the OPP pathway is operating in soybean embryo plastids and can provide reducing power for fatty acid synthesis. The presence of an active plastidial OPP pathway has also been demonstrated for plastids isolated from *Cuphea* seeds (Heise and Fuhrmann, 1994).

References

Burton, J.W. (1985) Breeding soybeans for improved protein quantity and quality. In Shibles (ed.), Proceedings of the 3rd World Soybean Research Conference. Westview Press, Boulder and London, pp. 361-367

Eastmond P.J. and Rawsthorne S. (2000) Coordinate changes in carbon partitioning and plastidial metabolism during the development of oilseed rape embryos (*Brassica napus* L.). Plant Physiology **122**, 767-774

Fox S.R., Hill I.M., Rawsthorne S. and Hills M.J. (2000). Inhibition of the glucose-6-phosphate transporter in oilseed rape (*Brassica napus* L.) plastids by acyl-CoA thioesters reduces fatty acid synthesis. Biochemical Journal **352**, 525-532

Heise K.P. and Fuhrmann (1994) Factors controlling medium-chain fatty acid synthesis in plastids from *Cuphea* embryos. Progress in Lipid Research **33**, 87-95

Kang F. and Rawsthorne S, (1996) Metabolism of glucose-6-phosphate and utilization of multiple metabolites for fatty acid synthesis by plastids from developing oilseed rape embryos. Planta **199**, 321-327

Tetlow I.J, Blissett K.J. and Emes M.J. (1993) A rapid method for the isolation of purified amyloplasts from wheat endosperm. Planta **189**, 597-600

EXPRESSION OF ANTISENSE ACYL CARRIER PROTEIN-4 (LMI-ACP) REDUCES LIPID CONTENT IN *ARABIDOPSIS* LEAF TISSUE

J.K. BRANEN, M. KWON, N.J. ENGESETH
University of Illinois, Department of Food Science and Human Nutrition
259 ERML, 1201 W. Gregory Drive, Urbana, IL 61801 USA

1. Abstract

Arabidopsis plants were transformed with ACP4 in antisense conformation and driven by the Cauliflower Mosaic Virus 35S promoter (CaMV 35S). Our hypothesis was that reduction of ACP4 in leaf tissue will result in a reduction in lipid biosynthesis and in addition may affect fatty acid composition and leaf physiology. Several transgenic lines have been generated with reduced ACP4 protein in leaf tissue. Large reductions in ACP4 resulted in a reduction in leaf lipid content (22-42%) based on fresh leaf weight, as well as a bleached appearance and reduced photosynthetic efficiency. In addition, a decrease in 16:3 as a percentage of the total fatty acid composition was noted. Preliminary data does not indicate any noticeable changes in the leaf lipid class distribution; however, the fatty acid composition of the individual lipids as well as the activity of glycerolipid assembly enzymes is under investigation.

2. Introduction

Acyl carrier protein (ACP) is a critical cofactor in fatty acid biosynthesis. This small (9 kDa) acidic protein carries growing acyl chains through the various enzymatic steps in fatty acid biosynthesis. All higher plants that have been studied contain several isoforms of ACP, some of which are expressed constitutively (in all tissues) and others that are expressed in a tissue specific manner (Battey and Ohlrogge, 1990). For example, *Arabidopsis* has at least five isoforms of ACP displaying a range of tissue specific and constitutive expression.

Little is known about the functional role of various ACP isoforms in plant lipid and fatty acid biosynthesis. One of the goals of our research is to further understand the role of ACP isoforms in fatty acid biosynthesis, particularly their role in determining fatty acid content and composition of plant lipids. In order to reach this goal, we are conducting a number of *in vivo* manipulations of *Arabidopsis* ACP isoforms through sense and antisense expression. This report represents the results obtained by decreasing ACP4, the most prominent ACP isoform in *Arabidopsis* leaf tissue, through antisense expression.

N. Murata et al. (eds.), Advanced Research on Plant Lipids, 73–76.
© 2003 *Kluwer Academic Publishers. Printed in the Netherlands.*

3. Materials and Methods

Arabidopsis thaliana ecotype Columbia was grown at 22°C with a 16 hr day/8 hr night cycle. To prepare the DNA construct for plant transformation, ACP4 cDNA was cloned in antisense conformation into p1079 behind the CaMV 35S promoter. The 1.6 Kb *Not I* fragment containing 35S-ACP4 (antisense) was subcloned into the binary vector pART27 (Gleave, 1992) that contains *NPTII* to confer kanamycin resistance.

Leaf tissue from transgenic plants was analyzed for alterations in ACP levels by Western blot analysis. Development was with an anti-spinach ACP antibody detected with goat-anti-rabbit IgG-alkaline phosphatase conjugate. Intensity of bands was used for determination of relative levels of various ACP isoforms.

Total lipid was extracted from leaves by the method of Bligh and Dyer (1959). An aliquot of total lipid extract was dried and derivatized using methanolic HCL (3N). Fatty acid methyl esters were determined by gas chromatography with pentadecanoic acid (15:0) as an internal standard. Lipid classes were either separated by thin-layer chromatography (TLC) on silica gel plates with chloroform:methanol:acetic acid:H_2O (170:30:20:7) or on NH_4SO_4-impregnated silica gel plates with acetone:toluene:water (91:30:8). Bands were visualized with iodine vapor and were identified by comparing to known standards. Fatty acid composition of individual lipid classes was determined by scraping the bands from the TLC plate and preparing fatty acid methyl esters as described above. To determine Fv/Fm values, leaves were dark adapted for two minutes before fluorescence measurements were taken using a Hansatech fluorescence monitoring system FMS2 (Hansatech Instruments, Norfolk, UK).

4. Results and Discussion

Thirty-two independent transgenic plants were generated from the transformation of *Arabidopsis* with antisense ACP4. These plants will henceforth be referred to as pLMIA. Western blot analysis of protein from pLMIA leaves indicates that varying reductions in leaf ACP-4 were achieved (Figure 1).

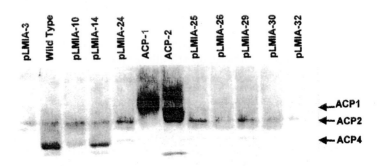

Figure 1 : Western blot from native PAGE of wild type and pLMIA leaf protein.

Plants that had the greatest reduction in ACP4 also displayed a bleached phenotype to varying degrees in leaf tissue (not shown). Transgenic plants with bleached phenotype and reduced ACP4 were found to have lower lipid content (22-42% lower lipid by fresh weight) compared to wild type plants (Figure 2).

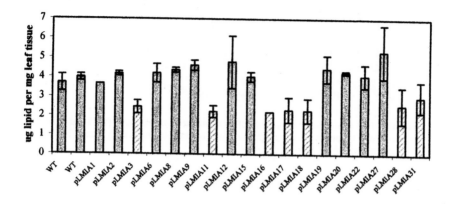

Figure 2: Lipid content of wild type (WT) and pLMIA leaf tissue. Values represent the average from two leaf samples from each plant. Hatched bars represent plants with bleached appearance.

In addition, those transgenic plants with lower lipid content also displayed reduced 16:3 as a percentage of the total fatty acids (Figure 3) compared to wild type leaves. Since 16:3 was lower in these transgenic plants, we hypothesized that there may be a decrease in monogalactosyl diacylglycerol in lipids from these plants. However, preliminary analysis of leaf lipid classes by TLC did not reveal any noticeable alterations in leaf lipids of transgenic plants compared to wild type plants (data not shown). Since plants with reduced ACP4 and lower lipid content were bleached, we measured the maximum quantum yield (Fv/Fm) of dark-adapted leaves to determine the impact of lowered leaf lipid on photosynthesis. Wild type *Arabidopsis* leaves had an average dark-adapted Fv/Fm value of 0.850, as did the pLMIA plants that were not dramatically reduced in ACP4 and lipid content. However, for those plants with low ACP4 lipid content and bleached phenotype the Fv/Fm average was 0.796 (data not shown).

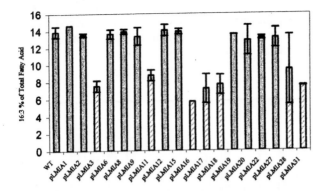

Figure 3: 16:3 content (as a percentage of total fatty acid) of wild type (WT) and pLMIA leaf tissue. Values represent the average from two leaf samples from each plant. Hatched bars represent plants with bleached appearance.

5. Conclusions

Antisense expression of ACP4 in *Arabidopsis* resulted in decreased ACP4 protein in the leaf tissue. As a result, there was a decrease in total leaf lipids. In addition, the fatty acid composition was altered, characterized mainly by a decrease in the percentage of 16:3. This may be due either to a decrease in monogalactosyl diacylglyceride (MGDG) content, which is the only leaf lipid with a significant portion of 16:3 in its fatty acids, or to an alteration in the ratio of MGDG packaged via the prokaryotic pathway vs the eukaryotic pathway. We are currently in the process of investigating microsomal and chloroplast glycerol-3-phosphate acyl transferase and lysophosphatidylcholine acyl transferase activity to further elucidate the effects of decreased fatty acid substrate on the activity of these enzymes. This may lead to a better understanding of the regulation of these enzymes in lipid biosynthesis.

Another consequence of reduction in leaf ACP4 was a bleached appearance and a decrease in photosynthetic efficiency. We are not yet sure whether this is due to the overall decrease in lipid or to the decrease in 16:3. We are planning to further investigate the photosynthetic capabilities of these transgenic plants to further understand the effect of altered lipid on the photosynthetic machinery.

6. References

Battey, J.F., Ohlrogge, J.B. 1990. Evolutionary and tissue specific control of expression of acyl carrier protein isoforms in plants and bacteria. Planta 180:352-360.

Bligh, E.G., Dyer, W.J. 1959. A rapid method of total lipid extraction and purification. Can. J. Biochem. Physiol. 37: 911-917.

Gleave, A.P. 1992. A versatile binary vector system with a T-DNA organizational structure conducive to efficient integration of cloned DNA into the plant genome. Plant Mol. Biol. 20: 1203-1207.

BIOSYNTHESIS OF 9,12,15–OCTADECATRIEN–6–YNOIC ACID IN THE MOSS *DICRANUM SCOPARIUM*

I.A. GUSCHINA [1,3], G. DOBSON [2] AND J.L. HARWOOD [3]

[1] *Institute of Ecology of the Volga River Basin RAS, Togliatti 445003, Russia*

[2] *Scottish Crop Research Institute, Invergowrie, Dundee DD2 5DA, Scotland, U.K.*

[3] *School of Biosciences, Cardiff University, PO Box 911, Cardiff CF10 3US, U.K.*

1. Introduction

In comparison to higher vascular plants, bryophytes are notable for their ability to synthesise very long chain (>18C) polyunsaturated fatty acids and, in some cases, triacylglycerols with acetylenic fatty acids. The acetylenic acids, 9, 12–octadecadien–6–ynoic acid (18:2A),
9, 12, 15–octadecatrien–6–ynoic acid (18:3A) and 11, 14–eicosadien–8–ynoic acid (20:2A) have been isolated from various mosses (e.g. Jamieson and Reid, 1976; Kohn *et al.*, 1987a) while 9–octadecen–6–ynoic acid (18:1A) has been identified in some liverwort species (Kohn *et al.*, 1987b).

Pathways for the formation of 18:2A and 18:3A have been examined in the moss, *Ceratodon purpureus* (Kohn *et al.*, 1994). In higher plants, the formation of crepenynic acid was first studied in seeds of *Crepis rubra* (Haigh *et al.*, 1968). Acetylenase enzymes have been isolated from *Crepis alpina* (Lee *et al.*, 1998) and a bifunctional desaturase/acetylenase from *Ceratodon purpureus* was found to be a member of the growing cytochrome b_5 family of fusion desaturases (Sperling *et al.*, 2000).

In mosses, the acetylenic acids are located exclusively in TAGs, although Kohn *et al.* (1994) were able to detect very small amounts of radiolabelled acetylenic acids in PtdCho following labelling from radioactive linoleate. This parallels the situation in higher plants where many unusual fatty acids are thought to be made on PtdCho and then transferred to TAGs (see Voelker and Kinney, 2001). However, it is possible that TAG–located desaturation also occurs in mosses (Kohn *et al.*, 1994; Beutelmann and Menzel, 1997).

As part of our research concerning environmental effects on lipid metabolism in mosses, we have been analysing species with acetylenic acids. Here we describe experiments on the biosynthesis of 18:3A in *Dicranum scoparium*.

2. Materials and Methods

N. Murata et al. (eds.), Advanced Research on Plant Lipids, 77–80.
© 2003 *Kluwer Academic Publishers. Printed in the Netherlands.*

2.1 Incubations

Moss samples were incubated with [1-^{14}C]acetate at 20°C under 100μE m^{-2} s^{-1} illumination. Pulse (30 min)–chase experiments were continued for up to 48h after rinsing the moss and continuing the incubation in water. After incubation, tissues were rinsed thoroughly and metabolism terminated with hot isopropanol.

2.2 Lipid analysis

Lipids were extracted by the method of Garbus et al. (1963). Lipid classes were separated by TLC and fatty acids by radio-GLC. For identification of acetylenic acids, FAME were converted to 4, 4–dimethyloxazoline derivatives (Christie, 1998) and analysed by GC-MS.

3   Results and Discussion

3.1 Acetylenic acids in D. scoparium

Dicranum scoparium contains 9, 12, 15–octadecatrien–6–ynoic acid (18:3A) in large quantities in its TAG fraction (up to 45% total fatty acids). Apart from DAG, we could not detect 18:3A in any other lipid class. The TAG fraction, itself, separated into two fractions on TLC. The faster–moving subfraction contained common fatty acids while the slower–moving subfraction contained >95% 18:3A.

3.2 Radiolabelling of lipid classes

When moss samples were pulse–labelled from [1-^{14}C]acetate, the polar lipids were rapidly labelled during the 30 min pulse. Increased labelling of neutrals was found during the chase period (up to 4h) whereas the total radioactivity in the polar lipids remained constant during this period. The label persisted in the total neutrals for up to 48h whereas that in polars gradually declined.

The total neutral and polar lipids were separated into classes by TLC. PtdCho was rapidly labelled during the pulse (45% total polars) and its radioactivity increased (to 65% of total polars) during 30 min of chase. After that, the radioactivity in PtdCho gradually declined. PtdGly was initially labelled well (30% total polars) and remained the second most labelled polar lipid after 48h. Other significantly labelled classes included PtdEtn, MGDG and DGDG.

The only neutral lipids which were well labelled during the pulse were DAG and the two TAG subfractions. During the chase period the radioactivity in DAGs declined from 40% total neutrals to around 10% within 4h. Labelling of the common–TAGs remained at around 25% of the total neutrals throughout the experimental period while the proportion of label in acetylenic TAGs rose from 24% after the pulse to 53% after 4h and 64% by 48h. Taking into consideration that the absolute neutral lipid labelling increased during the chase period (above), then the radioactivity in the acetylenic–TAGs rose about 15–fold. Thus, the total counts in acetylenic–TAGs increased from about 39x10^3 to about 600x10^3 d.p.m. per g fresh wt. This rise was almost equal to the decline in radioactivity in PtdCho (8.1x10^5 down to 3x10^5 d.p.m. per g fresh wt.).

3.3 Fatty acid radiolabelling

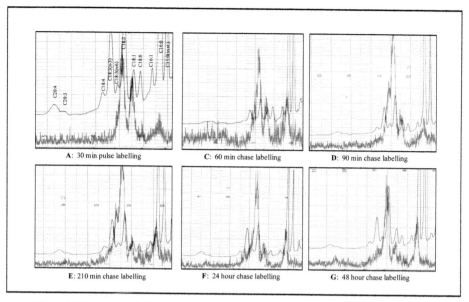

Figure 1. Pulse-chase labelling of phosphatidylcholine fatty acids in the moss *Dicranum scoparium*

The patterns of fatty acids labelled in different lipid classes were examined by radio–GLC. A typical time–course for the labelling of PtdCho is shown in Figure 1. In the initial pulse period, palmitate, oleate and linoleate were the main radioactive fatty acids and this was also true after 60 min of chase. Labelling of γ–linolenate was significant after 90 min and 210 min of chase but was barely detectable at longer times. These data are consistent with PtdCho being a substrate for Δ6–desaturation of linoleate to γ–linolenate, followed by transfer of the latter to other lipids.

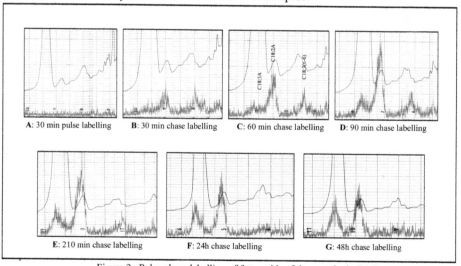

Figure 2. Pulse-chase labelling of fatty acids of the acetylenic-TAG subfraction in the moss *Dicranum scoparium*

When radiolabelling of the neutral lipid classes was examined, radiolabelled acetylenic fatty acids were only detected in TAGs, despite the presence of unlabelled 18:3A (and a trace of unlabelled 18:2A) in DAG. Traces for the time–course of radiolabelling of the acetylenic–TAG fraction are shown in Figure 2. As discussed before, radiolabelling of this fraction increased during the chase period when γ–linolenate and 18:2A were the first major labelled acids. After 60 min of chase, the first traces of radiolabelled 18:3A were detected and this increased with time as the radiolabelled γ–linolenate declined. After 48h chase, the amounts of radioactivity in 18:2A and 18:3A were approximately equal and no other fatty acids were significantly labelled.

3.4 Conclusions

Our data agree with the proposal of Kohn et al. (1994), using *C. purpureus*, that the main pathway for the synthesis of 18:3A is from γ–linolenate via 18:2A. However, our experiments clearly demonstrate the involvement of PtdCho only as a substrate for the first Δ6–desaturation of linoleate to γ–linolenate in *D. scoparium*. Although the exact nature of the substrate for the next step in the biosynthesis of 18:3A cannot be concluded from our results, the second Δ6–desaturation may occur on γ–linolenate after it has been removed from PtdCho and transferred to TAG. However, further experiments are necessary in order to establish unequivocally the details of these reactions.

Acknowledgement: supported by a Royal Society/NATO Fellowship to I.A.G.

4. References

Beutelmann, P. and Menzel, K. (1997) Degradation of acetylenic triacylglycerols and the inactivation of membrane preparations from moss protonema cells. In J.P. Williams, M.V. Khan and N.W. Lem (eds), Physiology, Biochemistry and Molecular Biology of Plant Lipids. Kluwer Academic Publishers, Dordrecht, pp. 253–255.

Haigh, W.C., Morris, L.J. and James, A.T. (1968) Acetylenic acid biosynthesis in *Crepis rubra*. Lipids 3, 307–312.

Jamieson, G.R. and Reid, E.H. (1976) Lipids of *Fontinales antipyretica*. Phytochemistry 15, 1731–1734.

Kohn, G., Demmerle, S., Vandekerkhove, O., Hartmann, E. and Beutelmann, P. (1987a) Distribution and chemotaxonomic significance of acetylenic fatty acids in mosses of the *Dicranales*. Phytochemistry 26, 2271–2275.

Kohn, G., Vierengel, A., Vanderkerkhove, O., Hartmann, E. and Beutelmann, P. (1987b) 9–Octadecen–6–ynoic acid from *Riccia fluitans*. Phytochemistry 26, 2101–2102.

Kohn, G., Hartmann, E., Symne, S. and Beutelmann, P. (1994) Biosynthesis of the acetylenic fatty acids in the moss *Ceratodon purpureus* (Hedw.) Brid. J. Plant Physiol. 144, 265–271.

Lee, M., Lenman, M., Banas, A., Bafor, M., Singh, S., Schweizer, M., Nilsson, R., Liljenberg, C., Dahlqvist, A., Gummeson, P.D., Sjodahl, S., Green, A. and Stymne, S. (1998) Identification of non–heme diiron proteins that catalyse triple bond and epoxy group formation. Science 280, 915–918.

Sperling, P., Lee, M., Girke, T., Zahringer, U., Stymne, S. and Heinz, E. (2000) A bifunctional Δ6–fatty acyl acetylenase/desaturase from the moss *Ceratodon purpureus*. Eur. J. Biochem. 267, 3801–3811.

Voelker, T. and Kinney, A.J. (2001) Variations in the biosynthesis of seed–storage lipids. Ann. Rev. Plant Physiol. Plant Mol. Biol. 52, 335–361.

THE ROLE OF THE PLASTIDIAL GLUCOSE 6-PHOSPHATE TRANSPORTER IN THE CONTROL OF FATTY ACID BIOSYNTHESIS OF DEVELOPING EMBRYOS OF OILSEED RAPE (*Brassica napus L.*)

P. C. NIELD, S. RAWSTHORNE, U-I FLÜGGE AND M. J. HILLS
Department of Metabolic Biology
John Innes Centre, Norwich Research Park, Norwich, Norfolk, UK

Abstract

The expression of the plastidial glucose 6-phosphate transporter was altered using RNAi to investigate the role that this transporter plays in the regulation of fatty acid biosynthesis in oilseed rape and Arabidopsis embryos. A number of transgenic lines of oilseed rape were produced, each showing a reduction in mature seed mass and germination rate. Two lines showed up to a 90% reduction in mature seed number and contain less than 50% oil (w/w) than wild type.   A reduction in total seed oil was also observed in transgenic Arabidopsis.

Introduction

The process of fatty acid biosynthesis is localized to the plastid of heterotrophic (non-photosynthetic) tissues (Murphy et al 1993).   Pyruvate is converted through acetyl-CoA, to acyl-ACP via the action of fatty acid synthase. The long chain fatty acid is then cleaved from the acyl-ACP and exported from the plastid as acyl-CoAs. These are then carried via acyl-CoA binding proteins (ACBP's) to the endoplasmic reticulum, for further processing into various lipids including storage as triacylglycerol (Hills et al 1994). Glc6P enters the plastid via the glucose 6-phosphate transporter (GPT) supplying substrate for starch biosynthesis, glycolysis, yielding pyruvate, and the oxidative pentose phosphate pathway (OPPP), which amongst other products, yields reducing power in the form of NADPH which is required by the fatty acid synthase. Pyruvate may also enter the plastid either directly or as phosphoenol pyruvate (PEP) via a PEP transporter (PPT). (Fischer et al 1997).

It has been found that in vitro if the concentration of long chain acyl-CoA exceeds that of ACBP then the action of the GPT is inhibited by up to 80% (IC_{50} = 200nM) (Fox et al 2000). This suggested that the GPT may be important in regulation of oil synthesis. To investigate the importance of

N. Murata et al. (eds.), Advanced Research on Plant Lipids, 81–84.

Glc6P in controlling flux to oil *in vivo* we have created RNAi lines of both *Brassica napus* L. and *Arabidopsis thaliana* to reduce expression of the plastidial GPT.

Experimental

The RNAi construct was produced in the binary vector pBin19 utilizing the seed specific oleosin promoter and the nos terminator, driving expression of a 420bp fragment of *Brassica napus* L. GPT DNA in the antisense orientation. This fragment was also >90% identical to GPT genes identified in *Arabidopsis thaliana* so therefore should be effective in reducing expression in both species. Plants of *Brassica napus* L. were transformed using the cotyledonary method (Maloney et al 1989) and Arabidopsis was transformed by floral dip (Clough and Bent 1998).

Seed mass was measured using a Mettler Toledo (AG104) balance using random sampling of each batch of seed to give an average result. Oil analysis was performed using an Oxford (QP20+) NMR machine on both Arabidopsis and oilseed rape according to the supplied instructions. Fatty acid composition of the oil was measured by FAMES analysis using gas chromatography (Perkin Elmer Autosystem) and analyzed using Turbochrom (Perkin Elmer) software.

Results

Seven lines of transgenic *Brassica* have been created using the RNAi construct. All these lines show a reduction in mature seed mass of up to 20% compared to wild type (fig 1).

RNAi line	seed mass (mg)		Wildtype (%)	germination rate (%)
A	3.24	± 7.87	81.8	7
B	3.13	± 7.60	79.0	8
C	3.06	± 7.44	77.4	49
D	3.19	± 7.75	80.6	52
E	3.38	± 8.21	85.3	59
F	3.54	± 8.60	89.4	57
G	3.38	± 8.20	85.2	61
H	3.55	± 8.62	89.6	63
I	3.69	± 8.97	93.2	55
Wild Type	3.96	± 9.62	100.0	92

Figure 1: *Brassica* T1 seed data

The first two lines created (A and B) also show a 90% reduction in the number of mature seeds that are produced. This severe phenotype is thought to be related to the copy number of the RNAi construct. Plant A contains 4 constructs and plant B contains

5. The remaining transgenic lines contain single inserts, and do not have any noticeable effect on mature seed number. Germination rates of the seeds harvested from all these plants shows a reduction in competence of up to 93% compared to wild type (fig 1). Oil measurement of the T1 seeds harvested from these lines shows a decrease in seed oil content of up to 62% (fig 2).

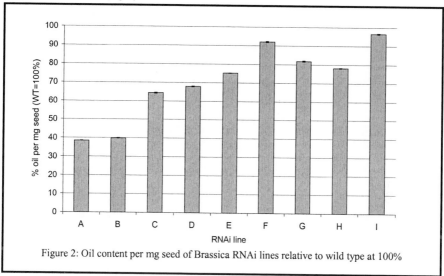

Figure 2: Oil content per mg seed of Brassica RNAi lines relative to wild type at 100%

All lines show a distinct phenotype where the germinating seedling has cotyledons which are reduced in size compared to wild type. In extreme cases the cotyledons are absent altogether (fig 3).

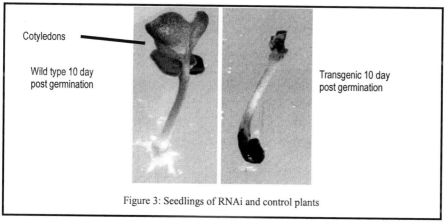

Figure 3: Seedlings of RNAi and control plants

This does not appear to affect the growth of the seedling and primary leaves develop 15 days post germination and the plant establishes at the same rate as a wild type. The mature plants of all lines are identical to wild type.

Preliminary results from Arabidopsis transformants indicate that the lines carrying the

RNAi construct contain less oil (w/w) than wild type (fig 4).

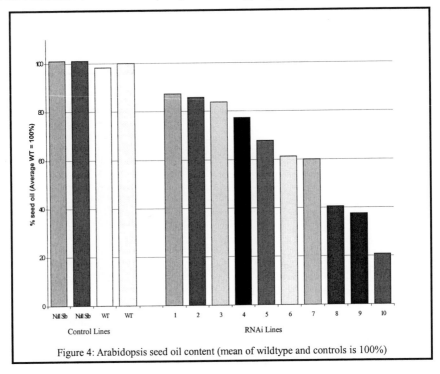

Figure 4: Arabidopsis seed oil content (mean of wildtype and controls is 100%)

Discussion

These results suggest that reducing the expression of the plastidial GPT results in reduced oil synthesis in the developing embryo of both *Brassica napus* L. and *Arabidopsis thaliana*, though lipid composition is unaffected (data not shown). Other storage products such as starch, sugar and protein may be affected as these all use glucose 6-phosphate as a part of their biosynthesis. This is to be investigated further.

References

Murphy D.J, Rawsthorne S. Hills M.J. (1993) Seed Science Research 3: 79-95

Hills M.J, Dann R, Lydiate D, Sharpe A. (1994) Plant Molecular Biology 25: 917-920

Fischer K et al (1997) The Plant Cell 9: 453-462

Fox S.R et al (2000) Biochem J. **352**:525-532

Maloney M.M., Walker J.M., Sharma K.K. (1989) Plant Cell Reports, Vol. 8, No. 4, pp 238-242

Clough S.J, Bent A.F (1998) The Plant Journal 16(6):735-743

COMPARATIVE ANALYSIS OF EXPRESSED SEQUENCE TAGS FROM DEVELOPING SEEDS OF *SESAMUM INDICUM* AND *ARABIDOPSIS THALIANA*

[1]M. C. SUH, [1]J. M. BAE, and [2]JOHN B. OHLROGGE

[1]Graduate School of Biotechnology, Korea University, Seoul 136-701, South-Korea; [2]Department of Plant Biology, Michigan State University, East Lansing, MI 48824, USA

Introduction

Sesame (*Sesamum indicum* L.) belongs to the Pedaliaceae family is an annual and self-pollinated non-green oilseed crop. The seeds produce approximately 50% oil of their dry weight and are mostly consumed as edible oil or an additive in food for special flavor. Although oil content and composition of cultivated sesame seeds was known, the metabolic pathways and the related genes responsible for accumulation of the seed-storage lipids from photosynthate, and the regulatory mechanism that determines the partitioning of seed reserves into the storage lipid are little understood.

Oilseed species could be divided into two groups depending on whether the seeds are green or non-green during development. Green seeds contain photosynthetic plastids to generate reducing power required for biosynthesis of storage lipid by photosynthesis. Whereas, non-green seeds have simple structure similar to proplastids (Browse and Slack, 1985). Despite the generation of reducing power independent on photosynthesis is important as well as carbon flow for the accumulation of storage lipid, little genetic studies for the generation of cofactors were reported in non-green seeds.

N. Murata et al. (eds.), Advanced Research on Plant Lipids, 85–88.

In this study, 3372 expressed sequence tags (ESTs) were generated from developing seeds of sesame and compared with *Arabidopsis* developing seed ESTs.

Results and Discussion

1. Generation and analysis of sesame developing seed ESTs

To investigate the regulatory mechanism of metabolic pathways of photosynthate into the storage lipid during sesame seed development, 3372 expressed sequence tags (ESTs) from 5- to 30-days-old immature seeds were developed by 5'end single-pass sequencing. Based on sequence homology analyses against NCBI BLASTX and *Arabidopsis* Genomic database systems, approximately 15% of sesame ESTs has currently no match or low similarity in both databases. Approximately 2,800 sesame ESTs corresponding to *Arabidopsis* genome sequence were classified in functional categories assigned by Munich bioinformatics center and comparatively analyzed with 10,240 *Arabidopsis* developing seed ESTs.

2. Comparative analyses of developing seed ESTs between sesame and *A. thaliana*

To provide novel and seed-specific gene pool during sesame seed development, the number of ESTs from sesame and *Arabidopsis* was compared by hit ID of *Arabidopsis* genome. As resulted in *Arabidopsis* developing seed ESTs (White et al., 2000), the most abundant ESTs from sesame developing seeds encoded for seed storage proteins including 12S cruciferin (previously identified as a 11S globulin), 2S albumin, and legumin-like protein. The number of ESTs encoding oleosin and translation elongation factor (eEF-1 alpha chain) was significantly high in both ESTs. In contrast, both major latex protein that is a laticifer-specific protein, and metallothionein-like protein that is involved in metal detoxification and homeostasis were highly abundant in sesame ESTs rather than *Arabidopsis* ESTs. In addition, ESTs encoding chlrophyll-binding proteins (At1g29930) occurred only 2 times, whereas 5 ESTs encoding Rubisco small subunits (At1g67090) were detected in sesame developing seed ESTs. Low frequency of ESTs encoding chlrophyll-binding proteins in sesame ESTs than *Arabidopsis* ESTs is consistent with that sesame seeds contain non-photosynthetic plastids. However, relatively high expression of Rubisco small subunits in sesame seeds suggests that it might be involved in the recapture of CO_2 released from the conversion of pyruvate to acetyl-CoA by the plastid pyruvate dehydrogenase.

One interesting aspect from this comparison was that ESTs encoding plastidic pyruvate kinase are highly abundantly in *Arabidopsis* developing seeds, but not detected in sesame developing seeds. In contrast, cytosolic malate dehydrogenase (At5g43330) is highly abundant in sesame ESTs, but completely missing in *Arabidopsis* ESTs. According to Browse and Slack, (1985) and Eastmond et al., (1997), malate was more preferable substrate than pyruvate for fatty acid synthesis in developing seeds of safflower (*Carthamus tinctorius*) and castor endosperms (*Ricinus communis*). These genetic and biochemical results indicate that malate is a major substrate and could be transported to provide a reducing power as well as carbon source for fatty acid synthesis in the leucoplasts of non-green seeds (Fig. 1).

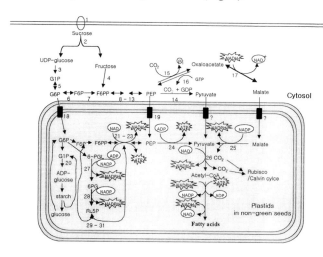

Fig 1. Simple diagram of metabolic pathway in the plastids of non-green oil seeds: the use of sucrose imported from photosynthetic tissues to generate cofactors as well as carbon source for fatty acid synthesis in the leucoplasts.

Conclusion

Although a variety of sesame cultivars have been developed by conventional breeding method, the low yield of sesame is the major obstacle to sesame cultivars' expansion over the world. The genetic improvement of sesame cultivars is strongly demanded for the generation of new varieties producing economically and nutritionally valuable oils. We summarized the similar or different gene expression between sesame and *Arabidopsis* developing seed ESTs as follows. 1. More than 15% of sesame ESTs might be sequence-specific genes or novel genes in sesame seeds. The seed storage proteins and oleosins are abundantly expressed in both sesame and *Arabidopsis* developing seeds, whereas most abundant ESTs encoding lipid transfer proteins, major latex protein, and metallothionein are significantly high in sesame rather than *Arabidopsis* seeds, implicating that expression of those genes is a specific characteristic of sesame developing seeds. 2. Although ESTs processing the same reaction are existed in both sesame and *Arabidopsis* ESTs, some specific isoforms (e.g. 12S cruciferin, oleosin, LTP, ACP, and so on) in multiple gene families are preferentially expressed during sesame or *Arabidopsis* seed development. 3. High expression of cytosolic malate dehydrogenase and no detection of plastidic pyruvate kinase in sesame ESTs suggest that malate is a major substrate for the generation of reducing power as well as carbon source required for storage lipid biosynthesis in non-green seeds. All the information of sesame developing seed ESTs including their nuclotide sequences, gene description annotated by BLASTX and *Arabidopsis* genome search, and comparative analyses of developing seed ESTs between sesame and *Arabidopsis* will be presented at our web site (http://plant.pdrc.re.kr).

References

Browse, J. and Slack, C.R. (1985) Fatty-acid synthesis in plastids from maturing safflower and linseed cotyledons. Planta 166, 74-80.

Eastmond, P.J., Dennis, D.T. and Rawsthorne, S. (1997) Evidence that a malate inorganic phosphate exchange translocator imports carbon across the leucoplast envelope for fatty acid synthesis in developing castor seed endosperm. Plant Physiol. 114, 851-856.

White, J.A., Todd, J., Newman, T., Focks, N., Girke, T., Martinez de Ilarduya, O., Jaworski, J.G., Ohlrogge, J.B. and Benning, C. (2000) A new set of *Arabidopsis* expressed sequence tags from developing seeds. The metabolic pathway from carbohydrates to seed oil. Plant Physiol. 124, 1582-1594.

SUBSTITUTION OF PHE[282] WITH SER IN LEA *B. napus* c.v. WESTAR RESTORES FAE1 ENZYME ACTIVITY

V. KATAVIC[1], E. MIETKIEWSKA[2,3], D. L. BARTON[1], E. M. GIBLIN[2], D. W. REED, D. C. TAYLOR[2]

[1]*Saskatchewan Wheat Pool Agricultural Research and Development, 201-407 Downey Road, Saskatoon, SK, S7N 4L8, Canada;*

[2]*National Research Council of Canada, Plant Biotechnology Institute, 110 Gymnasium Place, Saskatoon, SK, S7N 0W9, Canada;*

[3]*Plant Breeding and Acclimatization Institute, Mlochow, Poland.*

1. Abstract

We have isolated *fatty acid elongation 1* (*FAE1*) genomic clones from high erucic acid (HEA) B. *napus, B. rapa and B. oleracea,* and low erucic acid (LEA) B. *napus.* Nucleotide sequences corresponding to open reading frames of 1523 bp were translated and proteins of 506 a.a. were deduced. Comparative study of FAE1 protein sequences from HEA and LEA Brassicas revealed the one crucial amino acid difference: the serine residue at position 282 of the HEA FAE1 sequences is substituted by phenylalanine in LEA *B. napus.* Using site directed mutagenesis the phenylalanine[282] residue was substituted with a serine residue in FAE1 polypeptide from LEA *B. napus,* the mutated gene was expressed in yeast, and GC analysis showed presence of very long chain mono-unsaturated fatty acids indicating that the elongase activity was restored in the LEA FAE1 enzyme. Thus, for the first time, the low erucic acid trait in *B. napus* has been attributed to a single amino acid substitution which prevents the biosynthesis of very long chain mono-unsaturated fatty acids.

2. Introduction

In high erucic acid (HEA) *Brassicaceae,* a seed-specific fatty acid elongase 1 (FAE1) is the condensing enzyme (3-ketoacyl-CoA synthase; 3-KCS) that catalyzes the first of four enzymatic reactions of FAE complex, resulting in the synthesis of VLCMFAs which are the major constituents of their seed oil.

89

N. Murata et al. (eds.), Advanced Research on Plant Lipids, 89–92.

Intense research is ongoing by several groups to elucidate the mutations involved in the loss of 3-KCS activity in LEA *B. napus* cultivars. Han et al. (2001) speculated that the presence of serine at position 282 in all functional 3-KCS proteins instead of phenylalanine in non-functional LEA *B. napus* 3-KCS could be important for the activity of the condensing enzyme. Roscoe et al. (2001) hypothesized that the LEA phenotype could be the result of one or more lesions in the genes that encode or regulate 3-KCS activity. In order to clarify this controversy, we decided to examine the role of the amino acid serine at position 282 in the 3-KCS protein sequence to determine if this apparent mutation from serine to phenylalanine led to the LEA *B. napus* phenotype. We introduced a point mutation into the LEA *B. napus* cultivar Westar *FAE1* coding region to substitute phenylalanine with serine at position 282 in attempt to restore the FAE1 condensing enzyme activity.

Here we report and discuss the results of analyses of heterologous expression in yeast and site directed mutagenesis of LEA *Brassica napus* FAE1 KCSs.

3. Results

3.1. *Site directed mutagenesis of the LEA B. napus c.v. Westar FAE1 gene: expression in yeast shows the enzyme activity is restored in a mutated condensing enzyme*

In order to test the importance of serine[282] to 3-KCS function, using a site-directed mutagenesis (SDM) approach, we changed the phenylalanine[282] residue in LEA c.v. Westar FAE1 to the highly conserved serine residue. The sequence analyses of five different clones revealed that two of them (WS-SDM1 and WS-SDM18) had been successfully mutated with a Ser at position 282 (data not shown).

Due to the presence of hydrophobic phenylalanine in non-functional *B. napus* c.v. Westar wild-type 3-KCS (WS-wt) at position 282 instead of the hydrophilic serine in the mutated c.v. Westar 3-KCS (WS-SDM), the hydrophilicity values for

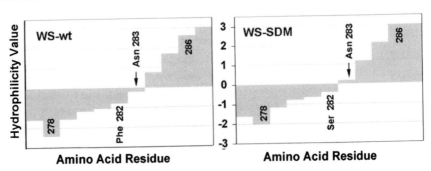

Figure 1. Hydrophilicity (Kyte-Doolittle) plot for the WS-wt and WS-SDM protein domains. Arrows indicate position 283 with asparagine (Asn) which is the last amino acid residue in the WS-wt hydrophobic domain and the first amino acid residue in the WS-SDM hydrophilic domain. The analyses were performed using Protean, Lasergene Biocomputing Software for Windows (DNAStar, Madison, WI). WS-wt, LEA *B. napus* c.v. Westar 3-KCS; WS-SDM, Westar site directed mutated 3-KCS.

amino acids in the transition region from highly hydrophobic to highly hydrophilic protein domains are significantly different in the non-functional WS-wt compared to the mutated WS-SDM. This resulted in different profiles of the hydrophilicity plots in the region spanning from alanine[278] to glycine[286] in WS-wt protein and the mutated WS-SDM, with asparagine[283] (hydrophilicity value −0.22) being shifted from the hydrophobic membrane domain in WS-Wt to the hydrophilic domain (hydrophilicity value +0.18) in WS-SDM (Figure 1).

We expressed wild-type c.v. Westar *FAE1* (WS-wt) and mutated clone WS-SDM1 in yeast cells. The results of fatty acid analyses of transformed yeast cell lysates by GC of the FAMEs revealed that 3-KCS activity was restored in mutated WS clone; yeast cells expressing mutated 3-KCS produced 20:1 Δ11 and Δ13 isomers, and 22:1 Δ13 and Δ15 isomers. In contrast, yeast cells expressing wild-type 3-KCS had fatty acid profiles typical of yeast, with no detectable mono-unsaturated VLCFAs present (Figure 2).

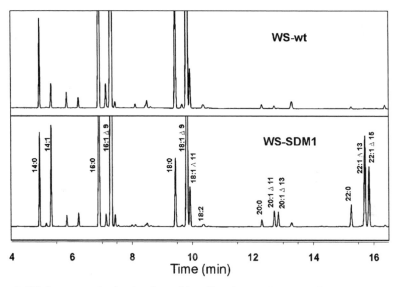

Figure 2. GC chromatographs showing fatty acid profiles of transgenic yeast cells. FAMEs were prepared from yeast cell lysates expressing FAE1 3-KCSs from LEA *B. napus* c.v. Westar wild-type and mutated FAE1 clone and analyzed by GC. WS-wt, LEA *B. napus* c.v. Westar wild-type; WS-SDM1, Westar site directed mutated clone 1.

3.2. *Both wild-type and mutated B. napus c.v. Westar FAE1 are translated in yeast*

In order to detect FAE1 3-KCS proteins in yeast cells expressing c.v. Westar wild-type FAE1 and two mutated c.v. Westar FAE1 clones western blot analyses were performed using microsomes isolated from yeast cells after FAE1 heterologous expression, and anti-FAE1 3-KCS antibodies raised against the *C*-terminus domain of the *Arabidopsis* FAE1 3-KCS protein. Protein bands corresponding to FAE1 KCS/V5-His fusion

92

were detected in all experimental samples except in the pYES2.1/V5-His-TOPO-only control (Figure 3).

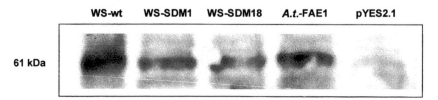

Figure 3. Immunoblot analysis of yeast microsomes expressing FAE1 3-KCSs. Proteins (100 μg per lane) from yeast expressing 3-KCS from wild-type LEA *B. napus* c.v. Westar, mutated c.v. Westar clones, *A. thaliana* and empty plasmid pYES2.1 were probed with Arabidopsis anti-FAE1 3-KCS antibodies. Detected protein bands correspond to the FAE1/V5-His fusion protein of *ca* 61kDa (56 kDa FAE1 + 5 kDa V5-His). WS-wt, LEA *B. napus* c.v. Westar 3-KCS; WS-SDM1, WS-SDM18, site directed mutated clones 1and 18; A.t-FAE1, *Arabidopsis thaliana* (positive control); pYES2.1, pYES2.1/V5-His-TOPO experimental negative control.

4.   Conclusions

The single-base change of nucleotide 845 from a T to a C residue, which resulted in substitution of a.a. residue phenylalanine (F) in FAE1 3-KCS from LEA c.v. Westar at position 282 with serine (S), led to successful restoration of elongase activity in a previously catalytically-inactive enzyme.

The analyses of translation rates in LEA *FAE1* 3-KCS demonstrate that the loss of activity is *not* due to reduced quantity or stability of the enzyme.

Based on changes in hydrophilicity values of amino acids in the vicinity of serine[282] in mutated (functional) Westar 3-KCS that affect the region in which the hydrophobic transmembrane domain (α-helix) ends and hydrophilic loop starts, we hypothesize that in LEA *B. napus* 3-KCS, phenylalanine at position 282 affects binding to the endoplasmic reticulum membrane and impairs function of the condensing enzyme in the fatty acid elongase complex.

To our knowledge, our study constitutes the first mutagenesis of catalytically inactive 3-ketoacyl-CoA-synthase from a low erucic acid *B. napus* (canola) cultivar and successful restoration of 3-KCS enzyme activity.

5.   References

Han, J., Lühs, W., Sonntag, K., Zähringer, U., Borchardt, D.S., Wolter, F.P., Heinz, E. and Frentzen, M. (2001) Functionl characterization of β- ketoacyl-CoA synthase genes from *Brassica napus* L. Plant Mol. Biol. 46, 229-239.

Roscoe, T.J., Lessire, R., Puyaubert, J., Renard, M. and Delseny, M. (2001) Mutations in the *fatty acid elongation 1* gene are associated with a loss of β-ketoacyl –CoA synthase activity in low erucic acid rapeseed. FEBS Letts. 492, 107-111.

Chapter 4:

Fatty Acid Desaturation

PRESENCE OF A PALMITOYL-CoA DELTA 6-DESATURASE IN THE PREPUTIAL GLAND OF THE MOUSE

M. MIYAZAKI*, F. E. GOMEZ* AND J. M. NTAMBI*#

Departments of *Biochemistry and #Nutritional Sciences,
University of Wisconsin, Madison, Wisconsin, 53706 USA

1. Introduction

Stearoyl-CoA desaturase (SCD) is a rate-limiting enzyme in the biosynthesis of monounsaturated fatty acids. It catalyzes the delta 9-cis desaturation of acyl-CoA substrates, the preferred substrates being palmitoyl (C16:0)-CoA and stearoyl(C18:0)-CoA, which are converted to palmitoleoyl (C16:1n-7)-CoA and oleoyl-CoA (C18:1n-9), respectively (1). These monounsaturated fatty acids are used as substrates for the synthesis of triglycerides, wax esters, cholesteryl esters and membrane phospholipids. The ratio of stearic acid to oleic acid has been implicated in the regulation of cell growth and differentiation through effects on membrane fluidity and signal transduction (2). Monounsaturated fatty acids also influence apoptosis (3) and may have some role in mutagenesis in some tumors (4, 5). Overall, SCD expression affects the fatty acid composition of membrane phospholipids, triglycerides and cholesterol ester, resulting in changes in membrane fluidity, lipid metabolism and obesity (2). Thus, the regulation of SCD is of considerable physiological importance and high SCD activity is associated with a wide range of diseases including cardiovascular disease, obesity, diabetes, neurological disease, skin disease and cancer.

Three mouse, two rat, and a single human gene are the best characterized SCD gene isoforms (6-9). The physiological role of each SCD isoform and the reason for having three or more SCD gene isoforms in the rodent genome are currently unknown. A clue as to the physiological role of the SCD, at least SCD1 gene and its endogenous products came from recent studies of asebia mouse strains that have a natural mutation in the SCD1 gene and a mouse model with a targeted disruption of the SCD1 gene (10, 11). The three mouse SCD genes are highly homologous at the nucleotide and amino acid level and encode the same functional protein (6, 7, 12). Most organs of different mouse strains express SCD1 and 2 with the exception of liver (7) and skin, which express mainly the SCD1 and SCD3 isoforms respectively (12). SCD2 is constitutively expressed in the brain (6). Despite the fact that the mouse SCD1, SCD2

N. Murata et al. (eds.), Advanced Research on Plant Lipids, 95–99.
© 2003 *Kluwer Academic Publishers. Printed in the Netherlands.*

and SCD3 genes are structurally similar, sharing ~87 % nucleotide sequence identity in the coding regions, their 5' flanking regions differ somewhat resulting in divergent tissue-specific gene expression (6, 7, 12). However, in some tissues such as the adipose and eyelid both SCD1 and SCD2 genes are expressed (11) whereas in the skin and Harderian gland all the three gene isoforms are expressed (12, 13).

1.1   *The mouse preputial glands express SCD1, SCD2 and SCD3:*

The preputial glands are large sebaceous glands rich in wax ester, alkyl-2,3-diacyl glycerols and triglycerides and are believed to play a role in behavioral interactions through the release of pheromones (14). Many hormones including sex steroids and pituitary and androgenic hormones control the holocrine product of the gland (14-16). We have analyzed total RNA from the preputial gland of the mouse and by using specific cDNA as radioactive probes have found that the mouse preputial glands (PG) express the three gene isoforms (SCD1, SCD2 and SCD3). However, mice with a targeted disruption in the SCD1 isoform (SCD1-/-) have very low expression of the SCD3 isoform while the expression of SCD2 isoform and several other lipogenic genes including FAS, C/EBP-α, PPAR-γ2, SREBP-1, SREBP-2 were not altered. Since androgens such as testosterone have been known to induce sebocyte differentiation and lipid synthesis in PG (14, 17), we examined whether testosterone affects the expression of SCD isoforms in PG. We found that testosterone administration to the wild-type mice increased SCD1, SCD2 and SCD3 mRNAs and increased delta 9-desaturase activities towards the palmitoyl-CoA and stearoyl-CoA substrates in the SCD1+/+ mice while in the microsomes of the SCD1-/- mice the delta 9-desaturase activity was increased only on the palmitoyl-CoA but not on stearoyl-CoA as substrate. These results indicate that SCD2 could not substitute for the SCD3 in the production of 16:1n-7 in the SCD1-/- mice and confirm that SCD3 prefers C16:0-CoA as a substrate over 18:0-CoA. These studies also strongly suggest that the SCD isoforms have different substrate specificity and may explain why there are several SCD isoforms in the mouse genome.

1.2   *Increased synthesis of C16:1n-10 in the preputial glands:*

When conducting the GC analysis to obtain the fatty acid composition in the lipid fractions of the PG we noticed that the chromatogram had a 50% increase in a peak with retention time very close to that of the C16:1n-7. We analyzed the pyrrolidide derivatives of the methyl esters by gas-chromatography mass-spectrometry and found that the unknown fatty acid was C16:1n-10. In addition, the fragmentation pattern of the unknown fatty acid was also consistent with that of C16:1n-10 from the seed oil of *Thunbergia alata*, which is composed of nearly 85-weight % of fatty acids (13).

1.3   *The preputial glands contain a delta 6-palmitoyl-CoA desaturase:*

To determine the metabolic origin of C16:1n-10 fatty acid the microsomes from the SCD1+/+ and SCD1-/- preputial gland were incubated with [^{14}C]16:0-CoA and delta 6 and delta 9 desaturase cofactors. The products of the reactions were analyzed using a TLC separation system at -20^{0}C. This TLC system has been used for the separation of 16:1n-7 from 16:1n-10 (18). We observed that there was no production of C16:1n-7 by the preputial gland of SCD1-/- mice. However, C16:1n-10 was produced in preputial gland of both the SCD1-/- and SCD1+/+ mice but with increased levels in the SCD1-/- mice. The greater than 50% increase in the levels of 16:1n-10 is consistent with an increase in the activity of the palmitoyl-CoA delta 6-desaturase activity because the rate of conversion of [^{14}C]-16:0-CoA to [^{14}C]-16:1n-10 was 2.6-fold higher in the microsome from PG of SCD1-/- mice than SCD1+/+ mice.

In plants a specific 16:0-ACP delta 6-desaturase has been identified (18) and this enzyme has extensive homology with the 18:0-ACP delta 9- and 16:0-ACP delta 9-desaturases (18). Our results now indicate that mammals also have a palmitoyl-CoA delta 6-desaturase that inserts a double bond between positions 5 and 6 in the fatty acyl-CoA chain to synthesize C16:1n-10. It is not presently known whether this palmitoyl-CoA delta 6-desaturase has homology to any of the mammalian or plant delta 9-SCD isoforms. We previously suggested that the differences in the catalytic selectivity of the delta 9-SCD isoforms might be to contribute to the establishing of the lipid composition of the cell. The existence of structurally related acyl-CoA desaturases with different substrate recognition and double bond-positioning properties would be to further refine changes in the fatty acid composition of various lipids. A finer control can be provided by regulated expression of several isoforms with differing selectivity than by expression of either one or two with the same substrate selectivity.

1.4   *Fate of palmitate in preputial glands*

We propose as depicted in figure 1 that in the PG, the palmitate that is synthesized de novo by fatty acid synthase complex (FAS) from acetyl-CoA can be acted on three enzyme systems. Palmitate can serves as a substrate for the microsomal malonyl-CoA dependent elongase to produce stearate, which then serves as the main substrate of SCD1 and SCD2 for the synthesis of oleate. Palmitate is the substrate of SCD1 and SCD3 isoforms for the synthesis of C16:1n-7 and by the palmitoyl-CoA delta 6-desaturase to generate C16:1n-10. The palmitoyl-CoA delta 6-desaturase is induced in the SCD1-/- mice. The mechanism of induction of this activity is not currently known at the present time. The palmitoyl-CoA delta 6-desaturase we have described in this study may be specific to the PG because we could not detect C16:1n-10 in other tissues.

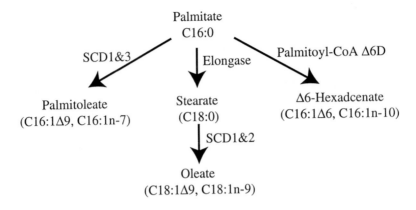

Figure 1. Palmitate can serves as a substrate for the microsomal malonyl-CoA dependent elongase to produce stearate, which then serves as the main substrate of SCD1 and SCD2. It is the substrate of SCD3 isoform to produce C16:1n-7 and by the palmitoyl-CoA delta 6-desaturase to generate C16:1n-10.

Acknowledgements. This work was supported in part from a grant from the American Heart Association and in part by funds from Xenon Biogenetics (Vancouver Canada) to JMN. FEG received partial support from the National Council of Science and Technology (CONACYT-Mexico)

References

1. Enoch, H.G., and P. Strittmatter. (1978) Role of tyrosyl and arginyl residues in rat liver microsomal stearylcoenzyme A desaturase. Biochemistry. 17, 4927-4932.

2. Ntambi, J.M. (1999) Regulation of stearoyl-CoA desaturase by polyunsaturated fatty acids and cholesterol. J Lipid Res. 40, 1549-1558.

3. Kasai, T., K. Ohguchi, S. Nakashima, Y. Ito, T. Naganawa, N. Kondo, and Y. Nozawa. (1998) Increased activity of oleate-dependent type phospholipase D during actinomycin D-induced apoptosis in Jurkat T cells. J Immunol. 161, 6469-6474.

4. Fermor, B.F., J.R. Masters, C.B. Wood, J. Miller, K. Apostolov, and N.A. Habib. (1992) Fatty acid composition of normal and malignant cells and cytotoxicity of stearic, oleic and sterculic acids in vitro. Eur J Cancer. 28A, 1143-1147.

5. Khoo, D.E., B. Fermor, J. Miller, C.B. Wood, K. Apostolov, W. Barker, R.C. Williamson, and N.A. Habib. (1991) Manipulation of body fat composition with sterculic acid can inhibit mammary carcinomas in vivo. Br J Cancer. 63, 97-101.

6. Kaestner, K.H., J.M. Ntambi, T.J. Kelly, Jr., and M.D. Lane. (1989) Differentiation-induced gene expression in 3T3-L1 preadipocytes. A second differentially expressed gene encoding stearoyl-

CoA desaturase. J Biol Chem. 264,14755-14761.

7. Ntambi, J.M., S.A. Buhrow, K.H. Kaestner, R.J. Christy, E. Sibley, T.J. Kelly, Jr., and M.D. Lane. (1988) Differentiation-induced gene expression in 3T3-L1 preadipocytes. Characterization of a differentially expressed gene encoding stearoyl- CoA desaturase. J Biol Chem. 263,17291-17300.

8. Mihara, K. (1990) Structure and regulation of rat liver microsomal stearoyl-CoA desaturase gene. J Biochem (Tokyo). 108,1022-1029.

9. Zhang, L., L. Ge, S. Parimoo, K. Stenn, and S.M. Prouty. (1999) Human stearoyl-CoA desaturase: alternative transcripts generated from a single gene by usage of tandem polyadenylation sites. Biochem J. 340, 255-264.

10. Zheng, Y., K.J. Eilertsen, L. Ge, L. Zhang, J.P. Sundberg, S.M. Prouty, K.S. Stenn, and S. Parimoo. (1999) Scd1 is expressed in sebaceous glands and is disrupted in the asebia mouse. Nat Genet. 23, 268-270.

11. Miyazaki, M., W.C. Man, and J.M. Ntambi. (2001) Targeted disruption of stearoyl-CoA desaturase1 gene in mice causes atrophy of sebaceous and meibomian glands and depletion of wax esters in the eyelid. J Nutr. 131, 2260-2268.

12. Zheng, Y., S.M. Prouty, A. Harmon, J.P. Sundberg, K.S. Stenn, and S. Parimoo. (2001) Scd3-a novel gene of the stearoyl-coa desaturase family with restricted expression in skin. Genomics. 71, 182-191.

13. Miyazaki, M., H.J. Kim, W. Man Chi, and J.M. Ntambi. (2001) Oleoyl-CoA is the major de novo product of stearoyl-CoA desaturase 1 gene isoform and substrate for the biosynthesis of the Harderian Gland 1-alkyl-2,3-Diacylglycerol. J Biol Chem. 276, 39455-39461.

14. Cooper, M.F., P.W. Bowden, D. Meddis, A.J. Thody, and S. Shuster. (1976) Effects of testosterone and alpha-melanocyte-stimulating hormone on preputial-gland (sebaceous) activity. Biochem Soc Trans. 4, 798-800.

15. Sansone, G., and J.G. Hamilton. (1969) Glyceryl ether, wax ester and triglyceride composition of the mouse preputial gland. Lipids. 4, 435-440.

16. Sansone-Bazzano, G., G. Bazzano, R.M. Reisner, and J.G. Hamilton. (1972) The hormonal induction of alkyl glycerol, wax and alkyl acetate synthesis in the preputial gland of the mouse. Biochim Biophys Acta. 260, 35-40.

17. Thody, A.J., M.F. Cooper, P.E. Bowden, D. Meddis, and S. Shuster. (1976) Effect of alpha-melanocyte-stimulating hormone and testosterone on cutaneous and modified sebaceous glands in the rat. J Endocrinol. 71, 279-288.

18. Cahoon, E.B., A.M. Cranmer, J. Shanklin, and J.B. Ohlrogge. (1994) delta 6 Hexadecenoic acid is synthesized by the activity of a soluble delta 6 palmitoyl-acyl carrier protein desaturase in Thunbergia alata endosperm. J Biol Chem. 269, 27519-27526.

LINOLINIC ACID ACCUMULATION IN DRAGONHEAD

MOHAMMED ABDEL-REHEEM, RESHAM BHELLA, AND DAVID HILDEBRAND

Plant Biochemistry/ Physiology/ Molecular Biology Program
University of Kentucky, N-122 Agri. Sci. Bldg-North, Lexington, KY 4056-0091, USA

1. Abstract:

Dragonhead (*Dracocephalum moldavica*) seeds accumulate linolenate (18:3) to one of the highest levels known. To elucidate the factors that contribute to the high 18:3 content of dragonhead compared to normal oilseeds, a pulse-chase study of radiolabeled $^{14}C^1$ linoleoyl (18:2)-CoA was carried out at an early and late stage of embryo triglyceride (TG) synthesis. In dragonhead ^{14}C-18:2 incorporation in PC increased and ^{14}C-18:3 decreased, meanwhile the opposite was observed in TG (the incorporation of ^{14}C-18:3 in TG is two-fold that of ^{14}C-18:2 in PC). This suggests that both ω-3 desaturase is effectively desaturating ^{14}C-18:2 in PC and acyltransferases are transferring ^{14}C-18:3 into TG. In contrast, soybean data indicate that ^{14}C-18:2 is rapidly removed from PC and incorporated directly into TG. Also ^{14}C-18:3 increased in dragonhead TG but no significant changes occurred in ^{14}C-18:3 levels in soybean. This indicates that desaturase activity decreased during the late stage in soybean embryos and most of ^{14}C-18:2 was released from PC and incorporated into TG without being converted into 18:3. In contrast most of the 18:2 was desaturated to 18:3 before transfer to the free fatty acid/acyl-CoA pool in dragonhead.

1.1 *Pulse-Chase study of radiolabeled 18:2 COA for Soybean and Dracocephalum:*

Seeds of two different cotyledons developmental stages from both soybean (7-9 mm& 9-11mm), and *Dracocephalum* (0.05-0.1mg & 0.1-0.2mg), were taken. 0.0033 µCi of radiolabeled 18:2 CoA / mg seed tissue, 1 µL/mg weight of embryo of 0.1 M potassium phosphate buffer, pH 7.2, 0.1% Tween 20 were added to clean test tubes. Seed tissues were added, and then Incubated for 90 min. in the presence of fluorescent light. Two replicates, from each stage of both soybean and *Dracocephalum* were taken for lipid analysis (zero time), and the rest were incubated for further chasing. Samples were collected for lipid analysis after 30, 60 (1h), and 120min. (2h), from the zero time. Total lipids were extracted in chloroform: methanol (2:1 v/v). Individual lipids were separated by one-dimensional thin layer chromatography by the method of (Kumar *et al.*, 1983). TLC plates were developed in two solvent systems. Phospholipids and glycerolipids were separate in chloroform: methanol: water (65:25:4, v/v) for10 cm. Neutral lipids were separated in hexane: diethyl ether: acetic acid (100: 100: 2, v/v) for 19.5cm (Miquel and Browse, 1992). Lipids were

N. Murata et al. (eds.), Advanced Research on Plant Lipids, 101–104.
© 2003 *Kluwer Academic Publishers. Printed in the Netherlands.*

located by spraying the plates with solution of 0.005% primulin in 80% acetone, followed by visualization under UV light. The silica gel from each lipid band were scrapped, transferred to a tube containing 2 mL of 2% (V/V) H2SO4 in methanol. To separate linoleate (18:2) and linolenate (18:3) reveres phase TLC approach was applied. Fatty acids methyl esters, for each lipid classes, obtained from the previous steps were loaded into TLC plates plates were developed in acetonitrile: acetic acid: water (70: 10: 10 v/v) for 19.5 cm ascending. Staining on this plate was carried out using iodine. 18:2 and 18:3 bands were scrapped, directed for scintillation counting using liquid scintillation analyzer.

1.2 Results and Discussions:

Fig. 1 shows the incorporation of ^{14}C in different lipid fractions specially in 18:2 and 18:3, Fig. 1A shows that the activity of ^{14}C -18:2 increased by the increment in the chasing time in the PC of the seconded stage seed. However in the first stage ^{14}C -18:2 activity was almost steady during different chase times at low level. In fig. 1B, ^{14}C -18:3 activity level in triacylglycerol was dramatically increased in the seeds of the second stage during the chasing time. In the other hand the ^{14}C -18:2 activity level in TG was slightly decreased in the first cotyledon developmental stage, while it slightly increased at 1h then decreased at 2h of the chasing time. Data from these two figures highly suggests that 18:2 is desaturated into 18:3 withω-3-desaturase at the PC level and then 18:3 is rapidly transferred from PC to TG by acyltransferases. 18:2 incorporated into PC during the chasing the times, but due to the high activity of ω-3 desaturase and acyltransferases it does not accumulate in PC and is converted to 18:3 which is rapidly transferred into TG during the chasing times (Fig. 1B). Fig. 1C showed that ^{14}C -18:3 activity levels at a lower scale but it decreased after 1/2h chase time. From the data in fig. 1D the level of ^{14}C -18:2 decreased in free fatty acids fraction in seeds of the second cotyledon developmental stage. The data emphasizes the important role of acyltransferases in the incorporation of 18:3 into TG in *Dracocephalum* plant and in the incorporation of ^{14}C -18:2 from free fatty acyl-CoA pool into DAG which is used up rapidly to produce ^{14}C -18:3 through the desaturation of ^{14}C -18:2 PC and then ^{14}C - 18:3is transported from PC into TG the final oil storage form. Also this suggested that acyltransferases are very active since the level of ^{14}C -18:3 in free fatty acid/acyl-CoA pool during the chasing times (Fig. 1E) is relatively lower than its level in TG during the same chasing times (Fig. 1B). The incorporation of ^{14}C -18:2 and ^{14}C -18:3 were found to be at a very low level in MAG in both cotyledon developmental stages (Fig. 1D). The % of ^{14}C incorporation into different lipid fractions in *Dracocephalum* cotyledon at two different developmental stages differed (Fig. 2). In Fig. 2A % of ^{14}C incorporation in PC at 0h and 1/2h is higher than in TG at the same chasing times. However ^{14}C incorporation in PC declined to a steady state at 2h, meanwhile ^{14}C incorporation in TG increased to steady state at 2h. ^{14}C % in TG is higher than that of PC in the second stage (data not shown). All other lipid fractions have very low ^{14}C percentage at both first and second stages. This data is highly suggesting that the ω-3 desaturase and acyltransferases as well as possibly phospholipases are playing an important role in 18:3 accumulations in *Dracocephalum* seeds.

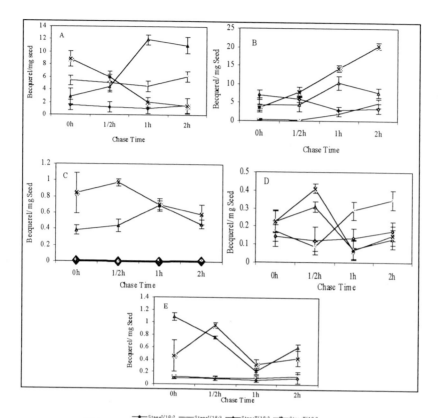

Figure 1: ^{14}C (18:2/ 18:3) distribution in *Dracocephalum* seeds **(A)** in phosphatidyl choline, **(B)** in triacylglycerol, **(C)** in diacylglycerol, **(D)** in monoacylglycerol, and **(E)** in FFA.

Fig. 4A shows that the incorporation of ^{14}C -18:2 in soybean PC at the first stage is increased at ½h of chasing time till the end of the chasing period, while ^{14}C -18:2 level in PC of the second stage decreased dramatically after 1/2h. The ^{14}C - 18:3 level in PC slightly increased by the end of chasing period. Fig. 4 B data showed that ^{14}C -18:2 levels in TG is increased dramatically in the first stage seeds after 1/2h and reached a maximum level at 2h. Data from Fig. 4A&4B highly suggest that ω-3 desaturase activities is very low in both stages and slightly converts a low amount of ^{14}C -18:2 from PC into ^{14}C -18:3. Fig. 4C data showed that ^{14}C incorporation level in soybean DAG is at a very low level in both stages but the incorporation of both ^{14}C -18:2 and ^{14}C -18:3 levels is dramatically decreased in DAG of soybean seeds at the first stage. From Fig. 4 D&4 E it is apparent that the ^{14}C incorporation level in MAG and free FA is very low. Also the ^{14}C -18:2 levels in both the first and second stages is much higher than ^{14}C -18:3 levels. Moreover Fig. 4D shows that ^{14}C -18:2 levels in MAG of first stage seeds is increased. However the percentage of ^{14}C incorporation in both TG and PC of the first stage of soybean cotyledon developmental stages is increased (data not shown), meanwhile ^{14}C percentage in both DG and FFA pools is dramatically decreased during the chasing time while ^{14}C level in MAG slightly increased. Also the ^{14}C level in PC at second stage decreased while it increased in TG at the same stage. And ^{14}C level in all other lipid fractions (MAG, DAG, and FFA) is almost steady during the chasing time (data not shown), only slightly increasing in FFA levels toward

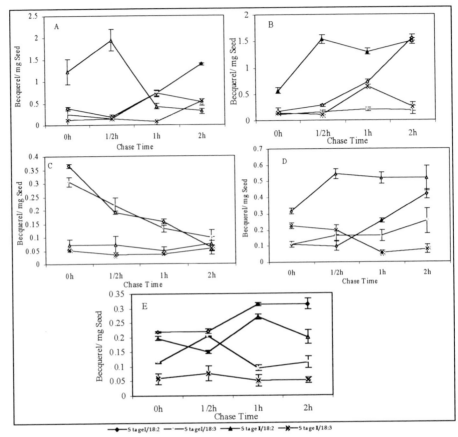

Figure 4 : ^{14}C (18:2/ 18:3) distribution in soybean (**A**) in phosphatidylcholine, (**B**) in triacylglycerol, (**C**) in diacylglycerol, (**D**) in monoacylglycerol, and (**E**) in free fatty acids.

the end of the chasing time, this may be due to that both PC and TG reaching a steady state by the end of the chasing time and therefore there is no use for the FFA pool. MAG showed a slight decrease toward the end of the second stage. This data emphasize that the lipid metabolism in soybean is in favor of accumulating 18:2 into TG (Dohmer *et. al*, 1991) also these data highly suggest that ω-3 desaturase activities is low compared with *Dracocephalum*. From radioactive 18:2 added exogenously, and other studies (Kumar, 1983) indicate that the synthesis of TG is via the glycerol-3-phosohate pathway.

References:

Dahmer, M.L., G.B. Collins and D.F. Hildebrand. (1991). Lipid content and composition of soybean somatic embryos. Crop Sci. 31:741-746.

Kumar D. Mukherjee, (1983). Lipid Biosynthesis in Developing Mustared Seed. Plant Physiol. (1983) 73, 929-934.

Miquel M., and Browse J. (1992). Arabidopsis mutants deficient in polyunsaturated fatty acids synthesis. J.Biol.Chem.267 (3): 1502-1509.

Stobart K., Styme S., Glad G., (1982). The control of triacylglycerol composition in the microsomes of developing oil seeds. In JFGM Wintermans. PJC Kuiper, eds, Biochemistry and Metabolism of Plant Lipids. Elsevier Biomedical, Amesterdam, pp 257-261.

SPIRULINA PLATENSIS MUTANTS DEFECTIVE IN γ-LINOLENIC ACID PRODUCTION : *Molecular characterization*

M. RUENGJITCHATCHAWALYA[1*], S. CHAMUTPONG[1], M. PONGLIKITMONGKOL[2], R. CHAIKLAHAN[1] and M. TANTICHAROEN[3]

[1]*School of Bioresources and Technology, KMUTT, Bangkok 10140, Thailand.* [2]*Faculty of Science, Mahidol University, Bangkok 10400, Thailand.* [3]*BIOTEC, NSTDA, Bangkok 10400, Thailand.*
*Author for correspondence; e-mail: marasri.rue@kmutt.ac.th

Introduction

Spirulina platensis, a filamentous cyanobacteria, has been a potential alternative source for a production of γ-linolenic acid (GLA, 18:3Δ6,9,12). This essential polyunsaturated fatty acid (PUFA) is of profound interest as its pharmaceutical values (see reviews in 1). Recently, *Spirulina* is commercially cultivated in several countries, mainly as a health food and feed, therefore choice of strains are required. In our laboratory, attempt to improve and select a high GLA producing strain(s) with high economic value, chemical treatment using SAN9785 (BASF 13-338, 4-chloro-5(dimethylamino)-2-phenyl-3(2H) pyridazinone) (2) or ethyl-methanesulfonate (EMS) have been investigated. However, we have obtained four mutants defective in the GLA production, namely I18, I22, I30 and I30/1, during a strain selection after EMS-mutagenesis. Reduction in the GLA level to half of that of wild type was found in the first three strains, but not in the I30/1 which completely lacked of the GLA. These mutants might be used as an alternative tools for clearly understand in the biosynthesis of the GLA. Since acyl-lipid desaturases are the enzymes that introduce double bond into fatty acids that are bound to glycerolipids in cyanobacteria (3). In this report, we thus examined the molecular characterization of *desD*, gene coding for Δ6-acyl lipid-desaturase, of the mutants in comparative to that of the wild type (WT), including the transcription level.

Materials and Methods

Cultivation condition and *Growth measurement*

The wild type (WT) *S. platensis* strain C1 (or *Arthrospira* sp. PCC 9438) and mutants were grown in Zarrouk's medium at 35 °C (or otherwise noted) under white fluorescent lamps at 80 μ mol photon $m^{-2} s^{-1}$ in an orbital incubator shaker at 130 rpm. Growth was measured as absorbance at 560 nm (cell density) and/or at 665 nm (chlorophyll a) (4,5).

Mutagenesis

The WT cells grown at 35 °C were collected by centrifugation at 1,000 × g for 10 min at the same growth temperature. The collected cells were suspended in 30 mM potassium

N. Murata et al. (eds.), Advanced Research on Plant Lipids, 105–108.
© 2003 Kluwer Academic Publishers. Printed in the Netherlands.

phosphate buffer (pH 7.0) at a cell density of 1×10^6 trichomes mL^{-1}. To 2.5 mL of the cell suspension, 0.4 M ethyl methanesulfonate (EMS) was added. The suspension was vortexed and incubated at 35 °C for 1 h in the dark. The EMS was inactivated by the addition of 5% sodium thiosulfate and the mutagenized cells were collected by centrifugation at 1,000 × g for 10 min. The cells were washed 3 times with 5 mL of Zarrouk's medium by centrifugation at 1,000 × g for 10 min and re-suspension. Then, the cell suspension was incubated at 25 °C for 2 days in the presence of ampicillin at 100 μg mL^{-1}. The treated cells were washed and re-suspended in 2.5 mL of Zarrouk's medium and then grown at 35 °C for 3 days. The cells were plated on Zarrouk's medium containing 1.5% agar and incubated under the dim light (10 μ mol photon $m^{-2}s^{-1}$) at 35 °C until colonies were formed.

Fatty acid analysis

Analyses were done using a modification of the method of Lepage and Roy (1984). The freeze-dried samples were direct transmethylated in 5% HCl in methanol at 85 °C for 1 h, and heptadecanoic acid (C17:0; Sigma Co.) was added as an internal standard. The fatty acid methyl esters (FAMEs) were analysed by gas chromatography (Fison instrument 8340), performed on a SP-2330 fused silica capillary column (60 m x 0.25 mm), and identified by co-chromatography with authentic standards (Sigma Co.) as described (5).

Isolation of RNA and Northern blot analysis

Cells were harvest by filtration and the cell pellets were immediately frozen in liquid nitrogen and broken by grinding in a cold mortar. Total RNA was extracted in TRIZOL® reagent (Gibco BRL, Life Technologies, UK) as described by the company. The *desD* specific probe (395 bp) was generated by PCR using the oligonucleotides 5'-TGGATGAAACTACTGGGTTGC-3' and 5'-TGCTATATCTACCCATGTCGG-3', as forward and reverse primers, repectively. Northern blot analysis was followed as described (6) using the *S. platensis* 16S rRNA as internal control.

Results and Discussion

The I18, I22, I30 mutants of *S. platensis*, obtained from mutagenesis by EMS, showed that the level of GLA (as % of total fatty acid, TFA) decreased about 32, 43 and 54%, respectively, from that of the WT. Another mutant, the I30/1, which was derived from the I30, showed completely lacking of the GLA (Table 1). Nucleotide sequence analysis revealed that *desD* gene encoding a Δ6-acyl-lipid desaturase of the I18, I22 and I30 were not altered. On the other hand, the I30/1 showed one base pair (T-656) deletion changing an amino acid residue from Leu-224 to a stop codon (data not shown), perhaps resulting in the loss of desaturase activity of this truncated protein. Transcriptional analysis (Figure 1) showed that 70% and 50% reduction in the level of the *desD* mRNA in cells grown at optimal temperature (35°C) of the I18 and I30, respectively, compared to that of the WT. This observation was in concordance with the level of GLA production. However, the *desD* mRNA of the I18 was remarkably increased to the same level as that of the WT after a temperature shift from 35°C to 22°C, whereas that of I30 showed only a 10% increase. This suggested that both I18 and I30, were defective in the GLA production at the transcription level. The defect in the I30 was probably involved with its promoter and/or regulatory gene(s) alterations, whereas that of the I18 was likely due to a defect in a temperature signaling factor(s). In contrast, the I22 showed the same level of the *desD* mRNA as that of the WT at both the optimal and low temperatures (Figure 1), but a much higher stability. Half-life of *desD* mRNA of the I22 cells grown at both 35°C and 22°C are 50 min, whereas those of the WT are 7 min and 20 min, respectively (Figure 2). These finding demonstrated that the low level of the GLA production might involve several changes including transcription or post-transcription level and/or other defect correlated to the fatty acid desaturation process.

Table 1. Fatty acid composition of *S. platensis* cells grown for 4 days at 35 °C under light intensity 120 μmol photon m^{-2}s^{-1}

Strain	Fatty acid composition (% of TFA[a])						TFA (% of dw[b])	GLA (% of dw)
	16:0	16:1	18:0	18:1	18:2	γ-18:3		
WT	44.8	5.0	1.2	8.1	21.5	19.3	4.8	0.9
I18	44.5	7.8	1.1	8.9	24.5	13.1	4.9	0.6
I22	49.9	9.1	0.8	4.0	24.6	11.0	6.0	0.7
I30	46.1	5.6	1.4	5.5	30.7	10.6	4.2	0.5
I30/1	43.3	3.4	0.7	5.9	42.6	0	5.8	0

a ,TFA: total fatty acid; b, dw: dry weight; c, GLA: γ-linolenic acid, 18:3Δ6,9,12

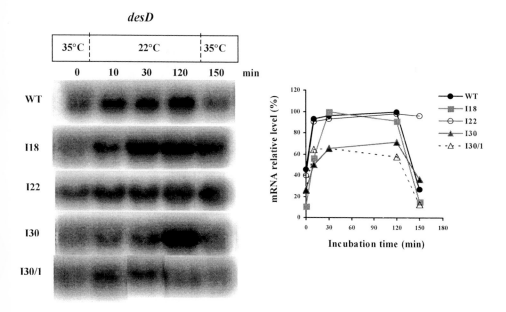

Figure 1. Levels of *desD* mRNA of *S. platensis* mutants V.S. WT after temperature shift from 35 °C to 22 °C

108

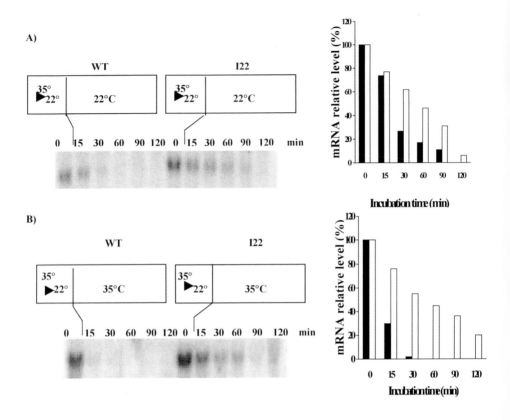

Figure 2. Half life analysis of *desD* mRNA of *S. platensis* mutant I22 (□) V.S. WT(■) after temperature shift from 35 °C to 22 °C (A) or vice versa (B).

Acknowledgments

This work was supported by a grant from National Center for Genetic Engineering and Biotechnology (BIOTEC), Bangkok, Thailand

References
1. Cohen, Z. (1997) The Chemicals of *Spirulina*. In A. Vonshak (ed.). *Spirulina platensis* (*Arthrospira*): Physiology, Cell-biology and Biotechnology. Taylor & Francis Ltd., pp. 175-212.
2. Tanticharoen, M. et al. (1994) Optimization of γ-linolenic acid (GLA) production in *Spirulina platensis*. J. Appl. Phycol. 6, 295-300.
3. Murata, N. et al. (1996) Biosynthesis of γ-linolenic acid in the cyanobacterium *Spirulina platensis*. In Y.S. Huang and D.E. Mill (eds.). γ-Linolenic Acid: Metabolism and its Roles in Nutrition and Medicine. AOCS Press, IL., pp. 22-32.
4. Lepage, G. and Roy, C.C. (1984) Improved recovery of fatty acid through direct transesterification without prior extraction or purification. J. Lipid Res. 25, 1391-1396.
5. Cohen ,Z, et al. (1993) Production and partial purification of γ-linolenic acid and some pigments from *Spirulina platensis*. J. Appl. Phycol. 5, 109-115.
6. Deshnium, P. et al. (2000) Temperature-independent and –dependent expression of deaturase genes in filamentous cyanobacterium *Spirulina platensis* strain C1 (Arthrospira sp. PCC9438). FEMS Microbiol. Lett. 164, 207-213.

OXYGEN AVAILABILITY REGULATES MICROSOMAL OLEATE DESATURASE (FAD2) IN SUNFLOWER DEVELOPING SEEDS BY TWO DIFFERENT MECHANISMS

J.M. MARTINEZ-RIVAS, A. SANCHEZ-GARCIA,
M.D. SICARDO and M. MANCHA
Instituto de la Grasa, CSIC. Avda. Padre García Tejero 4.
41012-Sevilla, Spain.

1. Introduction

The microsomal oleate desaturase (FAD2) catalyzes the desaturation of oleate to linoleate, which are the main fatty acids in storage triacylglycerols of oilseeds. The growth temperature modifies the relative contents of oleate and linoleate in the seed lipids, indicating a temperature regulation of the FAD2 activity.

Studies from our group using developing sunflower seeds have just shown that hull removing strongly increased FAD2 activity, and low oxygen concentration obtained by respiration of peeled seeds incubated in sealed vials brought about a decrease of the FAD2 activity. Our results indicated that, in addition to the direct effect of temperature on FAD2 activity, there is also an indirect effect. The enzyme seems to be regulated by oxygen availability, which is affected by its diffusion through the hull and the competition with respiration, being both factors temperature dependent (García-Díaz et al., 2002). This was the first time that a role for oxygen was demonstrated in the regulation of a plant desaturase.

2. Materials and Methods

Sunflower plants were grown in a growth chamber with a 16 h photoperiod at 25/15 °C (day/night). The capitula were collected at 18-19 DAF. Either detached achenes or peeled seeds (achenes without hull and seed membrane) were used for the experiments. Lots of 20 seeds were used to assure homogeneity.

The isolation of the microsomal fraction and the *in vitro* assay of the FAD2 activity were carried out as described by García-Díaz et al. (2002).

N. Murata et al. (eds.), Advanced Research on Plant Lipids, 109–112.
© 2003 *Kluwer Academic Publishers. Printed in the Netherlands.*

3. Results and Discussion

Further research has been done in order to confirm that oxygen availability regulates FAD2 activity in developing sunflower seeds and to characterize the involved mechanism.

After a fast and significant increase of FAD2 activity as a result of hull removing (Fig. 1), anoxia brought about a rapid decrease. Air reposition after 30 min of anoxia produced the recovery of initial activity level, confirming that oxygen was the responsible for the changes of FAD2 activity and showing that it was a reversible mechanism.

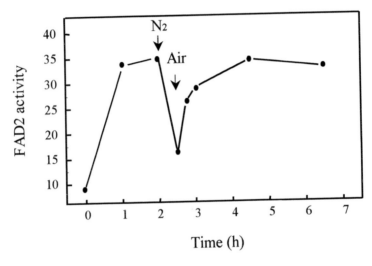

Figure 1. Effect of hull removing followed by anoxia and oxygen reposition on FAD2 activity in isolated microsomes from peeled developing sunflower seeds. Peeled seeds were incubated at 20°C in a stream of air. After 2h, they were incubated in nitrogen for 30 min and finally air was replaced. At the indicated times the seeds were homogenized and stored at –20°C. The homogenate was used to isolate the microsomal fraction and the FAD2 activity was measured in this fraction and expressed as nmol 18:2 $(g FW)^{-1} h^{-1}$.

When microsomes isolated from a homogenate obtained immediately after dehulling, were incubated for 2 h at 20°C instead of peeled seeds, no effect on FAD2 activity was observed. This result suggests that oxygen could initiate a signal transduction cascade *in vivo*, not being the direct responsible for the change in the activity level. Kinetic and thermal properties of the enzyme such as the Km for oxygen and the thermostability, decreased or increased respectively, when the low activity form was compared with the high activity one (results not shown). These data, together with the high speed of the process seem to indicate that an activation/inactivation mechanism could be involved, more than an induction/repression of the gene. A similar pattern of FAD2 activity changes that shown in Fig. 1 was obtained when detached achenes were subjected to low-high temperature transitions (García-Díaz et al., 2002), indicating that temperature changes indirectly control oxygen availability inside the achene, being a regulatory mechanism of physiological significance. A role for oxygen in the regulation of FAD2 activity was previously reported in the ameba *A. castellanii* (Thomas et al., 1998).

In order to determine the oxygen level that provoked a decrease in the FAD2 activity, peeled seeds were incubated for 30 min at different concentration of the gas (Fig. 2). Oxygen concentrations higher than 4% showed the maximal activity level, whereas when the oxygen level was reduced to 1-4% a strong decrease of the enzyme activity was observed.

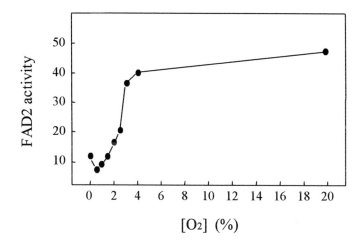

Figure 2. Effect of oxygen concentration on FAD2 activity in peeled developing sunflower seeds. Peeled seeds were incubated at 20°C in a stream of nitrogen with different oxygen concentrations. After 30 min the seeds were homogeneized and stored at –20°C. The homogenate was used to isolate the microsomal fraction and the FAD2 activity was measured in this fraction and expressed as nmol 18:2 (g FW)$^{-1}$ h^{-1}.

Surprisingly, an important increase in FAD2 activity was also noticed after 30 min when oxygen level was 0% (Fig. 2). To confirm this observation, sunflower achenes were incubated in anoxia for different times (Fig. 3). As expected, a relatively slow increase of the FAD2 activity was obtained, suggesting that possibly a *FAD2* gene induction occurred. The anoxic induction of the stearoyl-CoA desaturase (*OLE1*) gene from *S. cerevisiae* has been reported (Kwast et al., 1999).

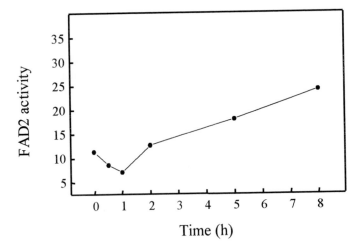

Figure 3. Effect of anoxia on FAD2 activity in sunflower detached achenes. Detached achenes were incubated at 20°C in a stream of nitrogen. At the indicated times the seeds were homogeneized and stored at –20°C. The homogenate was used to isolate the microsomal fraction and the FAD2 activity was measured in this fraction and expressed as nmol 18:2 $(g\ FW)^{-1}\ h^{-1}$.

In conclusion, these data indicate that oxygen availability regulates sunflower FAD2 activity in developing seeds by two different mechanisms. Currently, we are investigating whether these two different oxygen effects are due to any of the three *FAD2* genes/isoforms recently isolated from sunflower (Martínez-Rivas et al., 2001).

4. References

García-Díaz, M.T., Martínez-Rivas, J.M. and Mancha M. (2002) Temperature and oxygen regulation of oleate desaturation in developing sunflower (*Helianthus annuus*) seeds. Physiol. Plant. 114, 13-20.

Kwast, K.E., Burke, P.V., Staahl, B.T. and Poyton R.O. (1999) Oxygen sensing in yeast: Evidence for the involvement of the respiratory chain in regulating the transcription of a subset of hypoxic genes. Proc. Natl. Acad. Sci. USA, 96 5446-5451.

Martínez-Rivas, J.M., Sperling, P., Luehs, W. and Heinz E. (2001) Spatial and temporal regulation of three different microsomal oleate desaturase genes (*FAD2*) from normal-type and high-oleic varieties of sunflower (*Helianthus annuus* L.). Mol. Breed. 8, 159-168.

Thomas, K., Rutter, A., Suller, M., Harwood, J. and Lloid, D. (1998) Oxygen induces fatty acid (n-6) desaturation independently of temperature in *Acanthamoeba castellanii*. FEBS Lett. 425, 171-174.

IDENTIFICATION OF A Δ5-FATTY ACID DESATURASE FROM PHYSCOMITRELLA PATENS

P. SPERLING[1], J. M. LUCHT[2], T. EGENER[2], R. RESKI[2], P. CIRPUS[3], E. HEINZ[1]

[1]*Institut für Allgemeine Botanik, Universität Hamburg, Ohnhorststr. 18, 22609 Hamburg, Germany*
[2]*Universität Freiburg, Pflanzenbiotechnologie, Sonnenstr. 5, 79104 Freiburg, Germany*
[3]*BASF Plant Science GmbH, BPS-A 030, 67056 Ludwigshafen, Germany*

1. Introduction

Very long-chain polyunsaturated fatty acids (VLCPUFA) such as arachidonic (ARA, 20:4ω6) and eicosapentaenoic acid (EPA, 20:5ω3) are considered to be beneficial in the human diet and a sustainable source of VLCPUFA would be highly desirable. The biosynthesis of Δ5-unsaturated VLCPUFA occurs through sequential desaturation and chain-elongation in many eukaryotic organisms, but not in angiosperm plants. A good source to clone the required genes from a plant is the moss *Physcomitrella patens* with lipids containing up to 30 % ARA and some EPA (Grimsley et al., 1981). Recently we have identified genes encoding a Δ6-desaturase (Girke et al., 1998) and a Δ6-specific elongase from *P. patens* (Zank et al., in press).

In this paper, we report the functional identification of a Δ5-desaturase from *P. patens* by gene disruption and by heterologous co-expression with the moss Δ6-elongase (Zank et al., in press) leading to the production of ARA and EPA in *Saccharomyces cerevisiae*.

2. Material and Methods

An EST sequence database from the moss *P. patens* was used to search for cDNA clones showing sequence similarities to Δ5-, Δ6- and Δ8-fatty acid desaturases, all of which are cytochrome b_5-fusion proteins (Sperling and Heinz, 2001). One clone showing similarity to these front-end desaturases was characterized in more detail. For expression in yeast, the open reading frame (ORF) of the clone was amplified by PCR using primers introducing a 5'-KpnI and 3'-XhoI restriction site. The resulting DNA

N. Murata et al. (eds.), Advanced Research on Plant Lipids, 113–116.

fragment was cloned into the yeast expression vector pYES3CT (Invitrogen) downstream of the inducible GAL1 promotor yielding pΔ5DES. The PSE1 gene cloned into the BamHI site of vector pYES2 (Invitrogen) yielding pPSE1 was generated previously (Zank et al., in press).

S. cerevisiae strain UTL-7A (MAT a, leu2-3,112, trp1, ura3-52) was transformed with pPSE1 and either with pΔ5DES or the empty vector pYES3CT by the polyethylene gycol method (Dohmen et al., 1991). After growth (3d, 20°C) and immediate induction, total fatty acids of transgenic cells were converted into their methyl esters (FAMEs) and analysed by gas-liquid chromatography (GLC) as described before (Sperling et al., 2000). For feeding experiments the medium was supplied with 0.5 mM of the corresponding fatty acid substrate.

For an alternative proof of function, the Δ5-desaturase gene in *P. patens* was disrupted by homologous recombination. A cDNA fragment was replaced by a *npt*II cassette as a positive selection marker and the disrupted gene was tranformed into protoplasts as described by Strepp et al. (1998). Regenerated protonemata were selected on medium containing G418 for 14 d. Stable transformants were cultivated for mass production in non-selective liquid medium and FAMEs of isolated polar lipids were analysed by GLC as described before (Girke et al., 1998).

3. Results and Discussion

We could identify a cDNA clone from the moss *P. patens* by its homology to Δ5-, Δ6- and Δ-8-fatty aicd desaturases. The complete nucleotide sequence of the cDNA consisted of 1542 bp. It contained an ORF of 1443 bp encoding a protein of 480 amino acids with a calculated molecular mass of 54,3 kDa. The protein sequence showed the expected N-terminal cytochrome b_5-domain (Sperling and Heinz, 2001) and highest identity (33 %) and similarity (52 %) to the Δ5-desaturase from *Mortierella alpina* (Michaelson et al., 1998; Knutzon et al., 1998).

For a functional identification the ORF of Δ5DES was co-expressed with the moss elongase gene PSE1 in *S. cerevisiae* to provide enough C_{20}-PUFAs as substrates for Δ5-desaturation. Figure 1 shows the fatty acid profiles of yeast cells transformed with pPSE1 and with the empty vector pYES3CT and pΔ5DES, respectively, analysed by GLC. In the presence of γ-linolenic (γ-18:3), transgenic cells expressing the elongase and the empty vector produced 23 % 20:3ω6 of total fatty acids, whereas cells co-expressing the pΔ5DES converted 5 % of the provided 20:3ω6 to an additional peak showing the same retention time as ARA (Fig. 1A). Furthermore, transgenic cells expressing pPSE1 and fed with stearidonic acid (18:4ω3) produced 18 % 20:4ω3 of total fatty acids. Additional expression of pΔ5DES lead to a 6 % conversion of 20:4ω3 to a new peak co-eluting with EPA (Fig. 1B). On the other hand, saturated and mono-unsaturated fatty acids present in yeast cells as well as incorporated linoleic (18:2),

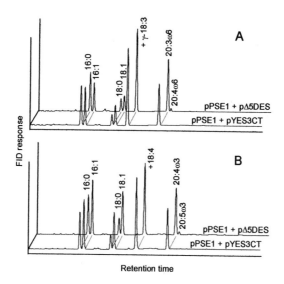

Figure 1. Fatty acid profiles of *S. cerevisiae* co-expressing the Δ6-elongase (pPSE1) and the Δ5-desaturase (pΔ5DES) from *P. patens* or the empty vector pYES3CT (control) in the presence of γ-linolenic (A) or stearidonic acid (B) yielding 20:4ω6 or 20:5ω3, respectively. FAMEs were prepared from whole cells and analysed by GLC.

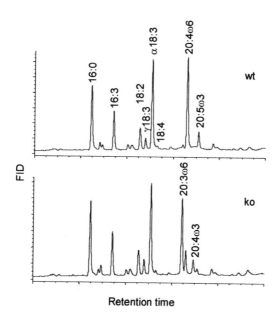

Figure 2. Disruption of a Δ5-desaturase gene from *P. patens*. FAMEs were prepared from polar lipids isolated from 14 d old protonemata of wild type (wt) and a knock-out line (ko) and analysed by GLC.

α-linoleic (18:3), 20:2-11,14 and 20:3-11,14,17 did not seem to serve as substrates for the moss Δ5-desaturase (data not shown).

For an alternative proof of function, we disrupted the Δ5DES gene in *P. patens* by homologous recombination. Figure 2 shows the fatty acid profiles of polar lipids isolated from wild-type and from a knock-out line of *P. patens* analysed by GLC. In contrast to the wild-type (wt), the knock-out line (ko) showed a strong decrease in ARA and EPA accompanied by a significant increase of the direct substrates 20:3ω6 and 20:4ω3 involved in Δ5-desaturation. Interestingly, the knock-out line still contained some ARA and EPA indicating that at least one other functional gene for a Δ5-desaturase should exist in *P. patens*. We did not detect any visibly altered phenotype in the knock-out plant, which maybe prevented by residual ARA and EPA.

Our data suggest, that two Δ5-desaturase genes are active on 20:3ω6 and 20:4ω3 in the moss *P. patens*. However, the identification of a Δ5-desaturase from *P. patens* completes the sequence of moss genes required for a successful reconstitution of VLCPUFA biosynthesis in oilseed crops.

4. References

Dohmen, R. J., Strasser, A. W. M., Höner, C. B., and Hollenberg, C. P. (1991) An efficient transformation procedure enabling long-term storage of competent cells of various yeast genera. Yeast 7, 691-692.

Girke, T., Schmidt, H., Zähringer, U., Reski, R., and Heinz, E. (1998) Identification of a novel Δ6-acyl-group desaturase by targeted gene disruption in *Physcomitrella patens*. Plant J. 15, 39-48.

Grimsley, N. H., Grimsley, J. M., and Hartmann, E. (1981) Fatty acid compositions of mutants of the moss *Physcomitrella patens*. Phytochemistry 20, 1519-1524.

Knutzon, D. S., Thurmond, J. M., Huang, Y.-S., Chaudhary, S., Bobik, E. G., Jr., Chan, G. M., Kirchner, S. J., and Mukerji, P. (1998) Identification of Δ5-desaturase from *Mortierella alpina* by heterologous expression in baker's yeast and canola. J. Biol. Chem. 273, 29360-29366.

Michaelson, L. V., Lazarus, C. M., Griffiths, G., Napier, J. A., and Stobart, A. K. (1998) Isolation of a Δ^5-fatty acid desaturase gene from *Mortierella alpina*. J. Biol. Chem. 273, 19055-19059.

Sperling, P. and Heinz, E. (2001) Desaturases fused to their electron donor. Eur. J. Lipid Sci. Technol. 102, 158-180.

Sperling, P., Lee, M., Girke, T., Zähringer, U:, Stymne, S., and Heinz, E. (2000) A bifunctional Δ^6-fatty acyl acetylenase/desaturase from the moss *Ceratodon purpureus*. A new member of the cytochrome b_5 superfamily. Eur. J. Biochem. 267, 3801-3811.

Strepp, R., Scholz, S., Kruse, S., Speth, V., and Reski, R. (1998) Plant nuclear knockout reveals a role in plastid division for the homolog of the bacterial cell division protein FtsZ, an ancestral tubulin. Proc. Natl. Acad. Sci. USA 95, 4368-4373.

Zank, T. K., Zähringer, U., Beckmann, C., Pohnert, G., Boland, W., Holtorf, H., Reski, R., Lerchl, J., and Heinz, E., Cloning and functional characterization of an enzyme involved in the elongation of Δ6-polyunsaturated fattty acids from the moss *Physcomitrella patens*. Plant J., in press

LOW-TEMPERATURE AND HYPOXIC SIGNALS REGULATING Δ9 FATTY ACID DESATURASE GENE TRANSCRIPTION ARE CONVERTED INTO MEMBRANE FLUIDITY SIGNAL IN YEAST

YOUJI NAKAGAWA, YOSHINOBU KANEKO and SATOSHI HARASHIMA

Department of Biotechnology, Graduate School of Engineering, Osaka University 2-1 Yamadaoka, Suita, Osaka 565-0871, Japan

1. Introduction

Poikilothermic organisms such as yeast, bacteria, plants and fish, respond to a decrease in temperature with the expression of a specific subset of proteins. Recently, two-component signal transducers consisting of histidine kinases and response regulators, as components of the pathway for perception and transduction of low-temperature signals, were identified in cyanobacteria (Suzuki et al., 2000) and bacilli (Aguilar et al., 2001). However, in eukaryotes, low-temperature signal sensors as well as low-temperature signal transducers have not yet been identified.

In yeast *Saccharomyces cerevisiae*, which is a unicellular eukaryote, fatty acid desaturation which requires molecular oxygen (O_2) as an electron acceptor is performed solely by Δ9 fatty acid desaturase encoded by *OLE1*. The *OLE1* transcription is regulated by nutrient fatty acids and molecular oxygen (O_2) (Nakagawa et al., 2001). Two functionally and genetically related *S. cerevisiae* membrane proteins, Spt23p and Mga2p, were found to be required for *OLE1* transcription (Zhang et al., 1999; Hoppe et al., 2000). Either Spt23p or Mga2p is sufficient for *OLE1* transcription because cells with *mga2 spt23* double mutation but not single mutation of either gene require unsaturated fatty acid (UFA) for growth (Zhang et al., 1999). Recently, Mga2p was shown to be essential for the hypoxic induction of *OLE1* expression (Jiang et al., 2001).

In this report, we demonstrate that *OLE1* transcription is transiently activated by temperature downshift. Furthermore, analysis of *OLE1* transcription in the Δ*mga2* cells revealed for the first time that Mga2p is a component of the low-temperature signal perception and transduction mechanism in eukaryotes. Since low-temperature reduces membrane fluidity, we suggest that hypoxic and low-temperature signals are converted to the membrane fluidity signal to activate *OLE1* transcription, therefore, Mga2p is the first identified eukaryotic sensor for recognizing membrane fluidity.

N. Murata et al. (eds.), Advanced Research on Plant Lipids, 117–120.

118

2. Results

2.1. OLE1 Transcription is Transiently Activated by Low-Temperature Signal

An important and widespread response of living organisms to environmental cold is an increase in the proportion of unsaturated fatty acids of membrane phospholipids (Hazel et al., 1990). This response offsets the direct cold-induced ordering of their membranes, termed 'homeoviscous adaptation' and is, perhaps, the clearest example of the homeostatic, cellular control of membrane lipid composition for structural and adaptive reasons. To clarify the effects of low temperature on *OLE1* transcription in *S. cerevisiae*, we examined the time course of *OLE1* transcription after temperature downshift.

Cells of wild-type strain were grown to exponential phase at 30°C and shifted to 10°C. Total RNAs were prepared from the cells before and after the shift and subjected to Northern blot analysis (Figure 1). Results revealed that the expression levels of *OLE1* and *OLE1*p-*PHO5* fusion gene transcripts are significantly higher 1h after the shift than at 0h (4.2- and 5.7-fold, respectively), indicating that *OLE1* transcription is activated by the temperature downshift.

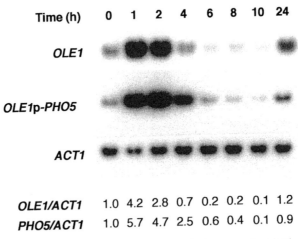

Time (h)	0	1	2	4	6	8	10	24
OLE1/ACT1	1.0	4.2	2.8	0.7	0.2	0.2	0.1	1.2
PHO5/ACT1	1.0	5.7	4.7	2.5	0.6	0.4	0.1	0.9

Figure 1. *OLE1* transcription is transiently activated by low-temperature signal. Northern analysis of the *OLE1*, *OLE1*p-*PHO5*, and *ACT1* transcripts in the wild-type cells after a temperature downshift from 30°C to 10°C. When the wild-type strain were grown at 30°C to exponential phase and shifted to 10°C, total RNAs were prepared from the cells immediately before the shift (0h) and 1h, 2h, 4h, 6h, 8h, 10h, and 24h after the shift. Equal amounts of RNA (10 μg) were electrophoresed on a 1.5% agarose gel in the presence of formaldehyde, transferred to a nylon filter, blotted, and hybridized with the following probes: ^{32}P-labelled DNA fragments containing *OLE1* and *PHO5* for detection of the transcripts of *OLE1*p-*PHO5* or *ACT1* as an internal control. The expression levels of all transcripts were normalized to those of *ACT1* mRNA and are presented as a decimal percent relative to that of the cells immediately before the shift.

2.2 Low-Temperature Signal is Transmitted to OLE1 through Membrane Protein Mga2p

To investigate whether the Spt23p and Mga2p play roles in transcriptional activation of *OLE1* by the low-temperature signal, Δ*spt23* and Δ*mga2* single disruptant, and Δ*spt23* Δ*mga2* double disruptant strains were grown to exponential phase at 30°C and shifted to 10°C. Total RNAs were prepared from the cells immediately before the shift (0h) and 1h after the shift, and subjected to Northern analysis (Figure 2). The degree of low-temperature-mediated induction of the *OLE1* transcript level in the Δ*mga2* single disruptant cells significantly decreased to one fifth compared with that in wild-type cells (lanes 1, 2, 5 and 6) and this induction was completely abolished in the Δ*mga2* Δ*spt23* double disruptant cells (lanes 9 and 10), indicating that Mga2p plays a vital role in the transcriptional activation of *OLE1* by the low-temperature signal.

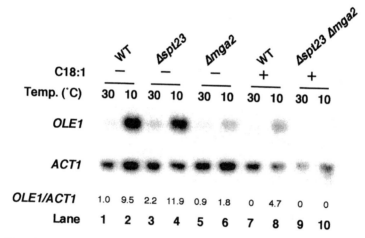

Figure 2. Mga2p is a vital component of low-temperature signal transduction pathway for inducing *OLE1* transcription. Northern analysis of the *OLE1* transcript in aerobically grown wild-type, Δ*spt23* and Δ*mga2* single, and Δ*spt23* Δ*mga2* double disruptant cells after a temperature downshift from 30°C to 10°C. The expression levels of *OLE1*, normalized to those of *ACT1* mRNA, are presented as decimal percentages relative to that for the wild-type cells before the shift.

3. Discussion

Based on the findings of this study, we propose an integrated model for *OLE1* transcriptional regulation by multiple signals, i.e., hypoxic, low-temperature and UFA signals. We assume that the hypoxic signal is recognized by cells as the intracellular UFA depletion signal. This idea is consistent with the facts that desaturation reactions require oxygen as an electron acceptor, and that the hypoxic activation of *OLE1* transcription is repressed by the presence of exogenously added UFA (Nakagawa et al., 2001). How then do cells recognize the intracellular UFA depletion? The decrease of intracellular UFA level is expected to reduce the biomembrane fluidity (Hazel et al., 1990). We found that *OLE1* transcription is transiently activated by a low-temperature signal (Fig. 1). As a low-temperature signal also reduces membrane fluidity (Hazel et al., 1990), it is possible that cells recognize low-temperature and hypoxic signals as a low-membrane-fluidity signal. We found that the membrane protein Mga2p is essential for low-temperature-mediated activation of *OLE1* transcription (Fig.2) (Nakagawa et al., 2002). Interestingly, Mga2p has also been shown to be essential for the hypoxic activation of *OLE1* transcription (Jiang et al., 2001). Therefore, it is highly possible that Mga2p is a sensor for recognizing alteration of membrane fluidity.

4. References

Aguilar, P. S., Hernandez-Arriaga, A. M., Cybulski, L. E., Erazo, A. C., and de Mendoza, D. (2001) Molecular basis of thermosensing: a two-component signal transduction thermometer in *Bacillus subtilis*. EMBO J. 20, 1681–1691.

Hazel, J. R., and Williams, E. E. (1990) The role of alterations in membrane lipid composition in enabling physiological adaptation of organisms to their physical environment. Prog. Lipid. Res. 29, 167-227.

Hoppe, T., Matuschewski, K., Rape, M., Schlenker, S., Ulrich, H. D. and Jentsch, S. (2000) Activation of a membrane-bound transcription factor by regulated ubiquitin/proteasome-dependent processing. Cell 102, 577-586.

Jiang, Y., Vasconcelles, M. J., Wretzel, S., Light, A., Martin, C. E. and Goldberg, M. A. (2001) *MGA2* is involved in the low-oxygen response element-dependent hypoxic induction of genes in *Saccharomyces cerevisiae*. Mol. Cell. Biol. 21, 6161-6169.

Nakagawa, Y., Sakumoto, N., Kaneko, Y. and Harashima, S. (2002) Mga2p is a putative sensor for low temperature and oxygen to induce *OLE1* transcription in *Saccharomyces cerevisiae*. Biochem. Biophys. Res. Commun. 291, 707-713.

Nakagawa, Y., Sugioka, S., Kaneko, Y. and Harashima, S. (2001) O2R, a novel regulatory element mediating Rox1p-independent O_2 and unsaturated fatty acid repression of *OLE1* in *Saccharomyces cerevisiae*. J. Bacteriol. 183, 745-751.

Suzuki, I., Los, D. A., Kanesaki, Y., Mikami, K., and Murata, N. (2000) The pathway for perception and transduction of low-temperature signals in *Synechocystis*. EMBO J. 19, 1327–1334.

Zhang, S., Skalsky, Y. and Garfinkel, D. J. (1999) *MGA2* or *SPT23* is required for transcription of the Δ9 fatty acid desaturase gene, *OLE1*, and nuclear membrane integrity in *Saccharomyces cerevisiae*. Genetics 151, 473-483.

FATTY ACID DESATURASES FROM THE DIATOM PHAEODACTYLUM TRICORNUTUM.

Reconstitution of polyunsaturated fatty acid biosynthetic pathways in transgenic yeast.

F. DOMERGUE[1], T.K. ZANK[1], A. ABBADI[1], P. SPERLING[1], A. MEYER[1], J. LERCHL[2,#] AND E. HEINZ[1].

[1] Institut für Allgemeine Botanik, Universität Hamburg, Ohnhorststr. 18, 22609 Hamburg, Germany;

[2] BASF Plant Science GmbH, BPS-A30, 67056 Ludwigshafen.

[#] present address: Plant Science Sweden AB, SE-26831 Svalöv, Sweden.

1. Introduction

Very long-chain polyunsaturated fatty acids (PUFAs) such as arachidonic acid (ARA, $20:4^{\Delta5,8,11,14}$) or eicosapentaenoic acid (EPA, $20:5^{\Delta5,8,11,14,17}$) have been shown to display many beneficial effects on human health. Today, the main commercial sources for these fatty acids are fish oil and micro-organisms. Since biotechnology may allow the production of specific fatty acids by genetic engineering of crops, the possibility of producing PUFAs in rapeseed or linseed has lead to the search for the genes encoding the enzyme activities involved in PUFA biosynthesis.

Phaeodactylum tricornutum is an unicellular silica-less diatom in which the EPA proportion reaches 30 % (Yongmanitchai and Ward, 1991). This marine diatom has been used as a model organism for PUFA biosynthesis, and labelling experiments have shown that EPA was synthesised by desaturations and elongation of fatty acids issued from the *de novo* fatty acid synthase (Arao and Yamada, 1994). We, therefore, decided to use this organism for the cloning of the different genes encoding the desaturases and the elongase involved in EPA synthesis.

2. Materials and methods

A combination of PCR, mass sequencing and library screening was used to identify the genes of interest and led to the identification of three putative fatty acid desaturases. The full-length clones were re-amplified by PCR with specific primers in order to insert specific restriction sites and subcloned in different yeast expression vectors (pYES2,

N. Murata et al. (eds.), Advanced Research on Plant Lipids, 121–124.

pVT102-U or pESC-LEU). For functional characterisation in *Saccharomyces cerevisiae*, the expressions were generally carried out for 48 hours at 20°C in the presence of 500 μM exogenously fed fatty acids. After expression, cells were harvested by centrifugation, washed and used to prepare FAMEs that were analysed by GLC as previously described (Zank et al., 2002).

3. Results

3.1 Isolation of putative desaturases

Three full-length sequences coding for putative fatty acid desaturases have been cloned. Using sequence homologies to already known desaturases, they have been annotated as Δ5-desaturase (PtD5, for *Phaeodactylum tricornutum* Δ5-desaturase), Δ6-desaturase (PtD6) and Δ12-desaturase (PtD12). The deduced protein sequence revealed that each sequence contains the 3 histidine clusters as well as the long hydrophobic stretches present in membrane-bound fatty acid desaturases. In addition, PtD5 and PtD6 contain a cytochrome b5 domain fused at their N-terminus, similarly to other front-end desaturases.

3.2 Functional characterisation

The enzymatic activities encoded by these sequences were determined by heterologous expression in *Saccharomyces cerevisiae*. The results obtained confirmed that we had isolated genes coding for fatty acid desaturases with different regioselectivities (Δ5, Δ6 and Δ12). The expression of PtD12 resulted in about 50 % conversion of oleic acid to $18:2^{\Delta9,12}$. For PtD6 and PtD5, the highest activities were measured with $18:2^{\Delta9,12}$ (28 % conversion) and $20:3^{\Delta8,11,14}$ (25 % conversion), respectively (Table 1).

TABLE 1: Substrate specificity of *Phaeodactylum tricornutum* desaturases expressed in *Saccharomyces cerevisiae*.

Yeast strain C13ABYS86 transformed with pYES2-PtD12, pVT102-U-PtD6 or pESC-LEU-PtD5) were grown for 48h at 20°C in the presence of different fatty acid substrates (500 μM or as indicated). Desaturation (%) was calculated as (product x 100) / (educt + product) using values corresponding to percent of total fatty acids.

Desaturase	Fatty acid substrate	Desaturation (%)	Desaturase	Fatty acid substrate	Desaturation (%)
PtD12	[a]$16:1^{\Delta9}$	13.4	PtD5	[b]$18:1^{\Delta11}$	2.0
	[a]$18:1^{\Delta9}$	50.1		[b]$20:1^{\Delta8}$	0
				[b]$20:1^{\Delta11}$	3.7
PtD6	[a]$16:1^{\Delta9}$	5.1		[b]$20:2^{\Delta11,14}$	10.5
	[a]$18:1^{\Delta9}$	4.9		$20:3^{\Delta11,14,17}$	11.1
	$18:2^{\Delta9,12}$	27.8		$20:3^{\Delta8,11,14}$	24.4
	$18:3^{\Delta9,12,15}$	23.5			

[a] in the absence of exogenously fed fatty acid, [b] in the presence of 1 mM exogenously fed fatty acid.

3.3 Reconstitution of PUFA biosynthetic pathways in yeast

Although no elongase was cloned from *Phaeodactylum tricornutum*, we reconstituted the biosynthetic pathway of arachidonic acid (ARA, $20:4^{\Delta5,8,11,14}$) in yeast using the elongase specific for $\Delta6$-PUFAs from *Physcomitrella patens*, PSE1 (Zank et al., 2002). The co-expression of PSE1 and PtD5 in the presence of γ-linolenic acid ($18:3^{\Delta6,9,12}$) or stearidonic acid ($18:4^{\Delta6,9,12,15}$) resulted in the accumulation of arachidonic acid ($20:4^{\Delta5,8,11,14}$) and eicosapentaenoic acid ($20:5^{\Delta5,8,11,14,17}$), respectively (data not shown). PSE1 elongated more than 50 % of each C_{18}-PUFA, whereas PtD5 converted 15 and 18 % of $20:3^{\Delta8,11,14}$ and $20:4^{\Delta8,11,14,17}$, respectively. The co-expression of PtD6, PSE1 and PtD5 in the presence of $18:2^{\Delta9,12}$ resulted in the synthesis of $18:3^{\Delta6,9,12}$, $20:2^{\Delta11,14}$, $20:3^{\Delta8,11,14}$ and $20:4^{\Delta5,8,11,14}$ in yeast (Figure 1).

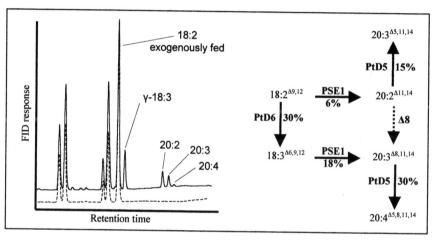

FIGURE 1: Arachidonic acid synthesis in transgenic yeast.

C13ABYS86 yeast strain transformed with either the empty vector (dashed line) or pVT102-U-PtD6 and pESC-LEU-PSE1-PtD5 (full line) was supplemented with 500 µM $18:2^{\Delta9,12}$ and grown for 48h at 20°C. FAMEs from the whole cells were prepared and analysed by GLC.

In detail about 30 % of the exogenously fed $18:2^{\Delta9,12}$ was converted to $18:3^{\Delta6,9,12}$ by PtD6, 18 % of it being further elongated to $20:3^{\Delta8,11,14}$ by PSE1. At the same time, PSE1 converted about 6 % of $18:2^{\Delta9,12}$ to $20:2^{\Delta11,14}$ due to its low activity on $\Delta9$-fatty acids. It should be noted that the presence of a $\Delta8$-desaturase like that found in *Euglena gracilis* (Wallis and Browse, 1999) could bring back this side product into the ARA pathway (Figure 1 right, broken arrow). PtD5 converted about 30 % of $20:3^{\Delta8,11,14}$ to $20:4^{\Delta5,8,11,14}$, which represented 0.5 % of the total fatty acids. In good agreement with its substrate specificity, PtD5 was also active on $20:2^{\Delta11,14}$, which resulted in the presence of $20:3^{\Delta5,11,14}$ (not shown). Although all these products could have been anticipated taking in consideration the substrate specificities of the different enzymes used, the rather low elongation of $18:3^{\Delta6,9,12}$ by PSE1 was unexpected as this enzyme had been shown to be highly specific for $\Delta6$-PUFAs (Zank et al., 2002).

3. Conclusions and perspectives

For the first time from a diatom, several fatty acid desaturases were cloned and functionally characterised. The three fatty acid desaturases described here display substrate specificities supporting their involvement in the biosynthesis of eicosapentaenoic acid ($20:5^{\Delta5,8,11,14,17}$). In good agreement with the favourite route for EPA biosynthesis in *Phaeodactylum tricornutum* described by Arao and Yamada (1994), the best activities of PtD6 and PtD5 were obtained with $18:2^{\Delta9,12}$ and $20:4^{\Delta8,11,14,17}$, respectively.

When the arachidonic acid pathway was reconstituted in yeast by co-expressing PtD6, PSE1 and PtD5 in the presence of $18:2^{\Delta9,12}$, a rather low elongation of $18:3^{\Delta6,9,12}$ was observed (Figure 1). This elongation differed considerably from the 50 % elongation obtained when $18:3^{\Delta6,9,12}$ was exogenously fed (Zank et al., 2002), suggesting that different lipid pools may be used for the desaturation and elongation steps of PUFA biosynthesis. Moreover, the fact that *Phaeodactylum tricornutum* accumulates rather exclusively EPA, whereas all the intermediates are barely detectable, suggests that additional activities may be involved in these organisms to achieve such a specific accumulation. Understanding the highly effective strategies to channel all biosynthetic intermediates towards the accumulation of a single end-product becomes a prerequisite to implement PUFA biosynthesis in oilseed crops.

Acknowledgement

This research has been supported by a Marie Curie Fellowship of the European Community programme Human Potential under the contract number HPMF-CT-1999-00148. We thank BASF Plant Science GmbH (Ludwigshafen, Germany) for providing the λ-ZAP Express library and for performing the random sequencing.

4. References

Yongmanitchai W. and Ward O. (1991) Screening of algae for potential alternative sources of eicosapentaenoic acid. Phytochemistry 30(9), 2963-2967.

Arao T. and Yamada M. (1994) Biosynthesis of polyunsaturated fatty acids in the marine diatom, *Phaeodactylum tricornutum*. Phytochemistry 35(5), 1177-1181.

Zank T.K., Zähringer U., Beckmann C., Pohnert G., Boland W., Holtorf H., Reski R., Lerchl J. and Heinz E. (2002) Cloning and functional characterisation of an enzyme involved in the elongation of Δ6-polyunsaturated fatty acids from the moss *Physcomitrella patens*. Plant J. (in press).

Wallis J.G. and Browse J. (1999) The Δ8-desaturase of *Euglena gracilis*: an alternate pathway for synthesis of 20-carbon polyunsaturated fatty acids. Arch. Biochem. Biophys. 365(2), 307-316.

CHARACTERIZATION OF TRANSGENIC MAIZE ENGINEERED WITH ANTISENSE FAD2 cDNA BY CHEMOMETRICAL ANALYSIS OF GLYCEROLIPIDS

IVELIN RIZOV, ANDREAS DOULIS*

AgroBioInstitute, 2232 Kostinbrod-2, Bulgaria,
* Mediterranean Agronomic Institute of Chania, Laboratory of Molecular Biology, Greece

Introduction

Chemometrics provides statistical methods for the study of joint relationships of intercorrelated biochemical data. Since several variables can be considered simultaneously, generalised interpretations can be made that could not be possible if univariate statistics were applied. Cluster analysis (CA), principal component analysis (PCA) and discriminant analysis (DA) have been widely used in lipid research (Lee D. et al., 1998; Matsumoto M et al., 1997; Stevens J.J. and Jones R.K., 1992; Tsimidou M. et al., 1987). However, we have seen no previous reports involving the use of a chemometrical approach to study the glycerolipid composition of photosynthetic plant tissues.

In this study, the content of seven glycerolipids: monogalactosyldiacylglycerol (MGDG), digalactosyldiacylglycerol (DGDG), sulphoquinovosyldiacylglycerol (SQDG) and phosphatidyletahnolamine (PEA), phosphatidylinositol (PI), phosphatidyl-choline (PC) , phosphatidilglycerol (PG), isolated from leaves, was used for characterization of eigththeen transgenic maize plants engineered with FAD2 cDNA in antisense orientation.

Material and Methods

Glycerolipid separation and identification was performed according to the method of Rizov and Doulis, 2001.

The chemometrical analysis of the biochemical data was performed with the SPSS/PC version 8.0 software. The content of lipid classes was presented as percentages .

The maize leave material was kindly provided by Prof. Paul Christou, Jhon Innes Center, UK (Gahakwa D. M., 2001). The used FAD2 cDNA was isolated from maize

N. Murata et al. (eds.), Advanced Research on Plant Lipids, 125–128.

(Mikkilineni, V., and T.R Rocheford, 2002) and with micro-projection bombardment was inserted in antisense orietation back to the maize plant under control of 35S promoter.

Results and Discussion

The glycerolipid composition of the leaves of maize plants engineered with FAD2 cDNA, in antisense orientation, as well as non-transgenic control tissue is presented in Table 1. The lipid profile of controls (C.1 and C.2) is in a good agreement with previously reported values. In transformed maize leaf tissues the content of SQDG was doubled in comparison with that of the control plants (Table 1). The contents of PEA and PI were increased approximately 1.5 fold (Table 1). The content of PC was also increased by 19.8%. Contrary to these increases, the levels of MGDG and PG were decreased approximately 20% for both. Nevertheless, the content of DGDG is not affected. For certain lipid classes some engineered maize plants (samples 5.0, 3.0, 4.10 vs ct.) displayed similarity to controls. On the other hand, there were other cases where transformed plants (samples 1.2, 5.0, 9.2 vs ct.) had overall lipid profile similar to this of controls.

Following these initial observations it became apparent that changes of seven glycerolipid classes across twenty samples cannot be analyzed by simple means comparisons. Consequently, a multivariate statistical approach was used to assess alterations of lipid profiles. Differences between plants were first assessed by means of CA, which was followed by PCA. The two analyses provided similar results.

The average linkage (UPGMA) clustering method showed that four transgenic maize plants (5.10, 2.0, 1.2 and 5.0) had lipid profiles similar to that of control (C.1 and C.2) and formed with them a single group. Five engineered maize plants (9.2, 6.12, 4.1, 1.0 and 3.0) were found to belong in a second group. A third group included the remaining nine maize plants (9.10, 9.0, 6.0, 4.0, 4.10, 9.13, 6.2, 9.11 and 4.3). The same grouping was maintained when additional similarity criteria (complete linkage and centroid method) were applied. The additional similarity criteria were applied in order to check for possible statistical artifacts in the grouping process.

The first two principal components (PC1 and PC2) of the PCA performed on the correlation matrix of the original data accounted for 68% of the total variability of the lipids. Additional components (3rd and 4th) increase this amount to 91.5%, but the corresponding figure adds little to the classification found in the space of PC1 and PC2. The first principal component is positively related with PEA, SQDG and PI and negatively with MGDG. The PG and the SQDG are the lipid classes which mainly determined variation in the second principal component. In the reduced lipid space (PC1 and PC2) the maize plants were clustered again into three groups. This grouping is similar to that produced by cluster analysis.

The increased distance between the group containing the controls and the two other

Table 1. Glycerolipid profile of fad2 antisense transgenic and non-transformed maize leaf tissues

No	Samples		Lipid classes, % of total						
		MGDG	DGDG	SQDG	PEA	PG	PC	PI
1	C.1	44,0	30,9	4,3	1,2	11,5	7,2	0,9
2	C.2	45,1	28,5	9,5	1,9	11,9	2,4	0,7
Mean of controls		44.6 ± 0.8	29.7 ± 1.7	6.9 ± 3.7	1.5 ± 0.5	11.7 ± 0.3	4.8±3.4	0.8 ± 0.1
3	4.1	34,8	30,4	13,0	6,5	10,9	2,2	2,2
4	4.0	33,8	27,0	16,9	4,1	6,8	10,1	1,4
5	6.0	30,4	29,4	14,0	5,6	9,4	9,4	1,9
6	6.2	30,0	32,5	22,5	0,8	5,0	5,8	3,3
7	4.10	31,1	29,8	17,4	1,7	7,8	7,8	4,4
8	1.0	39,1	34,2	11,0	2,4	9,8	2,4	1,0
9	9.10	26,8	28,7	15,3	6,9	10,7	9,6	1,9
10	9.0	26,5	28,6	16,3	6,1	10,2	10,2	2,0
11	4.3	24,2	25,3	16,3	9,0	11,9	10,8	2,5
12	3.0	31,3	33,0	6,3	7,1	15,2	4,5	2,7
13	9.13	34,0	24,3	19,4	5,8	10,2	2,4	3,9
14	9.2	41,2	30,9	15,5	1,1	5,2	5,2	1,0
15	6.12	38,0	30,4	17,1	1,1	7,6	4,6	1,1
16	9.11	30,8	33,6	21,0	1,0	8,4	3,9	1,4
17	5.10	53,6	31,2	2,8	0,7	7,1	3,5	1,1
18	2.0	51,0	30,3	2,0	1,0	10,2	4,1	1,4
19	1.2	47,2	30,2	3,4	1,0	10,1	6,7	1,4
20	5.0	44,4	31,0	7,8	2,3	13,0	0,3	1,3
Mean of transformants		36.0 ± 8.5	30.0 ± 2.6	13.2 ± 6.3	3.6 ± 2.8	9.4 ± 2.6	5.7±3.2	2.0 ± 1.0
Roughan and Batt, 1969		42	31	5	3	7	6	1
Leech et al., 1973	A*	29	23	12	10	3	23	N.D.[2]
	B*	36	26	12	7	3	15	N.D.[2]
	C*	43	25	16	4	3	9	N.D.[2]
	D*	47	27	15	3	3	6	N.D.[2]
	E*	51	28	13	1	3	3	N.D.[2]

* A to E refer to successive sections up the leaf. Section A is, therefore, at the base of the leaf and contains the youngest tissue; N.D.[2] – not determined

groups corresponded to increases in PEA, SQDG and PI and decreases in MGDG. This grouping is reflected in the mean values of these compounds in each group. Differences between the three plant groups (for each lipid class separately) were tested by means of ANOVA and Duncan tests. The most pronounced differences are exhibited by MGDG and SQDG. This is in agreement with the observation at the PCA part where MGDG and SQDG are two major glycerolipid classes determining the variation of the first principal component.

The engineered maize plants showing similarity with controls (belonging in the same group) exchibit very low expression of the transgene. On the other hand, engineered maize plants which were clustered in the third group are with the strongest expression of the engineered antisense ER oleate desaturase (FAD2). For the maize plants in the second group, the effect of the transgenic modification was significant in terms of altered lipid composition. This was especially obvious regarding the levels of SQDG, PEA and PI. Nevertheless, the alteration of the phenotype is in intermediate levels between control and third group.

Conclusion

The consensus between two different methods of multivariate analysis, one for ordination (PCA) and other for classification (CA) of FAD2 antisense maize plants in lipid space, provides strong evidence that the structure of the studied maize plant set results from FAD2 antisense engineering , i.e. is not merely a statistical artifact. In that respect is possible to conclude that the altering expression of a desaturase has affected the distribution of glycerolipid classes.

References

Lee D., Noh B., Bae S. and Kim K. (1998) Characterization of fatty acids composition in vegetable oils by gas chromatography and chemometrics, Analytica Chimica Acta 58, 163 – 175

Leech P.M., Rumsby M.G. and Thompson W.W. (1973) Plastid differentiation, acyl lipid, and fatty acid changes in developing green maize leaves, Plant Physiol. 52, 240 – 245

Matsumoto M., Furuya N. and Matsuyama N. (1997) Characterization of Rhizoctonia spp., causal agents of sheath diseases of rice plant, by total cellular fatty acid analysis, Ann. Phytopathol. Soc. Jpn. 63, 149 – 154

Mikkilineni, V., and T.R Rocheford (2002) Genomic Organization of Fatty Acid Desaturase-2 (*fad2*) and Fatty Acid Desaturase-6 (*fad6*) EST's in Maize. Theor. Appl. Genet. (in press).

Rizov I. and Doulis A. (2001) Separation of plant membrane lipids by multiple solid phase extraction, Journal of Chromatography A 922, 347-354

Roughan P.G. and Batt R.D. (1969) The glycerolipid composition of leaves, Phytochemistry 8, 363 – 369

Stevens J.J. and Jones R.K. (1992) Determination of whole-cell fatty acids in isolates of Rhizoctonia solani AG-1 IA, Phytopathology 82, 68 –72

Tsimidou M, Macrae R and Wilson I (1987) Authentication of virgin olive oils using principal component analysis of triglyceride and fatty acid profiles: part1– classification of Greek olive oils, Food Chem. 25, 251-256

BIOCHEMICAL STUDIES ON THE PRODUCTION OF DOCOSAHEXAENOIC ACID (DHA) IN *EUGLENA* AND *THRAUSTOCHYTRIUM*

A. MEYER, C. OTT AND E. HEINZ
Universität Hamburg, Institut für Allgemeine Botanik, Ohnhorststr. 18, 22609 Hamburg, Germany

1. Abstract

A strategy based on fatty acid α-oxidation was developed to discriminate $\Delta 4$-desaturation and the involvement of a polyketide synthase system in the production of docosahexaenoic acid ($22:6^{\Delta 4,7,10,13,16,19}$) from exogenously supplied docosapentaenoic acid ($22:5^{\Delta 7,10,13,16,19}$) in *Euglena gracilis* and *Thraustochytrium* sp. *Thraustochytrium* also seems to possess an $\omega 3$-desaturase activity working on C22 fatty acid substrates. This activity was apparently missing in *Euglena*.

2. Introduction

The inclusion of very long chain polyunsaturated fatty acids (VLC-PUFAs) such as docosahexaenoic acid ($22:6^{\Delta 4,7,10,13,16,19}$) in the diet is considered to be beneficial for human health. Attempts are now underway to implement the pathways of PUFA biosynthesis in oilseed crops (Abbadi et al., 2001). A variety of bacteria, algae and fungi, known for their ability to produce VLC-PUFAs, may prove as suitable gene sources (Russell and Nichols, 1999; Vazhappilly and Chen, 1998). Two different routes leading to PUFA biosynthesis have been described in microorganisms. First, a system of alternating desaturases and elongases referred to as the $\omega 6$- and $\omega 3$-pathways, and second, polyketide synthase (PKS) systems which are acting in some marine bacteria (Yazawa, 1996; Tanaka et al., 1999). PKS systems do not seem to accept intermediates of DHA synthesis as substrates. Since PKS systems might also be involved in PUFA synthesis in eukaryotes like the DHA-rich fungus *Schizochytrium* (Metz et al., 2001), it is advantageous to be able to discriminate between the two systems in order to evaluate the potential of these organisms as gene sources. *Thraustochytrium* sp., a fungus closely related to *Schizochytrium*, and *Euglena gracilis*, an alga with a very versatile fatty acid profile (Korn, 1964), were chosen as candidate organisms to test a method for discrimination.

N. Murata et al. (eds.), Advanced Research on Plant Lipids, 129–132.

3. Materials and methods

Euglena gracilis was grown at 23°C in a 16 h light / 8 h dark cycle. *Thraustochytrium* sp. ATCC 26185 was grown at 30°C in the dark. Cultures of both microorganisms were incubated for 12 or 24 h with docosatetraenoic acid ([2-^{14}C]22:4ω6) or docosapentaenoic acid ([2-^{14}C]22:5ω3), respectively. The radiolabelled fatty acids were kindly provided by BASF (Ludwigshafen, Germany). Fatty acid methyl esters (FAME) and free fatty acids (FFA) were prepared from cells or lipid extracts and analysed by RP-HPLC. The labelled and unlabelled DHA (22:6ω3, FFA) from 2-3 labelling experiments were collected, pooled, hydrogenated to behenic acid (22:0) and α-oxidised by potassium permanganate. The α-oxidation led to a chain shortening of the behenic acid beginning at the carboxyl end. The saturated fatty acids were converted into their bromophenacyl ester derivatives and analysed by RP-HPLC.

4. Results and discussion

In this presentation we demonstrate a method to discriminate between the desaturation/ elongation pathway and the involvement of a polyketide synthase system in the synthesis of docosahexaenoic acid (DHA) from the fatty acid substrate docosapentaenoic acid (22:5ω3). *Euglena gracilis* and *Thraustochytrium* sp. produced 22:5ω6 and 22:6ω3 from exogenously supplied 22:4ω6 and 22:5ω3, respectively. Radiolabelled C24 fatty acids could not be detected, indicating that the PUFA synthesis pathway occuring in mammals (Sprecher et al., 1995) probably does not operate in these microorganisms. *Thraustochytrium* was also able to introduce an ω3-double bond into C22 fatty acids. This activity could not be detected in *Euglena gracilis*. The radiolabel of the DHA isolated from both organisms could be confined to position C-1 or C-2, indicating that the fatty acid resulted from direct desaturation of docosapentaenoic acid. Only little radioactivity could be detected in other positions of the acyl chain from *Thraustochytrium*, which might originate from fatty acid β-oxidation and resynthesis by desaturases/elongases or a PKS system. *Thraustochytrium* is closely related to *Schizochytrium*, a fungus probably harboring a PKS system for the production of docosahexaenoic acid. *Thraustochytrium* possesses, in contrast to the alga *Euglena gracilis*, high proportions of endogenous DHA and lacks significant percentages of intermediate fatty acids (Weete et al., 1997), supporting the idea of an effective channeling by a PKS system, whereas the biochemical results in this work argue for the involvement of a Δ4-fatty acid desaturase in DHA production. The cloning of a Δ4-desaturase from *Thraustochytrium* (Qiu et al., 2001) demonstrated that the fungus may actually represent a useful gene source for both the desaturase/elongase and the polyketide synthase system.

TABLE 1. *In vivo* conversion of exogenously supplied docosatetraenoic acid ([2-^{14}C]22:4$^{\Delta 7,10,13,16}$) and docosapentaenoic acid ([2-^{14}C]22:5$^{\Delta 7,10,13,16,19}$) by *Eulgena gracilis* and *Thraustochytrium* sp.

Organism	[^{14}C]fatty acid supplied	[^{14}C]fatty acids detected*
Euglena gracilis	22:5$^{\Delta 7,10,13,16,19}$	22:5$^{\Delta 7,10,13,16,19}$
		22:6$^{\Delta 4,7,10,13,16,19}$
	22:4$^{\Delta 7,10,13,16}$	22:4$^{\Delta 7,10,13,16}$
		22:5$^{\Delta 4,7,10,13,16}$
Thraustochytrium sp.	22:5$^{\Delta 7,10,13,16,19}$	22:5$^{\Delta 7,10,13,16,19}$
		22:6$^{\Delta 4,7,10,13,16,19}$
	22:4$^{\Delta 7,10,13,16}$	22:4$^{\Delta 7,10,13,16}$
		22:5$^{\Delta 4,7,10,13,16}$
		22:5$^{\Delta 7,10,13,16,19}$
		22:6$^{\Delta 4,7,10,13,16,19}$

*Fatty acids were assigned by comparison with the retention times of known standards except for 22:5$^{\Delta 4,7,10,13,16}$ which was tentatively assigned due to its retention time comparable to that of a major fatty acid peak from *Thraustochytrium*.

TABLE 2. α-Oxidation of 22:6$^{\Delta 4,7,10,13,16,19}$ (DHA) that was produced from ^{14}C-labelled 22:5$^{\Delta 7,10,13,16,19}$ by *Euglena gracilis* and *Thraustochytrium* sp.

Euglena gracilis			
fatty acid peak	% radioactivity (R)	% UV absorption at 254 nm (A)	ratio (R/A)
22:0	77.8	5.0	15.6
21:0	21.1	1.3	16.2
20:0	n.d.*	3.4	-
19:0	n.d.	2.4	-
18:0	n.d.	8.6	-
17:0	n.d.	3.7	-
16:0	n.d.	8.3	-
15:0	n.d.	4.1	-

Thraustochytrium sp.			
fatty acid peak	% radioactivity (R)	% UV absorption at 254 nm (A)	ratio (R/A)
22:0	78.3	36.1	2.2
21:0	12.9	5.3	2.4
20:0	1.0	11.2	0.1
19:0	n.d.	5.4	-
18:0	1.6	8.5	0.2
17:0	n.d.	4.4	-
16:0	n.d.	7.0	-
15:0	n.d.	3.0	-

n.d.: not detected

132

5. References

Abbadi, A., Domergue, F., Meyer, A., Riedel, K., Sperling, P., Zank, T. K. and Heinz, E. (2001) Transgenic oilseeds as sustainable source of nutritionally relevant C20 and C22 polyunsaturated fatty acids? Eur. J. Lipid Sci. Technol. 103, 106-113

Korn, E. D. (1964) The fatty acids of *Euglena gracilis*. J. Lipid Res. 5, 352-62

Metz, J. G., Roessler, P., Facciotti, D., Levering, C., Dittrich, F., Lassner, M., Valentine, R., Lardizabal, K., Domergue, F., Yamada, A., Yazawa, K., Knauf, V. and Browse, J. (2001) Production of polyunsaturated fatty acids by polyketide synthases in both prokaryotes and eukaryotes. Science 293, 290-292

Qiu, X., Hong, H. and MacKenzie, S. L. (2001) Identification of a Δ4 fatty acid desaturase from *Thraustochytrium* sp. involved in the biosynthesis of docosahexaenoic acid by heterologous expression in *Saccharomyces cerevisiae* and *Brassica juncea*. J. Biol. Chem. 276, 31561-31566

Russell, N. J. and Nichols, D. S. (1999) Polyunsaturated fatty acids in marine bacteria – a dogma rewritten. Microbiology 145, 767-779

Sprecher, H., Luthria, D. L., Mohammed, B. S. and Baykousheva, S. P. (1995) Reevaluation of the pathways for the biosynthesis of polyunsaturated fatty acids. J. Lipid Res. 36, 2471-2477

Tanaka, M., Ueno, A., Kawasaki, K., Yumoto, I., Ohgiya, S., Hoshino, T., Ishizaki, K., Okuyama, H. and Morita, N. (1999) Isolation of clustered genes that are notably homologous to the eicosapentaenoic acid biosynthesis gene cluster from the docosahexaenoic acid–producing bacterium *Vibrio marinus* strain MP-1. Biotechnology Letters 21, 939-945

Vazhappilly, R. and Chen, F. (1998) Eicosapentaenoic acid and docosahexaenoic acid production potential of microalgae and their heterotrophic growth. JAOCS 75, 393-397

Weete, J. D., Kim, H., Gandhi, S. R., Wang, Y. and Dute, R. (1997) Lipids and ultrastructure of *Thraustochytrium* sp. ATCC 26185. Lipids 32, 839-845

Yazawa, K. (1996) Production of eicosapentaenoic acid from marine bacteria. Lipids 31, 297-300 (Supplement)

NONTHERMAL FACTORS AFFECTING PRODUCTION AND UNSATURATION OF ALKENONES IN *Emiliania huxleyi* AND *Gephyrocapsa oceanica*

J.M. SORROSA[1], M. YAMAMOTO[2] and Y. SHIRAIWA[1]

[1]*Institute of Biological Sciences, University of Tsukuba, Tsukuba, 305-8572, Japan,* [2]*Graduate School of Environmental Earth Science, Hokkaido University, Sapporo, 060-0810, Japan*

1. Introduction

Alkenones are long-chain (C_{37}, C_{38}, C_{39}) unsaturated methyl and ethyl ketones, and they occur widely in marine and lacustrine sediments. The recognized major source of the observe alkenones are the coccolithophorids, *Emiliania huxleyi* and the closely related *Gephyrocapsa oceanica*. Alkenones tend to retain the degree of unsaturation fixed at the time of synthesis at the sea surface and some of them were preserved at the sediment surface. $U^{K'}_{37}$ was used as an index of the unsaturation degree of C_{37} alkenones and has been widely used as a paleoceanographic tool (Volkman et al., 1980). $U^{K'}_{37}$ was calculated based on the previous report as: $U^{K'}_{37} = C_{37:2} / (C_{37:2} + C_{37:3})$ (Brassell et al., 1986; Prahl and Wakeham, 1987).

To assess the factors affecting the changes in the $U^{K'}_{37}$ and the production and the stability of alkenones and alkenoates during growth from lag to stationary phase, *Emiliania huxleyi* and *Gephyrocapsa oceanica* were grown in a batch culture at 10, 15, 20 and 25°C.

2. Materials and Methods

Pre-experimental cultures of *Emiliania huxleyi* (strain EH2) and *Gephyrocapsa oceanica* (strain GO1) were grown in an artificial seawater, Marine Art SF (Senju Pharmaceutical Co., Osaka, Japan) enriched with ESM supplement in which soil extract was replaced with 10 nmol/L selenite (Danbara and Shiraiwa, 1999) at various temperatures from 10°C to 25°C under continuous illumination at the intensity of 30μmol/m²/s. After acclimating cells at respective temperature for about 8 days, preculture cells were transferred to the experimental culture for analyzing alkenones and alkenoates. Parameters of cell growth and separation and analytical procedures of alkenones were according to Danbara and Shiraiwa (1999) and Yamamoto et al. (2000).

N. Murata et al. (eds.), Advanced Research on Plant Lipids, 133–136.

134

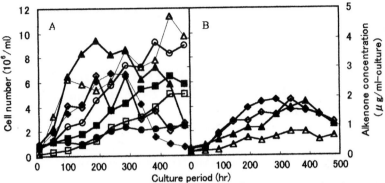

Figure 1. Changes in cell number (closed symbols) and alkenone concentration (open symbols) of *Emiliania huxleyi* (A) and *Gephyrocapsa oceanica* (B) cultured at varying temperatures such as 10°C (●○), 15°C (■□), 20°C (▲△) and 25°C (♦◇).

3. Results and Discussion

3.1 . *Alkenones and alkenoates*

G. oceanica did not grow at 15°C, indicating it is very sensitive to low temperature. Cells of *E. huxleyi* and *G. oceanica* contained the common alkenone components (C_{37}, C_{38}, C_{39}) at all growth temperatures, as reported previously (Volkman et al., 1980). Alkenone production was observed in accordance with growth status of cells and maximized in late logarithmic and linear phases (Fig. 1). The amount of alkenones remained unvaried even when cells started senescence, thus suggesting that alkenones are chemically and/or biologically stable compounds that can still be preserved even in dead cells.

Total cell volume of *G. oceanica* was several times bigger than *E. huxleyi* because it has larger cells and more coccoliths (data not shown). However, the alkenone content per cell in *E. huxleyi* was slightly higher than that in *G. oceanica* (Fig. 2). It indicates that the amount of alkenone synthesized in each cells is not influenced by cell size, as suggested by Conte et al. (1998) but rather controlled by genetics. Alkenone content per cell in *E. huxleyi* was significantly higher at 10°C than at the other temperatures, suggesting that alkenones might be necessary at such critical temperature (Fig. 2).

3.2. *Unsaturation index of alkenones and ethyl alkenoate ratio to total C_{37} alkenones*

Unsaturation index ($U^{K'}_{37}$) is useful for the estimation of paleo-sea-surface temperature from the analysis of alkenones preserved in oceanic sediments. $U^{K'}_{37}$ was slightly variables especially at lag and logarithmic phases during isothermal culture in both species at all growth temperatures (Fig. 3A,B). The ratio between ethyl alkenoate to total C_{37} (EE/K_{37}) might be useful in understanding relationship between the algal growth phase and the corresponding type of marine environment (Yamamoto et al.

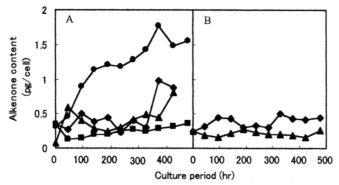

Figure 2. Changes in alkenone content per cell of *Emiliania huxleyi* (A) and *Gephyrocapsa oceanica* (B) cultured at varying temperatures such as 10°C (●), 15°C (■), 20°C (▲) and 25°C (◆).

2000). EE/K$_{37}$ changed slightly (Fig. 3C,D), suggesting that growth phase influences the unsaturation degree of alkenones as well as the relative proportion of the concentrations of alkenoate to alkenones.

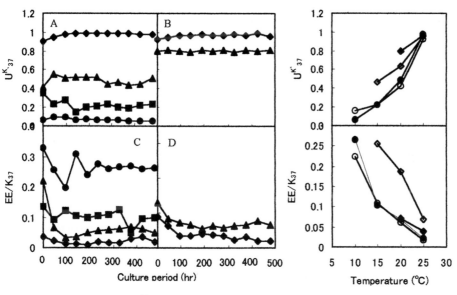

Figure 3 (Left). Changes in U$^K_{37}$ and the ratio of ethyl alkenoate to C$_{37}$ alkenones in *Emiliania huxleyi* (A and C) and *Gephyrocapsa oceanica* (B and D) grown at various temperatures such as 10°C (●), 15°C (■), 20°C (▲) and 25°C (◆).

Figure 4 (Right). U$^K_{37}$ and EE/K$_{37}$ values of *Emiliania huxleyi* (●○) and *Gephyrocapsa oceanica* (◆◇) plotted against growth temperature. The closed symbols (●◆) are the values obtained in this study and the open symbols (○◇) are the data by Sawada et al. (1996) in which the same strains were used.

$U^{K'}_{37}$ increased with temperature in both *E. huxleyi* and *G. oceanica,* while EE/K_{37} decreased with increasing temperature (Fig. 4), as previously reported (Prahl and Wakeham, 1987; Conte et al., 1998; Sawada et al. 1996). Interestingly, differences in those values were observed when results were compared to Sawada et al. (1996) (Fig. 4) where both studies used the same strains. Furthermore, Sawada et al. (1996) were able to grow *G. oceanica* at 15°C. This implies that maintenance of strains in a constant environment for a long period developed some physiological changes in temperature dependency of lipid metabolism.

In conclusion, alkenones produced remained undegraded at all temperatures even when cells started senescence. It indicates that alkenones are chemically and/or biologically stable compounds that can still be preserved without modification of $U^{K'}_{37}$ in damaged or dead cells. The synthesis of alkenones by *E. huxleyi* is stimulated especially at 10°C. Amount of alkenones was influenced not only by temperature but also by genetics and other nonthemal factor such as growth stage but independent on cell size. $U^{K'}_{37}$ values changed at early stage of growth even during isothermal culture.

4. References

Brassell, S.C, (1993) Applications of biomarkers for delineating marine paleoclimatic fluctuations during the Pleistocene. *in* M.H. Engel and S.A. Macko (eds.), Organic Geochemistry. Plenum Press, New York, pp. 699-738.

Conte, M.H., Thompson, A., Lesley, D., Harris, R.P. (1998) Genetic and physiological influences on the alkenone/alkeoate versus growth temperatures relationship in *Emiliania huxleyi* and *Gephyrocapsa oceanica.* Geochim. et Cosmochim. Acta 62, 51-68.

Danbara, A., Shiraiwa, Y. (1999) The requirement of selenium for the growth of marine coccolithophorids, *Emiliania huxleyi, Gephyrocapsa oceanica* and *Helladosphaera* sp. (Prymnesiophyceae). Plant Cell Physiol. 40, 762-766.

Prahl, F.G., Wakeham, S.G. (1987). Calibration of unsaturation patterns in long-chain ketone compositions for paleotemperature assessment. Nature 330, 367-369.

Sawada, K., Handa, N., Shiraiwa, Y., Danbara, A., Montani, S. (1996) Long-chain alkenones and alkyl alkenoates in the coastal and pelagic sediments of the northwest North Pacific, with special reference to the reconstruction of *Emiliania huxleyi* and *Gephyrocapsa oceanica* ratios. Org. Geochem. 24, 751-764.

Volkman, J.K., Eglinton, G., Corner, E.D.S., Forsberg, T.E.V. (1980) Long-chain alkenes and alkenones in the marine coccolithophorid *Emiliania huxleyi.* Phytochem. 19, 2619-2622.

Yamamoto, M., Shiraiwa, Y., Inouye, I. (2000) Physiological responses of lipids in *Emiliania huxleyi* and *Gephyrocapsa oceanica* (Haptophyceae) to growth status and their implications for alkenone paleothermometry. Org. Geochem. 31, 799-811.

Chapter 5:

Lipid Biosynthesis

DETERMINATION OF THE X-RAY CRYSTALLOGRAPHIC STRUCTURE OF *E.COLI* BUTYRYL-ACP.

J.W.SIMON[1], J. GILROY[1], A.ROUJENIKOVA[2],J. RAFFERTY[2],
C.BALDOCK[2], P.J.BAKER[2],D.W. RICE[2] ,A.R.STUITJE[3] AND A.R.SLABAS
[1]*Department of Biological and Biomedical Sciences, University of Durham,
South Rd., Durham DH1 3LE UK.,*
[2]*Krebs Institute for Biomolecular Research, Department of Molecular Biology
and Biotechnology, The University of Sheffield, Sheffield S10 2TN, UK,*
[3]*Department of genetics, Institute of Molecular Biological Sciences [IMBW],
VrijeUniversiteit, Biocentre Amsterdam, 1081 HV Amsterdam, Netherlands.*

1.Introduction

Acyl Carrier Protein [ACP] is a small acidic protein which is an essential component of the type 2 , dissociable fatty acid synthetase [1] . ACP is synthesised as an apo-protein and post-translationally modified, by the addition of a 4'phosphopantetheine [4'PANT] prosthetic group, which is linked to a serine residue by a phosphodiester bond. The amino acid sequence around the modified serine residue is highly conserved [2]. The post–translational modification is catalysed by holo-ACP synthetase [HAS], and this enzyme can also catalyse the direct conversion of apo-ACP to acyl-ACP using acyl-CoA as a source of the both the 4'PANT and the acyl group [3]. Apart from the role of ACP in fatty acid biosynthesis it is an important component of a number of other reactions which require acyl-transferase steps, these include: membrane derived oligosaccharides, polyketide antibiotics, biotin precursor, and acyl-transfer to glycerol-3-phosphate [*E.coli* and plants] and lysophosphatidic acid. Additionally acyl-ACPs are the substrates for both thioesterases and soluble desaturases. NMR structures exist for holo-ACP and acyl-ACPs [4] and recently a X-ray crystallographic structure has been reported for a HAS-ACP complex from *Bacillus subtilis* [5]. In this study

N. Murata et al. (eds.), Advanced Research on Plant Lipids, 139–142.

we are interested in determining the x-ray structure for an acyl-ACP and have used a combination of recombinant and non-recombinant proteins. Our eventual aim is to obtain structural data on a variety of different acyl-ACPs as well as acyl-ACP's complexed to enzymes of lipid metabolism.

1.1 Materials and methods.
ACP was isolated from *E. coli* by the method of Majerus [6] . cDNA clones for *E.coli* ACP and HAS were cloned and over-expressed in *E.coli* using the pET expression system. Synthesis of butyryl and octonoyl ACP was carried out using either the appropriate N-acylimidazole or by use of a synthetic system using acyl-CoA and HAS [3]. Apo, holo and acyl-ACPs were verified using MALDI-tof mass spectroscopy. Crystals of acyl-ACP were grown as previously described [7].

1.2 Wild type E.coli ACP and crystals.
Crystals for the butyryl derivative which defracted to 1.2Å and were stable in the X-ray beam were obtained. We call these crystals type A. A heavy metal derivative could not be obtained so a recombinant route was chosen to allow growth on selenomethionine .

1.3 Recombinant E.coli apo, holo and butyryl-ACP and the first heavy metal derivative.
Apo-ACP was prepared following over-expression of the *E.coli* ACP gene in *E.coli*. The protein was virtually homogeneous following a simple freeze thaw isolation from the induced cells and an anion exchange chromatography step. Apo-ACP was converted to holo-ACP or butyryl-ACP using HAS and either CoASH or butyryl-CoA. The products were verified following MALDI-Tof mass spectroscopy [Fig 1]. It can be seen that the major species in these spectra correspond to the exact molecular weights predicted. This is a far more convenient method of following derivatisation than the rather cumbersome conformational gel analysis. The total time from sample application to analysis is a matter of a few minutes.

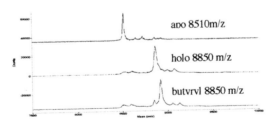

Figure 1. MALDI-tof spectra of E.coli apo, holo and butyryl-ACP

The recombinant protein crystallised under the same conditions as the wild type protein and had excellent defraction properties. *E.coli* ACP contains two methionine residues. One is the initiating methionine, which is lost following translation, and a second at position 45 within the protein. Cells, containing the *E.coli* ACP pET plasmid, were grown on selenomethionine and the ACP isolated and purified. These were verified to contain selenomethionine via mass spectroscopy. The butyryl selemomethionine ACP crystallised but in a slightly different from to the type A crystals obtained with wild type butyryl-ACP, we termed these crystals type B. Whilst the crystals defracted it was not possible to solve the structure with this single heavy metal derivative. Accordingly we decided to obtain a second heavy metal derivative by two different strategies involving site directed mutagenesis, these involved the introduction of, [a] cysteine and [b] methionine residues. The former could be used as "anchors" for heavy metals and the latter as potentially new second derivatives following growth on selenomethionine.

1.4 Recombinant cysteine and methionine mutant proteins of E.coli butyryl-ACP allows the x-ray structure to be determined.

The mutations produced were chosen following interrogation of the NMR structure of holo-ACP. The following were made T23C, A25C, S26C, D51C, I11M, T39M, I54M, I62M and I69M. All the genes expressed well and produced large quantities of the protein. The mutant proteins containing cysteine ran as dimers, on SDS-gels, in the absence of reducing agents, which prevented their utility as candidates for derivatisation with heavy metal. The selenomethionine I62M protein crystallised in the B form, it was in the space group $P2_12_12_1$, and it was possible to solve the structure, using the MAD method. The model was refined at 1.2Å resolution to a crystallographic R-factor of 0.160 and R_{free} of 0.193.

1.5 Features of the structure.

The structure consists of 4 α-helices that are connected by loop regions. Figure 2 compares the structure, determined in this investigation for the butyryl selenomethionine I62M, with the NMR structure determined for *E.coli* holo-ACP. Notably there is a difference in the angle of the α1-helix and a big difference in the position of the α3-helix. In the X-ray structure the α1 and α3 helices form a hydrophobic cavity into which the butyryl group can enter and it is highly possible that the butyryl group is "shielded' in this way and possibly protected from removal by thioesterases.

We are currently investigating the structure of other acyl-ACPs to determine what chain length acyl group the "hydrophobic pocket" can accommodate.

142

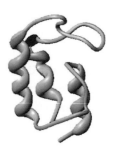

Figure 2. Ribbon traces of the fold of E.coli ACP as determined by X-ray crystallography (dark grey) and NMR (light grey). An extra α helix (3) and the bound acyl group (ball and stick) are seen in the crystal structure

References

1   Prescott, D. J. & Vagelos, P.R. (1972). PR Acyl carrier protein. *Adv. Enzymol.* 36, 269-311.
2   Slabas, A.R.& Fawcett, T. (1992). The biochemistry and molecular biology of plant lipid biosynthesis. *Plant Mol. Biol.* **19**, 169-191
3   Carreras, C.W., Gehring, A.M., Walsh, C.T., and Khosla, C., (1997). Utilization of enzymatically phosphopantetheinylated acyl carrier proteins and acetyl-acyl carrier proteins by the actinorhodin polyketide synthase. *Biochem.* 36:11757-11761.
4   Holak, T.A., Nilges, M., Prestegard, J.H., Gronenborn, A.M. & Clore, G.M. (1988). Three-dimensional structure of acyl carrier protein in solution determined by nuclear magnetic resonance and the combined use of dynamical simulated annealing and distance geometry. *Eur. J. Biochem.* **175**, 9-15.
5   Parris, K.D., Lin, L., Tam, A., Mathew, R., Hixon, J., Stahl, M., Fritz, C.C., Seehra, J. & Somers, W.S. (2000). Crystal structures of substrate binding to *Bacillus subtilis* holo-(acyl carrier protein) synthase reveal a novel trimeric arrangement of molecules resulting in three active sites. *Structure* **8**, 883-895.
6   Majerus P.W., Alberts A.W., and Vagelos P.R. (1964). The acyl carrier protein of fatty acid synthesis: purification, physical properties and substrate binding site. *PNAS.* (USA) 51:1231-1238.
7   Roujeinikova A, Baldock C, Simon WJ, Gilroy J, Baker PJ, Stuitje AR, Rice DW, Rafferty JB, Slabas AR (2002) Crystallization and preliminary X-ray crystallographic studies on acyl-(acyl carrier protein) from *Escherichia coli* Acta Crystallographica Section D-Biological Crystallography 58: 330-332.

DIACYLGLYCEROL ACYLTRANSFERASE 1 OF *ARABIDOPSIS THALIANA*

M.J. HILLS, D.H. HOBBS, J. KANG* AND C. LU*

John Innes Centre, Colney, Norwich, NR4 7UH, U.K.
* now at Washington State University, Pullman, U.S.A

Abstract

The activity of the DGAT1 promoter in Arabidopsis was assessed using promoter-reporter fusions in transgenic plants. Analysis showed that the promoter was active in embryos, pollen, germinating seeds and seedlings. It was found in root tips of older plants but not in mature leaves. In order to determine whether the DGAT played a role in the germination and growth of seedlings we studied the physiology and development of the DGAT1 mutants. This showed that the DGAT1 mutant seeds are more sensitive to 5% glucose than their parent lines. Lipid degradation is similar in the mutants and parents but the mutants have much higher activity of hexokinase. Whether this phenotype is due to a direct effect of lack of DGAT in the seedlings on sugar sensing systems or is due to an after-effect on sugar metabolism in the developing seed remains to be seen.

Introduction

Oilseeds such as *Arabidopsis thaliana*, accumulate large quantities of storage triacylglycerols during embryo development. The triacylglycerol is synthesized from diacylglycerol and long chain acyl-CoAs in a reaction catalysed by diacylglycerol acyltransferase (DGAT). Two sequence-unrelated DGAT genes have been cloned in Arabidopsis, DGAT1 (Hobbs et al 1999) for which mutants have been identified (Zou et al 1999; Routaboul 1999) and DGAT2 (Lardizarbal et al 2001) which was recently reported but for which no mutants have yet been described. The DGAT1 gene consists of 15 exons which when transcribed produce a messenger RNA, which is 1988 bases in length. The message has 5' untranslated region of 230 bases, which is relatively long for Arabidopsis. It was previously reported that the DGAT1 gene is expressed in a wide range of tissues in *Brassica* and *Arabidopsis* (Hobbs et al 1999; Zou et al 1999). This is a curious finding since although triacylglycerols are accumulated in large amounts in developing seeds and in pollen; it is present at quite low levels in other plant tissues such as roots, leaves and stems. In order to investigate the expression of DGAT1 with more precision

143

N. Murata et al. (eds.), Advanced Research on Plant Lipids, 143–146.

we have made a more detailed study of the pattern of expression of the DGAT1 gene during Arabidopsis growth and development. We have also studied the DGAT1 mutants in more detail and in particular have investigated the effect of the mutation on seed germination and early development of the seed. The data suggest that growth of the mutant seedlings is affected in ways that cannot be solely accounted for by lack of carbon due to the reduced level of storage oil in the seed.

Methods

We have made pBIN19 binary constructs that contain 2,100 bp of the DGAT1 promoter including the 5' UTR placed behind the GUS reporter gene followed by the DGAT1 gene 3' untranslated region. The constructs were transformed into Arabidopsis (Col) using the floral dipping method and progeny selected on agar plates containing kanamycin. The tissues of the kanamycin resistant progeny were stained for glucoronidase activity. For the germination assays, seeds of the DGAT1 mutants (AS11 – Zou et al 1999, and ABX45 – Routaboul et al 1999) and their respective parent lines Col and Ws were obtained from plants grown under identical conditions. They were imbibed for 3 days on 50% MS-agar plates containing various amounts of glucose or other components described in the text and then transferred to a growth room. Germination and seedling development were then assessed.

Results

A number of tissues from the transgenic Arabidopsis plants containing the DGAT::GUS::DGAT constructs were stained for GUS activity. As expected strong GUS staining was observed in developing seeds and in pollen grains. This was expected since these tissues accumulate large amounts of oil. We also observed GUS staining in other tissues, but this changed during growth and development of the plant. We found that germinating seeds stained strongly, as did 2 d old seedlings. The staining was found in the hypocotyls and cotyledons but not the root. This concurs with the previous observation that DGAT mRNA was found in both the tissues at levels similar to siliques and floral tissues (Zou et al 1999). However, as the seedlings developed the regions of GUS staining moved such that at 5 days only the root tips and stem apex were stained. Staining was no longer observed in the hypocotyls or cotyledons. We also determined the DGAT1 mRNA concentration in germinating seeds and seedlings at a number of stages up until 15 days after germination. The mRNA level was very similar at all stages – even though the GUS staining showed the location of the DGAT1 promoter activity moves during seedling growth.

In order to determine whether 2 d old seedlings were actually making TAG we fed [^{14}C]acetate and measured incorporation into lipids. The lipids were separated by thin layer chromatography, scraped and counted. In the wild-type, the amount of label in TAG was about 6% of the total incorporated into lipids, whereas in the mutants it was about 3% of total lipids. In leaves, only about 1% of the label was incorporated into TAG in both mutants and their wild type parents. This suggests that as in the developing seeds DGAT1 plays a role in TAG synthesis, but rates of synthesis are very low, considering that the DGAT1 mRNA concentration is as high in seedlings as in developing seeds.

In order to determine whether the DGAT1 plays a role in the developing seeds, other than the synthesis of storage TAG, we investigated germination and growth of the mutants compared to wild-type. It was known that the mutants germinated more slowly than wild type (Routaboul et al 1999). It was possible that this may have been due to a lack of carbon due to a decreased content of TAG in the seeds. However, we found that increasing the sugar content of the germination media actually decreased germination potential. Furthermore, seedling growth and development of the mutants was slowed compared to the wild type. We also found that the seedlings had a much greater sensitivity to abscisic acid compared to wild type, suggesting that the lack of DGAT1 activity had an impact on the sugar sensing system in the Arabidopsis seed

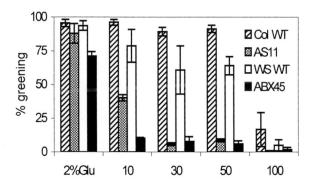

Figure 1. Arabidopsis seeds of the DGAT1 mutants AS11 and ABX45 were germinated on 50%MS agar plates containing 2% glucose and a range of concentrations of abscisic acid. At 7 days after germination the proportion of seedlings that had developed green cotyledons and produced a primary leaf was assessed.

The rate of lipid degradation as determined by total fatty acid content of the seedlings during development is similar in the mutants and parents. However, the seeds of the mutants contain over 100% more sucrose than the parents as might be expected given the reduce accumulation of TAG. We also found that the seedlings of the mutants have a 3-4 fold higher activity of hexokinase, whereas none of the other enzymes of glycolysis were significantly affected. Hexokinase has been implicated as being involved in the sugar sensing system in plants and animals. However, whether this phenotype of increased hexokinase activity and increase sensitivity of the seedlings to glucose and ABA is due to a direct effect of lack of DGAT in the seedlings on sugar sensing systems or is due to an after-effect of changes in sugar metabolism in the developing seed remains to be seen.

Acknowledgements

The authors acknowledge the Biotechnology and Biological Sciences Research Council who provided funding for this project through its competitive strategic grant to the John Innes Centre and through a grant funded under the Genome Analysis of Agriculturally Important Traits' initiative.

References

Hobbs DH, Lu C and Hills MJ (1999) Cloning of a cDNA encoding diacylglycerol acyltransferase from *Arabidopsis thaliana* and its functional expression. F.E.B.S. Letts **452**, 145-149.

Zou J., Wei Y., Jako C., Kumar A., Selvaraj G. and Taylor D.C. 1 (1999) The *Arabidopsis thaliana TAG1* mutant has a mutation in a diacylglycerol acyltransferase gene. Plant J, 19, 645-653;

Routaboul J., Benning C., Bechtold N., Caboche M. and Lepiniec L. 1999. The *TAG1* locus of *Arabidopsis* encodes for a diacylglycerol acyltransferase. Plant Physiol. Biochem. 37: 831-840.

Lardizabal K.D., Mai J.T., Wagner N.W., Wyrick A., Voelker T. and Hawkins D.J. 2001. *DGAT2* is a new diacylglycerol acyltransferase gene family. Purification, cloning, and expression in insect cells of two polypeptides from Mortierella ramanniana with diacylglycerol acyltransferase activity. J. Biol. Chem. 276: 38862-38869.

OLEOSINS AND PLASTID-LIPID-ASSOCIATED PROTEINS IN *ARABIDOPSIS*

HYUN UK KIM AND ANTHONY H. C. HUANG
Center for Plant Cell Biology
Department of Botany and Plant Sciences
University of California, Riverside, CA 92521, USA

1. Introduction

Some organelles in plant cells contain abundant neutral lipids, which are associated with specific amphipathic proteins. The proteins stabilize the lipids and may perform additional functions. Two types of such proteins, the oleosins and the plastid-lipid-associated proteins (PAPs), have been recognized.

Oleosins are present in seeds, the pollen interior, and the floral tapetum. In seeds, the storage oil body contains a matrix of triacylglycerols (TAGs) enclosed by a layer of phospholipids (PLs) and oleosins. In pollen, the intracellular storage oil bodies are similar in structure to those in seeds, by having a matrix of TAGs enclosed by a layer of PLs and oleosins. In the floret tapetum cells, TAGs and oleosins are housed in the organelles called tapetosomes. At a late stage of floral development, the tapetum cells lyze, and the oleosins but not the TAGs of the tapetosomes are deposited onto the adjacent maturing pollen as constituents of the coat.

PAPs are amphipathic proteins associated with neutral lipids in special structures in the chloroplasts and non-green plastids. These special structures include fibrils, tubules, crystalloids, and globuli. The neutral lipids may be carotenoids, TAGs, or steryl esters. In the chloroplasts, the carotenoids function as light receptors in photosynthesis and quenchers of excess energy in photoprotection, and the TAGs may provide temporary storage of energy or harmful free fatty acids from damaged lipids. In the chromoplasts in petals and fruits, the carotenoids function as attractants for animals in pollination and fruit dispersion. In the elaioplasts in the floral tapetum cells, the steryl esters are present in globuli; when the tapetum cells lyze, the steryl esters are deposited onto the pollen surface for waterproofing and other purposes.
We have explored the recently available *Arabidopsis* genome sequences and studied all the genes encoding the oleosins and the PAPs.

N. Murata et al. (eds.), Advanced Research on Plant Lipids, 147–150.

148

2. Oleosins and their genes in *Arabidopsis*

A search of the *Arabidopsis* genome database for genes containing the conserved hairpin sequence of oleosins (a stretch of 72 uninterrupted hydrophobic or neutral residues) has revealed 16 oleosin genes. About half of these genes have been reported previously as studied clones or expressed sequence tag sequences. They are scattered throughout the five chromosomes (Fig. 1). None of them is a pseudogene. The expression of these 16 genes in various tissues has been probed by RNA blot hybridization and RT-PCR (Kim, et al., 2002).

The 16 oleosin genes can be divided into three groups on the basis of their tissue-specific expressions (Fig. 1). The first group consists of eight genes, which are expressed specifically in the florets (tapetum). All the eight genes are on chromosome V, with seven of them in tandem. The second group consists of five genes, which are expressed only in maturing seeds (siliques). The third group consists of three genes, which are expressed in both maturing seeds and microspores (pollen).

The presence of oleosins inside the pollen, presumably associated with the storage oil bodies, has not been previous described. It is shown through the identification of an oleosin by N-terminal sequencing and immunoblotting with the use of isoform-specific antibodies. Collaborative evidence includes the detection of GUS activity in the microspores but not in other cells in the anthers in plants that have been transformed with a *GUS* gene driven by a pollen oleosin gene promoter.

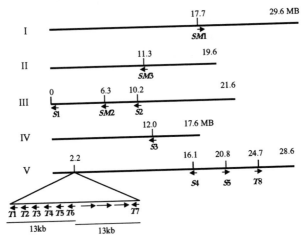

Figure 1. Location of the 16 oleosin genes on the five chromosomes of *Arabidopsis thaliana*. Arrows indicate the 5'- to 3'- direction of the genes. Simplified gene names are used: *T*, *S*, and *SM* genes are those expressed in tapetum, seeds, and seed plus microspores, respectively.

3. PAPs and their genes in *Brassica* and *Arabidopsis*

PAPs (also termed fibrillin and chromoplast-specific proteins) from several diverse species are highly similar in their amino acid sequences regardless of the type of neutral lipids that they interact with. Their deduced sequences have a putative plastid-targeting N-terminal peptide and a mature polypeptide of about 30 kD. The tertiary structure of the protein is unknown, and the amino acid sequence does not have a long hydrophobic stretch. It is predicted that the polypeptide resides on the surface of but does not penetrate into the neutral-lipid structures (e.g., fibrillin and globuli). PAP genes are expressed in leaves and other organs. They are especially active in specific organs during certain phases of development or in response to applied hormones and abiotic stresses. In all the studies of PAP and their genes in diverse species, only one gene from each species has been previously investigated. The uncertainty exists that the studied gene is not the one that encodes the studied PAPs because of the possible presence of multi-genes in an individual and the highly similar amino acid sequences.

A search of the *Arabidopsis* genome database for genes encoding proteins related to the *Brassica* PAPs has revealed 3 *Pap* genes. *AtPap*1 and *AtPap*2 are widely separated on chromosome IV, and *AtPap*3 is present on chromosome II. The characteristics of these three genes have been studied in *Brassica* with use of the corresponding *Brassica* genes (Kim et al., 2001). *Pap*1 and *Pap*2 are more similar to each other (79% identity) than to *Pap*3 (43% and 44% identity, respectively) in the sequences encoding the mature proteins.

The expression of the three *Pap* genes in various *Brassica* organs has been probed by RNA blot hybridization. *Pap*1 and *Pap*2 have much higher levels of expression than *Pap*3. *Pap*1 is expressed most abundantly in the anthers, presumably within the tapetum, where the elaioplasts house massive steryl esters in the globuli. *Pap*2 is expressed actively only in the petals that have ample chromoplasts (in *Brassica*, the petals are yellow). *Pap*3 is expressed in diverse organs at a relatively low level. The quantities of PAP in the various organs as probed by immunoblotting reflect the above-described expression patterns of the genes.

Abiotic stresses applied to the plants alter the expression of the *Pap* genes. Drought and ozone reduce the levels of expression the three *Pap*, whereas mechanical wounding and altering the light intensity enhance their expression.

Overall, the small *Pap* gene family has only three members in *Arabidopsis*. The DNA and the deduced amino acid sequences of *Pap* genes from diverse plant species are highly conserved, much more so than those of oleosin genes. This high conservation suggests a rigid requirement in the structures of the proteins to perform their functions, which include at least structural maintenance. The three PAPs may have distinct subplastid locations and associate with different lipids, and the expression of the three *Pap* genes are controlled by different mechanisms.

4. Comparison of oleosins, PAPs, and other proteins associated with neutral lipids

Amphipathic proteins associated with neutral lipids are present in diverse organisms. They include oleosins and PAPs in plants; apolipoproteins, perilipin and adipocyte-differentiation-related proteins in mammals and insects; and phasin in bacteria. Of all these proteins and others purported to be associated with neutral lipids, only oleosins have a long hydrophobic stretch of 72 uninterrupted residues that allow the bending of the stretch into a hairpin and penetrating of the hairpin beyond the surface PL (or glycolipid) layer into the hydrophobic lipid core. The longest hydrophobic stretches in these other proteins (e.g., 16, 10, and 16 residues in PAP, phasin, and caleosin, respectively) are insufficient to form a hairpin structure (of either α-helix or β-strand structure) penetrating beyond the acyl moieties of the surface PLs into the hydrophobic lipid core. The presence of proline residues at the center of a short hydrophobic stretch in proteins cannot be equated to the occurrence of three proline residues in the highly conserved center of the long hydrophobic stretch in oleosins; the genome database has more than a thousand different proteins with a short hydrophobic stretch containing proline residues at the center.

References

Kim, H. U, Wu, S. S. H., Ratnayake, C., and Huang, A. H. C. (2001) *Brassica rapa* has three genes that encode proteins associated with different neutral lipids in plastids of specific tissues. Plant Physiol. 126, 330-341.

Kim, H. U., Hsieh, K., Ratnayake, C., and Huang, A.H. C. (2002) A novel group of oleosins is present inside the pollen of *Arabidopsis*. J. Biol. Chem. (in press)

MEMBRANE-BOUND *SN*-2-MONOACYLGLYCEROL ACYLTRANSFERASE (MGAT) IS INVOLVED IN TRIACYLGLYCEROL SYNTHESIS

The presence and characterization of a novel function and its' role in triacylglycerol assembly in plants and fungi.

A.D.WATERS, T.C.M.FRASER, S.CHATRATTANAKUNCHAI & A.K.STOBART
School of Biological Sciences, University of Bristol
Woodland Road, Bristol, BS8 1UG, United Kingdom

1.Abstract

Sunflower microsomal preparations catalysed the incorporation of radioactivity from [^{14}C]oleoyl-CoA in the presence of *sn*-2-monoacylglycerol (2-MAG). Similar results were obtained using *sn*-2-mono-[^{14}C]oleoylglycerol and non-radioactive acyl-CoA. The observations were consistent with the activity of a monoacylglycerol acyltransferase (MGAT, E.C. 2.3.1.22.) catalysing the acylation of MAG yielding diacylglycerol (DAG), and the rapid utilization of DAG in TAG synthesis. The MGAT was particularly efficient with the 2-MAG substrate and was selective for mono-unsaturated acyl species. Similar membrane-bound MGAT was also present in linseed (*Linum usitatissimum* L.), *Mortiella alpina*, *Candida curvata* D. and *Saccharomyces cerevisiae*.

2.Materials and methods

2.1 *Chemicals*
Fine chemicals and solvents were purchased from Sigma. [1-^{14}C]Oleoyl-CoA (2kBq/nmol) and Tri-[U-^{14}C]oleoylglycerol (5.7 kBq/mol) were obtained from Amersham UK. *sn*-2 mono[^{14}C]oleoylglycerol was formed from TAG with type XI lipase from *Rhizopus arrhizus* from Sigma.

2.2 *Plant material and microsomal preparation*
Developing seeds of sunflower (*Helianthus annuus*) was harvested 14-16 days after flowering and microsomal membrane preparations were the same as described previously (Jackson *et al.*, 1998).

151

N. Murata et al. (eds.), Advanced Research on Plant Lipids, 151–154.
© 2003 Kluwer Academic Publishers. Printed in the Netherlands.

2.3 *Microsomal incubations and Analytical procedures*

All incubations were at 25°C. Reactions were terminated and the lipids extracted in chloroform/methanol (Bligh and Dyer, 1959). Complex lipids were purified by thin layer chromatography as described previously (Chatrattanakunchai *et al*, 2000). Incorporation of *sn*-2 mono[^{14}C]oleoyl-glycerol or [^{14}C]oleoyl-CoA into products was measured by imaging (Instant Imager, Packard).

3.Results

3.1 *Utilisation of MAG in TAG synthesis*

Microsomal membranes from sunflower were incubated with [^{14}C]oleoyl-CoA and *sn*-2-monooleoylglycerol. At regular intervals the radioactivity in complex lipids was determined. The results (Figure.1) show that the presence of MAG stimulated the incorporation of activity accumulating in TAG.

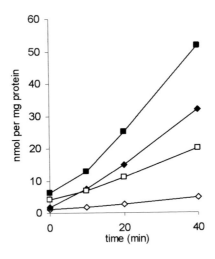

Figure 1. Incorporation of radioactivity from [^{14}C]oleoyl-CoA in the presence of MAG. Incorporation of [^{14}C]oleate from radioactive oleoyl-CoA into TAG (■) and phosphatidyl choline (PC) (♦).Occluded and open symbols plus minus MAG, respectively. Reaction mixtures contained 160 nmol of [^{14}C]oleoyl-CoA, 100µl of 0.1M phosphate buffer (pH 7.2), 100µl 0.1M MgCl$_2$, 50µl of 100 mg/ml BSA solution, +/- 100 nmol *sn*-2-monooleoylglycerol dissolved in 35µl ethanol in a volume of 1ml. The reaction was initiated by the addition of 0.50mg of microsomal protein.

Similar experiments were carried out using *sn*-2-mono[^{14}C]oleoylglycerol and non-radioactive acyl-CoA. Again efficient incorporation of label into DAG, PC and TAG was observed (results not shown).

These observations are consistent with the presence of a microsomal MGAT catalysing the acylation of MAG with fatty acids from acyl-CoA.

3.2 *Utilisation of MAG isomers*

Microsomal membranes were incubated with stereo specific acyl MAG species in the presence of [^{14}C]oleoyl-CoA. The results (Figure 2) show that *sn*-2-monooleyolglycerol was the most efficient substrate and that the unsaturated species was preferred.

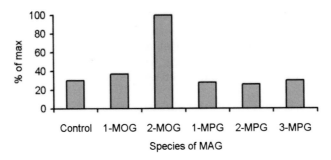

Figure 2. Substrate specificity of MGAT Radiolabel was determined in TAG after 2 hours incubation of microsomal preparations with a variety of MAG isomers and labelled oleoyl-CoA. Results are expressed relative to sn-2-monooleoylglycerol (2-MOG), the most active substrate, the control is minus MAG. 1-,2- and 3- refers to Fischer projection position of fatty acid on the glycerol backbone. Abbreviations, MPG monopalmitoylglycerol; MOG monooleoylglycerol.

Further experiments were carried out using a range of *sn*-2-MAG species. The results (figure 3) show that the enzyme was active only with monounsaturated *sn*-2-MAG.

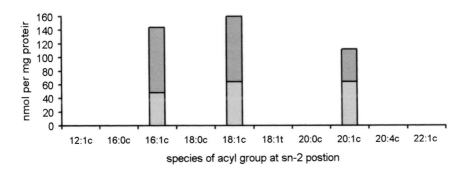

Figure 3. Radiolabel was determined in DAG and TAG after 2 hours incubation of microsomal preparations with a variety acyl group at the *sn*-2 position of MAG and labelled oleoyl-CoA.
Abbreviations: c – *cis*, t – *trans*.

Each reaction mixture contained 18 nmol of [^{14}C]oleoyl-CoA, 10 μl of 0.1 M phosphate buffer (pH 7.2), 10 μl 0.1 M MgCl$_2$, 5 μl of 100 mg/ml BSA solution, and 5 nmol of MAG. The reaction was initiated by the addition of 0.02 mg of microsomal protein.

4.Discussion

The primary route of TAG synthesis is considered to involve the reactions of the so-called Kennedy Pathway. Previously we reported the acyl-CoA independent formation of TAG from a pool of *de novo* synthesised endogenous DAG (Fraser *et al*, 2000). The diacylglycerol transacylase (DGTA) (Stobart *et al*, 1997) catalysed reaction has been shown to form TAG minus acyl-CoA, the other product of the reaction being *sn*-2-monoacylglycerol.

We have shown here that a membrane-bound acyltransferase in developing cotyledons of sunflower can recycle MAG, which in conjunction with the DAG:DAG transacylase could yield a net production of TAG (figure 4).

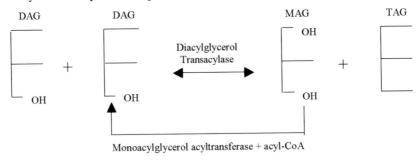

Figure 4. The role of MGAT and DGTA in TAG synthesis.

A similar enzyme activity was found in microsomal membranes from in linseed (*Linum usitatissimum* L.), *Mortiella alpina*, *Candida curvata* D. and *Saccharomyces cerevisiae* (results not shown).

Tumaney *et al* (2000) have recently characterised a soluble, non-membrane bound, MGAT from peanut. The peanut enzyme, unlike the activities reported above, was specific for the *sn*-1-MAG substrate and hence its' significance in TAG synthesis is unclear.

5.References

Bligh, E.G. and Dyer, W.J. (1959) *Can. J. Biochem. Physiol.* **37**, 911-917.

Chatrattanakunchai, S., Fraser, T.C.M., Stobart, A.K. (2000) *Biochem. Soc. T.* **28**, 707-709.

Jackson, F.M., Fraser, T.C.M., Smith, M.A., Lazarus, C., Stobart, A.K. and Griffiths, G. (1998) *Eur. J. Biochem.* **252**, 513-519.

Stobart, A.K., Mancha, M., Lenman, M., Dahlqvist, A. and Stymne, S. (1997) *Planta* **203**, 58-66.

Tumaney, A.W., Shekar, S. and Rajasekharan, R. (2001) *J. Biol. Chem.* **276** 10847-10852.

CLONING AND EXPRESSION OF *Vernonia* AND *Euphorbia* DIACYLGLYCEROL ACYLTRANSFERASE cDNAs

T. HATANAKA[1], K. YU[2] and D. F. HILDEBRAND[2]
[1]*Kobe University, Kobe, 657-8501, Japan*
[2]*University of Kentucky, Lexington, KY 40546-0091, USA*

1. Introduction

Plants are known to accumulate a wide diversity of unusual fatty acids, some of which have industrial uses. Epoxy fatty acids, such as vernolic acid, are an example of uncommon fatty acids that accumulate in triacylglycerol of a few plant species that is valuable for various industrial uses. *Vernonia galamensis* is known as one of the highest natural accumulators of vernolic acid (Krewson et al., 1966). In the seeds, vernolic acid can constitute 80% of triacylglycerol (TAG) fatty acids.

TAG is the main storage reserve in oilseeds. It is mainly synthesized during seed development before the embryo enters into the period of desiccation and dormancy. Although some researchers have suggested other possible pathways in recent years (Dahlqvist et al., 2000, Voelker et al., 2001), acyl-CoA:diacylglycerol acyltransferase (DGAT) can be an important enzyme catalyzing the final step of TAG synthesis. The first cDNA encoding DGAT was cloned from mice (Cases et al., 1998) and this led to other DGAT cloning from many different organisms including plants (Hobbs et al., 1999, Bouvier-Nave et al., 2000, Cahoon et al., 2000, Jako et al., 2001, Nykiforuk et al., 2002). We here report a cloning and functional expression of cDNAs encoding DGAT from *Vernonia galamensis* and *Euphorbia lagascae* which is another known natural accumulator of vernolic acid.

2. Materials and Methods

2.1. *cDNA cloning*

Degenerated primers were designed from the sequence information of DGAT and related genes. Partial cDNA fragments of *Vernonia galamensis* and *Euphorbia lagascae* were obtained by PCR using an RT-PCR strategy using RNA from developing seeds as a template. For determination of the full-length cDNA sequence, we then designed two

N. Murata et al. (eds.), Advanced Research on Plant Lipids, 155–158.

primers from the sequence information of the partial cDNA fragment, and both 5'- and 3'-RACE procedures were carried out using the Smart RACE cDNA amplification kit (Clontech).

2.2. *Gene testing*

The expression in Sf9 cells was tested with the Bac-to-Bac expression system (Gibco BRL), and the recombinant baculovirus was prepared following their instruction manual. Then Sf9 cells were infected by the baculovirus possessing *Vernonia* or *Euphorbia* DGAT and cultured for 4 days and the cells were collected. Another set of cultured cells were infected by the baculovirus without cloned genes as a control. Their lipids were extracted with chloroform:methanol (2:1), the TAG fractions were separated with thin layer chromatography (TLC), and the fatty acids were analyzed with gas chromatography.

2.3. *Yeast microsome assays*

Vernonia and *Euphorbia* DGATs were cloned into yeast vector pYES2 (Invitrogen, CA). The constructs along with the void vector were used to transform yeasts (*Saccharomyces cerevisiae*) strain INVScl (Invitrogen, CA). Transformed yeasts were cultured and the microsome fractions were prepared according Dahlqvist et al. (2000).

The reaction mixture (100 μL) contained 20 mM radiolabeled linoleic acid CoA or epoxy fatty acid CoA, 300 mM dioleyl diacylglycerol, 0.02% Tween 20, 100 mM Tris-HCl (pH 7.1), 1 mM $MgCl_2$, 0.5 mM CoASH, 0.5 mM ATP and microsomes (corresponding to 50 μg protein). The suspension was incubated at 30 °C with shaking (100 rpm) for 1 hour. The reaction was stopped by first placing the test tubes with the reaction mixture in ice and followed by adding 100 μg of soybean triacylglycerol (to linoleic acid CoA fed) or *Vernonia* oil (to epoxy fatty acid CoA fed) as carrier. The lipid was extracted with chloroform:methanol (2:1, v/v) . Samples were loaded on TLC plates and the radioactive bands were detected by phosphorimaging and scintillated.

For identification of radioactive products, methylated fractions were analyzed by TLC with a hexane:MTBE:acetic acid (85:15:1, v/v/v) solvent system and methyl vernoleate and methyl coronate were used as standards.

3. Results and Discussion

The cDNAs from *Vernonia galamensis* and *Euphorbia lagascae* are 1.8 kb and 2.3 kb, and they encode 523 and 508 amino acids, respectively. These sequences possess a motif consisting of 3 or 4 consecutive arginine residues typical of other DGATs (Bouvier-Nave et al., 2000). These genes therefore appear to be a DGAT, not an acyl CoA:cholesterol acyltransferase (ACAT). The similarities of the peptide sequences from *Vernonia* and *Euphorbia* with DGATs from other plants are between 66 to 79%.

The cDNA from *Vernonia* was tested in a yeast, *Saccharomyces cerevisiae,* expression system first. The yeast transformed with the *Vernonia* DGAT-gene accumulated TAG three times more than the control. When Sf9 insect cells were infected by the baculovirus possessing *Vernonia* DGAT, they accumulated a 30 times higher amount of TAG than the control (Table 1). For this excellent performance of the baculovirus system, we decided to use this system to examine the *Euphorbia* cDNA. However, the cells infected by the baculovirus with *Euphorbia* DGAT accumulated TAG only 30% higher than the control.

Table 1. TAG contents of virus-infected Sf9 cells.

Treatment	Average mg / g D.W. (Standard Deviation)
Control	1.34 (0.16)
Vernonia DGAT	40.23 (5.36)
Euphorbia DGAT	1.75 (0.06)

Numbers represent the mean and standard deviation of 3 replicates.

The results of linoleic acid and epoxy fatty acid incorporation into their corresponding triacylglycerides by *Euphorbia* and *Vernonia* DGATs transformed yeasts were shown in Table 2. The void vector transformed yeasts were able to incorporate linoleic acid and epoxy fatty acid into triacylglycerides. This complicated some of the results. *Vernonia* DGAT had higher activity than *Euphorbia* DGAT in yeasts. *Vernonia* and *Euphorbia* DGAT had no difference in incorporating linoleic acid and vernolic acid into triacylglycerides.

Table 2. Incorporation of epoxy fatty acid and linoleic acid into triacylglycerol in plant DGAT transformed yeasts

	Percentage of incorporation (% of radiation fed) into triacylglycerol Mean (standard deviation)					
	Linoleic acid		Coronaric acid		Vernolic acid	
Control	1.3	(0.4)	1.4	(0.2)	2.2	(0.6)
Euphobia DGAT	2.6	(0.7)	1.5	(0.4)	3.4	(0.1)
Vernonia DGAT	5.0	(0.6)	1.7	(0.5)	4.7	(0.3)

Note: the results are from two separate assays.

The in vitro results showed that the DGATs from the high epoxy fatty acid accumulators, *Vernonia* and *Euphorbia*, did not show substrate specificity on epoxy fatty acids over their precursor, linoleic acid. Thus, given the opportunity, linoleic acid should be equally incorporated into triacylglycerides in these high epoxy fatty acid accumulators. The high performance of *Vernonia* DGAT shows great promise in improving seed oil contents and also can be compared to other DGATs in the same expression system.

4. Acknowledgements

Funding was provided by United Soybean Board, CPBR and the Kentucky Soybean Promotion Board.

5. References

Bouvier-Nave, P., Benveniste, P., Oelkers, P., Sturley, S.L., Schaller, H. (2000) Expression in yeast and tobacco of plant cDNAs encoding acyl CoA:diacylglycerol acyltransferase. Eur. J. Biochem. 267: 85-96.

Bradford, M.M. (1976) A rapid and sensitive method for the quantitation of microgram quantities of protein

Cases, S., Smith, S.J., Zheng, Y.W., Myers, H.M., Lear, S.E., Sande, R., Novak, S., Collins, C., Welch, C.B., Lusis, A.J., Erickson, S.K., Farese, R.V. (1998) Identification of a gene encoding an acyl CoA: diacylglycerol acyltransferase, a key enzyme in triacylglycerol synthesis. PNAS 95: 13018-13023.

utilizing the principle of protein-dye binding. Anal. Biochem. 72, 248-254.

Cahoon, E.B., Cahoon, R., Kinney, A.J. (2000) Plant diacylglycerol acyltranseferases. World Patent WO0032756.

Dahlqvist, A., U. Stahl, M. Lenman, A. Banas, M. Lee, L. Sandager, H. Ronne, and S. Stymne. (2000) Phospholipid:diacylglycerol acyltransferase: An enzyme that catalyzes the acyl-CoA-independent formation of triacylglycerol in yeast and plants. PNAS, 97: 6487-6492.

Hobbs, D.H., Lu, C., Hills, M.J. (1999) Cloning of a cDNA encoding diacylglycerol acyltransferase from *Arabidopsis thaliana* and its functional expression. FEBBS Letters 452: 145-149.

Jako, C., Kuar, A., Wei, Y., Zou, J., Barton, D.L., Giblin, E.M., Covello, P.S., Taylor, D.C. (2001) Seed-specific over-expression of an *Arabidopsis* cDNA encoding a diacylglycerol acyltransferase enhances seed oil content and seed weight. Plant Physiol. 126: 861-874.

Krewson, C.F., Riser, G.R., Scott, W.E. (1966) *Euphorbia* and *Vernonia* seed oil products as plasticizer-stabilizers for polyvinyl chloride. J. Am. Oil Chem. Soc. 43: 171-174.

Nykiforuk,C.L., Furusawa-Stoffer,T.L., Huff, O.W., Sarna, M., Lorache, A., Moloney, M.M., Weselake, R.J. (2002) Characterization of cDNAs encoding diacylglycerol acyltransferase from cultures of *Brassica napus* and sucrose-mediated induction of enzyme biosynthesis. Biochem. Biophys. Acta 1580: 95-109.

Voelker, T., Kinney, A. (2001) Variations in the biosynthesis of seed-storage lipids Annu. Rev. Plant Physiol Plant Mol. Biol. 52: 335-361.

REUTILIZATION OF ARACHIDONYL MOIETIES OF TRIACYLGLYCEROLS IN THE MICROALGA *PARIETOCHLORIS INCISA* FOLLOWING RECOVERY FROM NITROGEN STARVATION

P. SHRESTHA, I. KHOZIN-GOLDBERG AND Z. COHEN

The Albert Katz Department for Desert Biotechnologies, The Jacob Blaustein Institute for Desert Research, Ben Gurion University of the Negev, Sde-Boker Campus 84990, Israel

I. Introduction

We have recently shown that the freshwater alga *Parietochloris incisa* (Chlorophyceae) is unique in its ability to accumulate triacylglycerols (TAG), rich in arachidonic acid (AA, 20:4ω6). Under nitrogen starvation conditions, TAG accounted for 90% of total acyl lipids and AA constituted up to 50-60% of the fatty acids of TAG (Bigogno et al., 2002). We have previously proposed that polyunsaturated (PUFA)-rich TAG may have a role as a depot of PUFAs that can be mobilized for the construction of chloroplastic membranes under certain environmental conditions (Bigogno et al., 2002; Khozin-Goldberg et al., 2000). In the present work, in an attempt to unravel the role of AA-rich TAG, other than its use as a source of energy, the changes in fatty acid compositions of both neutral lipids and polar lipids were studied during recovery from nitrogen starvation at 24 °C and 12 °C.

2. Materials and methods

Cultures of *P. incisa* at the early stationary phase were resuspended and maintained in nitrogen-free BG11 medium for 14 days (Bigogno et al., 2002). Recovery was initiated by resuspending the cells in four volumes of full medium. Cells were batch cultivated at 24 °C and 12 °C and sampled every 24 h over a 6-d period. In another series of experiments, after 13 days of N- starvation, the cultures were labeled for 24 h with the ammonium salt of [1-^{14}C]18:1 (5 μCi, specific activity 52 mCi/mmol, Amersham) and four times diluted in full medium and chased for 48 h. Lipids were extracted and analyzed as previously described (Bigogno et al., 2002).

N. Murata et al. (eds.), Advanced Research on Plant Lipids, 159–162.

3. Results and Discussion

Nitrogen starvation resulted in growth arrest of *P. incisa* and significantly altered the lipid content and composition. Net synthesis of fatty acids, and particularly AA, continued, reaching 29.2% and 14.4% (of dry wt.), respectively, after 14 d. Cells accumulated TAG (up to 86% of lipids) in extraplastidial lipid bodies. During the recovery period at room temperature, after a 1-day lag, cell division was resumed and significant chlorophyll and biomass synthesis commenced. At low temperature, however, pigment and biomass synthesis as well as cellular divisions significantly slowed down (data not shown).

The lipid distribution and fatty acid composition were determined periodically. In table 1 we compare the results obtained after 2d recovery at 24 °C and 4d recovery at 12 °C. These time points were chosen on the basis of similar increase in chlorophyll and dry wt. In both cases recovery was characterized by a decrease in the total fatty acid content as well as in the proportion of TAG (from 86 to 78% of total lipids). The content (μg/mL) of MGDG and DGDG increased, especially at 24 °C. The major molecular species of MGDG, during starvation, were of the 18/16 group. During recovery at 24 °C, these molecular species were more desaturated, consisting mostly of 18:3ω3/16:3ω3, but their share decreased in favor of 18/18, 20:4/18 and even 20:4/20:4 molecular species. However, after 4 days at low temperature, there was a lower increase in C_{20}-containing molecular species while other molecular species contained mostly 18:3ω3 together with 16:3ω3 and 20:4ω6. Similar, though less pronounced, results were obtained after 2 days. These findings were in keeping with those obtained in radiolabelling experiments. The label, which initially accumulated in TAG (mostly as AA), was transferred to chloroplastic lipids at 24 °C, but not at 12 °C (Figure 1).

Recovery from N-starvation at room temperature requires a swift synthesis of PUFAs. *P. incisa* was isolated from the slopes of a snow mountain, an environment which is characterized by rapid changes in environmental conditions. Under such conditions, the de novo synthesis of PUFAs could be too slow. We hypothesize that *P. incisa* can mobilize AA, accumulated in TAG, to temporarily provide galactolipids, predominantly MGDG, with the required level of unsaturation. At low temperatures, the growth rate is much lower and apparently the supply of 18:3ω3 by desaturation of preexisting molecular species is adequate and import of AA from TAG is much less needed. We have previously shown that the red alga *Porphyridium cruentum* can also utilize AA moieties for the construction (Khozin Goldberg et al., 2000). These conclusions are in keeping with the suggested role of lipid bodies in intracellular trafficking of lipids during the cycle of lipid body formation and turnover (Murphy, 2001).

TABLE 1. Changes in the molecular species distribution in MGDG of *P. incisa* following recovery (Rec) from N-starvation (2 days at 24 $^{\circ}$C or 4 days at 12 $^{\circ}$C). Positional analysis was not performed. TFA – total fatty acids, tr – traces.

Molecular species	Molecular species distribution & content					
	N-starvation 14 days		Rec 24 $^{\circ}$C 2 days		Rec 12 $^{\circ}$C 4 days	
	% TFA	µg/mL	%TFA	µg/mL	%TFA	µg/mL
C18/C16						
18:2/16:2	31.3	5.0	2.6	0.8	tr	tr
18:3ω3/16:2	16.0	2.6	4.9	1.5	1.2	0.2
18:2/16:3	12.5	2.0	4.3	1.3	0.6	tr
18:3/16:3	24.2	3.9	25.4	7.7	53.1	9.7
Sum	**83.9**	**13.4**	**37.1**	**11.3**	**54.9**	**10.0**
C18/C18						
18:1/18:2	tr	tr	3.8	1.2	tr	tr
18:2/18:2	2.0	0.3	6.4	1.9	tr	tr
18:1/18:3ω3	0.8	tr	tr	tr	tr	tr
18:2/18:3ω3	2.1	0.3	2.8	0.8	1.6	0.3
18:3ω3/18:3ω3	1.3	0.2	2.4	0.7	8.0	1.5
Sum	**6.1**	**1.0**	**15.3**	**4.7**	**9.6**	**1.7**
C20/C18						
20:4/18:1	2.4	0.4	10.1	3.1	1.8	0.3
204/18:2	4.4	0.7	21.2	6.5	6.7	1.2
20:4/18:3ω3	2.1	0.3	7.2	2.2	20.2	3.7
Sum	**8.9**	**1.4**	**38.5**	**11.8**	**28.6**	**5.2**
C20/C20						
20:4/20:4	**1.0**	**0.2**	9.0	2.8	6.9	1.3

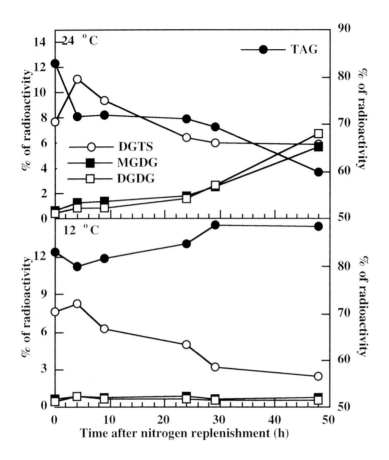

Figure 1. Redistribution of radioactivity in *P. incisa* following labeling with [1-^{14}C]18:1. Cultures were N-starved, labeled, resuspended in full medium and cultivated at 24 °C or 12 °C.

4. References

Bigogno, C., Khozin-Goldberg, I. and Cohen, Z. (2002). Accumulation of arachidonic-acid rich triacylglycerols in the microalga *Parietochloris incisa* (Trebuxiophyceae, Chlorophyta). Phytochemistry (in press).

Khozin-Goldberg, I., Hu, Z. Y., Adlerstein, D., Didi Cohen, S., Heimer, Y. M. and Cohen, Z. (2000) Triacylglycerols of the red microalga *Porphyridium cruentum* participate in the biosynthesis of eukaryotic galactolipids. Lipids 35, 881-889.

Murphy, D.J. (2001). The biogenesis and functions of lipid bodies in animal, plants and microorganisms. Progress in lipid research 40, 325-438.

Lipid Composition in Yeast Mutant Lacking Storage Lipids Synthesis Capacity

M. H. GUSTAVSSON[1], L. SANDAGER[2], A. DAHLQVIST[3], H. RONNE[4], AND S. STYMNE[1]

[1]*Department of Crop Science, Swedish University of Agricultural Sciences, Box 44, SE-230 53 Alnarp, Sweden*

[2]*Biology Department, 463 Brookhaven National Laboratory, Upton, NY 11973-5000, USA*

[3]*Scandinavian Biotechnology Research AB, Box 166, SE-230 53 Alnarp, Sweden*

[4]*Department of Plant Biology, Swedish University of Agricultural Sciences, Box 7080, SE-750 07 Uppsala, Sweden*

Abstract

Steryl esters and triacylglycerol (TAG) are the main storage lipids in eukaryotic cells. In the yeast *Saccharomyces cerevisiae*, these storage lipids accumulate during stationary growth phase within organelles known as lipid bodies. We have used multiple gene disruptions to study storage lipid biosynthesis in yeast. Four genes, *ARE1*, *ARE2*, *DGA1* and *LRO1*, were found to contribute to TAG synthesis. A yeast strain that lacks all four genes is viable and has no apparent growth defects under standard conditions. The strain is devoid of both TAG and steryl esters, and fluorescence microscopy revealed that it also lacks lipid bodies [1]. We have now further investigated the lipid content and found differences between the two strains. The levels of diacylglycerol and fatty acids as well as the composition of acyl groups were altered in the disrupted strain compared to the wild type. We conclude that even though storage lipids are non-essential for growth in yeast the lack of synthesis of storage lipids affects the acyl group composition of different lipid groups.

Introduction

The most common lipid storage molecule is triacylglycerol (TAG). The biosynthesis of TAG has been suggested to occur mainly in an acyl-CoA-dependent manner via acyl-

N. Murata et al. (eds.), Advanced Research on Plant Lipids, 163–166.
© 2003 *Kluwer Academic Publishers. Printed in the Netherlands.*

CoA:diacylglycerol acyltransferases, DGAT. There are two DGAT1-like genes present in yeast, *ARE1* and *ARE2*, and both these genes participate in both TAG [1, 2] and steryl ester biosynthesis [3, 4]. In addition, we recently identified a DGAT2 encoding gene, *DGA1*, in *S. cerevisiae* [1]. Further, an additional TAG biosynthetic enzyme, phospholipid:diacylglycerol acyltransferase (PDAT), was recently discovered in the yeast *S. cerevisiae* and in plants [5]. PDAT is acyl-CoA-independent and is encoded by *LRO1* in yeast.

A yeast strain disrupted in *ARE1*, *ARE2*, *DGA1*, and *LRO1* was found to lack storage lipid biosynthetic capacity, storage lipids, and lipid bodies [1]. Since the quadruple-disrupted strain was viable under standard growth conditions and showed a normal growth rate, we conclude that neither storage lipid biosynthesis nor lipid body formation is essential for vegetative growth in yeast. It has previously been suggested that lipid bodies may play a role during vegetative growth as storage buffers for phospholipids [6]. Our previous results suggest that if the latter is true, then *de novo* lipid biosynthesis must be able to provide those membrane lipid precursors during vegetative growth that otherwise would be generated from the storage lipids. Earlier work on *S. cerevisiae* has also shown that TAG metabolism is exclusively coupled to phospholipid biosynthesis and that TAG might serve to regulate the fatty acid species found in the membrane phospholipids [7]. In this paper we show that disrupted storage lipid biosynthesis affects both the amounts of individual lipids as well as their acyl group composition. We conclude that even though storage lipids are non-essential for growth in yeast the lack of synthesis of storage lipids affects the acyl group composition.

Materials and methods

Yeast strains used are congenic to W303-1A [8]. The wild type control strain SCY62 (*MATa ADE2*) and the quadruple-disrupted strain H1246 (*MATα are1-Δ::HIS3 are2-Δ::LEU2 dga1-Δ::KanMX4 lro1-Δ::TRP1 ADE2*) are as described in [1].

The yeast cells were cultivated at 30ºC on a rotary shaker in liquid YPD complete medium 2% glucose, diluted to OD_{600}=0.2 and grown for 32 h prior to being harvested. Cells were harvested and analyses of individual lipid species were performed as previously described by Sandager et al. [1].

Results and discussion

The lipid content of the quadruple-disrupted yeast cells compared to the wild type cells was investigated at stationary growth phase. As previously described by us the cells disrupted in *ARE1*, *ARE2*, *DGA1*, and *LRO1* completely lacks storage lipids, i.e., steryl esters and TAG [1]. The total acyl group content is 107 nmol / mg dry weight in the storage lipid deficient strain while in the wild type strain it is 278 nmol / mg dry weight. Notably, the entire difference is due to TAG and steryl ester accumulation, since the

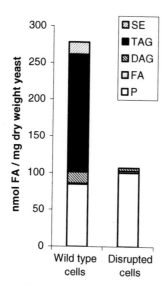

Figure 1. Lipid content of wild type and quadruple-disrupted yeast cells at stationary phase. The values shown are the means of four measurements. Abbreviations: SE, steryl ester; TAG, triacylglycerol; DAG, diacylglycerol; FA, fatty acids; and P, polar lipids.

total amount of other lipids remains the same in both yeast strains (Figure 1). However, the amount of polar lipids increases with 16% and the amount of diacylglycerol decreases with 73% in quadruple-disrupted yeast cells compared to wild type. The quadruple-disrupted yeast cells also contain a minor amount of free fatty acids, which we do not find in the wild type strain (Figure 1). This might reflect that the disruption of the synthesis of storage lipids affects the synthesis of fatty acids.

Analysis of the acyl group composition of the total lipid content revealed a higher accumulation of the uncommon fatty acid, vaccenic acid, while palmitic acid was less accumulated in the quadruple-disrupted yeast cells compared to the wild type cells [1]. The reason for these alterations in acyl group composition is not known but we suggest that by disrupting the storage lipid biosynthesis we affect the acyl-CoA pool in the cells. To further characterize the changes in lipid metabolism in the yeast strain deficient in storage lipid biosynthesis we continued to analyze the acyl group composition of the individual lipid groups of stationary phase yeast cells (Table 1). In all the groups of lipids analyzed we see a higher percentage of unsaturated acyl groups mainly on the expense of palmitic acid. The polar lipids have less alteration in the compositions of acyl groups compared to diacylglycerol and fatty acids. The reason for these alterations is not known, but we note that palmitoyl-CoA is converted to palmitoleoyl-CoA and then to vaccenoyl-CoA in two consecutive reactions catalyzed by a desaturase and an

Table 1. Acyl group composition of individual lipids in stationary phase yeast cells

Acyl group	Wild type cells					Quadruple-disrupted cells				
	SE	TAG	DAG	FA	P	SE	TAG	DAG	FA	P
16:0	6.2	17.0	23.0	45.8	18.7	-	-	5.7	15.1	6.8
16:1Δ^9	37.8	29.0	22.0	2.2	26.8	-	-	28.5	5.3	31.6
18:0	4.5	12.3	21.5	44.9	7.2	-	-	10.0	30.9	8.4
18:1Δ^9	51.5	40.2	32.9	7.1	47.3	-	-	43.7	27.2	42.6
18:1Δ^{11}	-	1.6	-	-	-	-	-	12.0	21.5	10.6

Results are expressed as mol percent of the acyl group composition of individual lipid groups and are the mean of four measurements. Abbreviations: SE, steryl ester; TAG, triacylglycerol; DAG, diacylglycerol; FA, fatty acids; and P, polar lipids.

elongase, respectively. Reducing or eliminating the acyl-CoA-dependent TAG synthesis is likely to affect both the size and the composition of the acyl-CoA pool. The effect of this change is more pronounced on diacylglycerol and fatty acids, since these lipids in wild type have 44.5 % and 90.7 % saturated fatty acids, respectively. These results show that in yeast the lack of synthesis of storage lipids affects the *de novo* biosynthesis of fatty acids, how it is affected remains to be revealed.

References

1. Sandager, L., Gustavsson, M. H., Stahl, U., Dahlqvist, A., Wiberg, E., Banas, A., Lenman, M., Ronne, H., and Stymne, S. (2002) Storage lipid synthesis is non-essential in yeast. J Biol Chem 277, 6478-82

2. Sandager, L., Dahlqvist, A., Banas, A., Stahl, U., Lenman, M., Gustavsson, M., and Stymne, S. (2000) An acyl-CoA:cholesterol acyltransferase (ACAT)-related gene is involved in the accumulation of triacylglycerol in *Saccharomyces cerevisiae*. Biochem Soc Trans 28, 700-2.

3. Yu, C., Kennedy, N. J., Chang, C. C., and Rothblatt, J. A. (1996) Molecular cloning and characterization of two isoforms of *Saccharomyces cerevisiae* acyl-CoA:sterol acyltransferase. J Biol Chem 271, 24157-63.

4. Zweytick, D., Leitner, E., Kohlwein, S. D., Yu, C., Rothblatt, J., and Daum, G. (2000) Contribution of Are1p and Are2p to steryl ester synthesis in the yeast *Saccharomyces cerevisiae*. Eur J Biochem 267, 1075-82.

5. Dahlqvist, A., Stahl, U., Lenman, M., Banas, A., Lee, M., Sandager, L., Ronne, H., and Stymne, S. (2000) Phospholipid:diacylglycerol acyltransferase: an enzyme that catalyzes the acyl-CoA-independent formation of triacylglycerol in yeast and plants. Proc Natl Acad Sci U S A 97, 6487-92.

6. Schneiter, R., and Kohlwein, S. D. (1997) Organelle structure, function, and inheritance in yeast: a role for fatty acid synthesis? Cell 88, 431-4.

7. Taylor, F. R., and Parks, L. W. (1979) Triaglycerol metabolism in *Saccharomyces cerevisiae*. Relation to phospholipid synthesis. Biochim Biophys Acta 575, 204-14.

8. Thomas, B. J., and Rothstein, R. (1989) Elevated recombination rates in transcriptionally active DNA. Cell 56, 619-30.

SYNTHESIS OF CHLOROPLAST GALACTOLIPIDS IN APICOMPLEXAN PARASITES (*TOXOPLASMA GONDII* AND *PLASMODIUM FALCIPARUM*)

E. MARÉCHAL[1], N. AZZOUZ[2], C. MERCIER[3], C. SANTOS de MACEDO[2], M. A. BLOCK[1], J.-F. DUBREMETZ[4], J.E. FEAGIN[5], M.-F. CESBRON-DELAUW[3], R.T. SCHWARZ[2] & J. JOYARD[1]

1: Laboratoire de Physiologie Cellulaire Végétale, UMR 5019 CNRS - CEA - Université J. Fourier, CEA-Grenoble, 17 rue des Martyrs, 38054 Grenoble Cedex 9, France 2: Philipps-Universität Marburg, Germany 3: CNRS FRE 2383 Pathogenèse des Sporozoaires, Université J. Fourier, Grenoble, France 4: Laboratoire de Dynamique Moléculaire des Interactions Membranaires, UMR 5539 CNRS - Université Montpellier II, France 5: Seattle Biomedical Research Institute, Seattle, USA
e-mail: emarechal@cea.fr

Abstract

Apicomplexan parasites, including major pathogens like *Toxoplasma*, *Plasmodium* and *Eimeria*, contain a vestigial chloroplast named the apicoplast. We report that in *T. gondii* and *P. falciparum* lysates, tritiated UDP-galactose is incorporated into monogalactosylcerebrosides (MGCB), and chloroplastic galactolipids, *i.e.* monogalactosyldiacylglycerol (MGDG) and possibly digalactosyldiacylglycerol (DGDG). Syntheses of MGCB and MGDG in *T. gondii* exhibit distinct enzymological features. Additionally, we show that apicoplast proteins have related epitopes with land plant chloroplast envelope proteins. This paper therefore reports the first enzymological and immunological tools to investigate the compositions of apicoplast membranes and introduces new targets for interventions against malaria and toxoplasmosis.

Introduction

Apicomplexan parasites contain a vestigial chloroplast DNA in a small organelle (~0.1-1 μm) named the apicoplast (for review Maréchal and Cesbron, 2001). There is only one apicoplast per cell and sub-cellular fractionation is still technically impossible. Apicoplast proteins, encoded by the nucleus, exhibit bipartite targeting sequences (a signal peptide upstream of a chloroplast-like transit peptide). The outermost membrane of the organelle is contiguous with the ER as in euglenoids and dinoflagellates. The envelope membranes that are shared between green and non-green plastids might be among the inner membranes limiting the apicoplast but tools to investigate the similar-

167

ity to plastid envelope membranes have heretofore been lacking. The plant and algal plastid envelope is a major contributor to the biogenesis of plastids: in particular, it is the site of synthesis of monogalactosyldiacylglycerol (MGDG) and digalactosyldiacylglycerol (DGDG). With the exception of plants lacking an envelope phosphatidate phosphatase, all chloroplasts contain the complete pathways for galactolipids biosyntheses, starting from stromal acyl-ACP and glycerol-phosphate. Because the FA biosynthetic machinery has been shown to localize to the apicoplast (review Maréchal and Cesbron, 2001), we asked whether galactolipids are synthesized in apicomplexans. We were unable to detect any homologue of the higher plant MGD and DGD genes in *T. gondii* and *P. falciparum* sequence databases. Consequently, we addressed the occurrence of both enzymes by enzymological assays.

Results and discussion

After incubation of *T. gondii* and *P. falciparum* lysates with tritiated UDP-galactose, lipids were analyzed by thin layer chromatography. We identified 3 types of labeled lipids: type 1 co-migrates with MGDG from spinach chloroplasts, type 2 with MGCB (1'-*O*-β-D-galactopyranosyl-*C*-ceramide) from bovine brain and type 3 with DGDG.

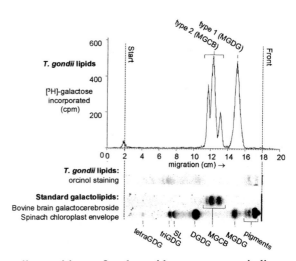

Fig. 1. Synthesis of MGDG and MGCB in *T. gondii* lysates after one hour incubation with UDP-[4,5-³H]-galactose in the absence of Mg^{2+}. Lipids were separated by TLC. Incorporation of tritiated galactose was monitored with a TLC-Scanner controlled by phosphorimager detection after a 2-week exposure. SL, sulfolipid.

Types 1 and 2 (see Fig. 1) were equally sensitive to β-galactosidase treatments, indicating that galactose was incorporated with a β-linkage as in MGDG and MGCB. Type 1 was completely deacylated by mild alkaline hydrolysis, demonstrating that its hydrophobic moiety was a diglyceride. In addition it co-migrated with standard MGDG in 2D-TLC. These points indicate that it is identical to plant MGDG. By contrast, type 2 lipids were not affected by alkaline treatment, showing that they were galactocerebrosides. The three peaks of MGCB in *T. gondii* correspond to the three complex ceramide structures of the bovine brain standard. Type 3 lipids appear in the presence of

Mg^{2+}. Type 3 was not affected by β-galactosidase treatment and decreased after an α-galactosidase incubation and mild alkaline hydrolysis, like plant DGDG. We injected a rabbit with spinach chloroplast DGDG, and obtained a polyclonal antibody that react with DGDG from spinach chloroplast envelope and from cyanobacteria. When lipid extracts from *T. gondii* were analyzed by TLC, no glycolipid could be detected by orcinol staining at the level of DGDG. However, after a blind scraping of the TLC plate at the R*f* level of DGDG and blotting on nitrocellulose, the anti-DGDG antibody reacted with the corresponding spots. In addition, when the scraped lipids from *T. gondii* or from spinach chloroplast DGDG were hydrolyzed by mild alkaline treatment and the remaining lipids re-extracted, the anti-DGDG antibody failed to react with any spot, confirming that the hydrophobic moiety of the spotted lipid was a diacylglycerol.

Diacyglycerol and ceramide, substrates for formation of MGDG and MGCB, exhibit free hydroxyl-residues with a close steric hindrance. Hence *in vitro*, the chloroplast envelope UDP-galactose:diacylglycerol galactosyltransferase (MGDG synthase) activity has a very low specificity for ceramides (Jorasch *et al*, 2000). *Vice versa*, the endoplasmic reticulum CGalT can transfer limited amounts of galactose onto diacylglycerol (van der Bijl *et al*, 1996). In *T. gondii* lysates, we showed that MGDG synthetic activity was two times more sensitive to EDTA inhibition than MGCB synthetic activity. In addition, only MGDG synthesis was enhanced by Mg^{2+} whereas only MGCB synthesis was inhibited by Ca^{2+}. An MGDG/MGCB synthesis ratio of 0.48 ± 0.01 was measured in control conditions, decreasing to 0.21 ± 0.06 in presence of 1 mM EDTA and increasing to 0.8 (0.71 ± 0.08 in average) in presence of 5 mM Mg^{2+}. From these data, we conclude that MGDG and MGCB synthetic activities in *T. gondii* are likely distinct.

Besides the fact that MGDG and DGDG are obvious building-blocks for plastid membranes, these components are strongly conserved from cyanobacteria to all plastids analyzed to date. The importance of MGDG in plastid biogenesis extends to the protein components of the organelle. In plants, MGDG binds specifically to the chloroplast transit peptides of nuclear-encoded proteins (Bruce, 1998). Since the bipartite targeting sequences of apicoplast proteins include chloroplast-like transit peptides, it is possible that MGDG plays a similar role in protein sorting in apicomplexans. Additionally, MGDG is an important cofactor for enzymatic activities associated with plastid membranes and contributes to the proper conformational stability of membrane proteins. We searched *T. gondii* and *P. falciparum* databases but found no sequences having a statistically reliable similarity with the primary sequences of major chloroplast envelope membrane proteins (E10, E24, E30, E37). Still, we speculated that apicoplast and chloroplast envelope proteins should share three-dimensional structures. Polyclonal antibodies (anti-ceHMWP and anti-ceMMWP) were raised against chloroplast envelope high- or medium-molecular-weight polypeptides. The anti-ceHMWP antiserum specifically labels a spherical organelle, close to the nucleus, with an apical localization and containing DNA, *i.e.* the apicoplast (Fig. 2). The range of protein sizes recognized in the *T. gondii* lanes is similar to that in the chloroplast envelope lanes, *i.e.*, higher MW for anti-ceHMWP and medium MW for anti-ceMMWP. This indicates that the shared epitopes between the samples are harbored by polypeptides of generally similar sizes. Therefore, in addition to plant chloroplast envelope lipids in apicomplexans, apicoplast proteins share structural features with chloroplast envelope membrane proteins.

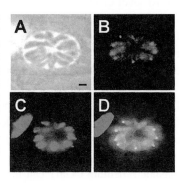

Fig. 2. Detection of apicoplast proteins sharing epitopes with spinach chloroplast envelope proteins. (A-D), Immunofluorescence staining of the apicoplast from *T. gondii* with antibodies raised against high-molecular-weight (MW) polypeptides from spinach chloroplast envelope membranes (anti-ceHMWP). (**A**) Phase-contrast image corresponding to the immunofluorescence image. bar: 2 μm. (**B**) Hoechst detection of DNA (nucleus, large ovoid body; apicoplast, small spherical body) in false colors (red in place of the blue staining of the Hoechst coloration). (**C**) Formaldehyde-fixed and detergent permeabilized intracellular parasites treated with ceHMWP anti-serum. (**D**) Colocalization of the immunostained organelle with the Hoechst-stained apicoplast.

In conclusion, plant chloroplast galactolipids are reported in apicomplexans, formed by synthetic activities that also occur in algae and land plants. Future prospects include the dissection of the full galactoglycerolipid biosynthesis pathway. In addition to stromal FA biosynthetic machinery, the apicoplast may contain a complete Kennedy-like pathway to synthesize glycerolipids. To date, no pure apicoplast subfractions have been isolated; GFP fusion proteins used as markers are lost after breakage of the organelle during the purification procedure. The apicoplast enzymological and immunological envelope markers we describe here may be useful to isolate at least the membrane fraction from the apicoplast, allowing analysis of its components, particularly its lipids and proteome, and eventually unraveling new functions for this intriguing and vital organelle.

References

Bruce, B.D. (1998) The role of lipids in plastid protein transport. Plant Mol. Biol. 38, 223-246.

Jomaa, H., Wiesner, J. , Sanderbrand, S., Altincicek, B., Weidemeyer, C., Hintz, M., Turbachova, I., Eberl, M., Zeidler, J., Lichtenthaler, H.K., Soldati, D. and Beck., E. (1999) Inhibitors of the nonmevalonate pathway of isoprenoid biosynthesis as antimalarial drugs. Science. 285, 1573-1576.

Jorasch, P., Warnecke, D.C., Lindner, B., Zahringer, U. and Heinz., E. (2000) Novel processive and nonprocessive glycosyltransferases from Staphylococcus aureus and Arabidopsis thaliana synthesize glycoglycerolipids, glycophospholipids, glycosphingolipids and glycosylsterols. Eur. J. Biochem. 267, 3770-3783.

Maréchal, E, Azzouz, N., Santos de Macedo, C., Block, MA, Feagin, JE, Schwarz, RT and Joyard, J. (subm.) Synthesis of chloroplast galactolipids in Apicomplexan parasites

Maréchal, E., and Cesbron-Delauw, M.F. (2001) The apicoplast: a new member of the plastid family. Trends Plant Sci. 6, 200-205

van der Bijl, P., Strous, G.J., Lopes-Cardozo, M., Thomas-Oates, J. and van Meer, G. (1996) Synthesis of non-hydroxy-galactosylceramides and galactosyldiglycerides by hydroxy-ceramide galactosyltransferase. Biochem J. 317, 589-597.

DETERMINATION OF THE EFFECTS OF TEMPERATURES ON PHOSPHATIDYLGLYCEROL BIOSYNTHESIS IN THYLAKOID MEMBRANES BY ANALYSIS OF MOLECULAR SPECIES

Y. N. XU, Z. N. WANG, G. Z. JIANG, L. B. LI, T. Y. KUANG

Key Laboratory of Photosynthesis and Environmental Molecular Physiology, Institute of Botany, Chinese Academy of Sciences
20 Nanxincun, Xiangshan, Beijing 100093, China

1. Introduction

Phosphotidylglycerol (PG) in thylakoid membranes is synthesized in the plastids. The incorporation of fatty acids into the glycerol backbone and fatty acid desaturation are involved in the processes of PG biosynthesis. Palmitic acid (16:0) and oleic acid (18:1) participate in the PG synthesis when they are synthesized in the acyl carrier protein bound (ACP-bound) form in the plastid by the fatty acid synthase. 18:1 is always esterified to the *sn*-1 position of glycerol backbone catalyzed by glycerol-3-phosphate acyltransferase and 16:0 is esterified to both *sn*-1 and *sn*-2 positions of glycerol catalyzed by glycerol-3-phosphate acyltransferase and 1-acylglycerol-3-phosphate acyltransferase, respectively. 18:1 is desaturated into linoleic acid (18:2) and then linolenic acid (18:3), and 16:0 at the *sn*-2 position of glycerol is desaturated into Δ^3-*trans*-hexadecenoic acid (16:1(3*t*)) after PG is formed (Browse and Somerville, 1991). PG in thylakoid membranes has been reported to be related to the susceptivity of plants to low temperature (Murata, 1983; Murata *et al.*, 1992). Low temperature induced an increase in fatty acid unsaturation degree in PG (Xu and Siegenthaler, 1996a). In this work, we analyzed the PG molecular species in cotyledons of squash and spinach grown at different temperatures to investigate how the temperature affected the PG biosynthesis briefly described above.

2. Materials and Methods

Squash (*Cucurbita pepo* L. cv Quintal) and spinach (*Spinacia oleracea* L. cv Novel) plants were grown from seeds in soil in a controlled environmental growth chamber. Seeds were germinated in darkness at 15°C, 20°C and 30°C for spinach and 20°C, 30°C and 35°C for squash, respectively. Plants were grown at the same temperature of

N. Murata et al. (eds.), Advanced Research on Plant Lipids, 171–174.
© 2003 *Kluwer Academic Publishers. Printed in the Netherlands.*

172

germination under continuous lights (350 μmol m^{-2} s^{-1}). Spinach cotyledons illuminated for 3, 9, 24, 48 and 96 h, and squash cotyledons illuminated for 3, 6, 12, 24, 48 and 96 h were collected. Thylakoid membrane isolation and lipid extraction were carried out according to Xu and Siegenthaler (1996a). PG purified by TLC was treated with phospholipase C to produce diacylglycerol (Xu and Siegenthaler, 1996b). The diacylglycerol was reacted with 3,5-dinitrobenzoyl chloride to give the dinitrobenzoyl derivatives which were separated by reversed-phase HPLC (Kito *et al.* 1985).

3. Results and Discussion

When plants were exposed to light after seed germination under all conditions, major changes were a decrease in the molecular species containing 18:3 and an increase in those containing 16:0 at the *sn*-1 position of glycerol backbone (Figure 1). After 24 h and 9 h illumination for squash and spinach, respectively, each molecular species reached a plateau (we do not show the spinach data here). The levels of molecular species were significantly affected by temperature, the lower the temperature, the higher content of the molecular species containing 18:3 and the lower content of those containing 16:0 at the *sn*-1 position of glycerol backbone were observed (Figure 1).

To avoid the effects of plant growth, the effects of temperature on the relative content of molecular species containing the same fatty acid at the *sn*-1 position of glycerol backbone were further investigated only when these molecular species levels had reached a plateau. The relative content of each molecular species (mean values from 48 and 96 h illuminated squash plants, and 9, 24, 48 and 96 h illuminated spinach plants) was linearly correlated with the growth temperatures (Figure. 2 and 3). The correlation between

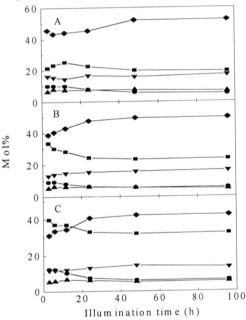

Figure 1. Effects of growth temperature on the composition of phosphatidylglycerol molecular species containing the same fatty acid at the *sn*-1 position of glycerol backbone in thylakoid membranes of squash cotyledons. Seedlings were exposed to continuous light (350 μmol m^{-2} s^{-1}) at 35°C (A), 30°C (B) and 20°C (C) after seeds were germinated in darkness at the same temperature. Molecular species were analyzed from seedlings illuminated for 3, 6, 12, 24, 48 and 96 h. ■ 18:3/16:0 + 18:3/16:1(3*t*); ● 18:2/16:0 + 18:2/16:1(3*t*); ▲ 18:1/16:0 + 18:1/16:1(3*t*); ▼ 18:0/16:0 + 18:0/16:1(3*t*); ◆ 16:0/16:0 + 16:0/16:1(3*t*).

molecular species content and the temperatures can be described by a straight line: $y=ax+b$. Here, y is the relative content of molecular species, the slope (a) indicates the molecular species change rate induced by temperature, x is the growth temperatures (°C), and b represents theoretically the relative content of molecular species at 0°C. The molecular species change rate (a) induced by the growth temperatures allows us to estimate the effects of temperature on the incorporate rate of fatty acids 16:0, 18:0 and 18:1 into glycerol, as well as the desaturase activities. As shown in Figure 2A, when squash cotyledons were grown at the temperatures in the range of 20 to 35°C, decreasing 1°C induced a decrease about 0.60 mol% in molecular species containing 16:0 at the sn-1 position of glycerol backbone, and 0.17 mol% in those containing 18:0. The sum of the decrease rate of these two molecular species must correspond to a same increase rate (0.77 mol%) of other molecular species that represented an increase in the

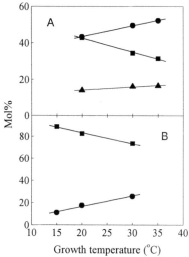

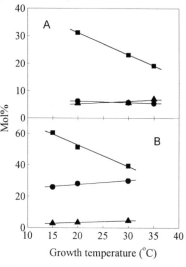

Figure 2. The relationship between growth temperatures and the relative content of the phosphatidylglycerol molecular species containing unsaturated fatty acids (■, $y = -0.77x + 57.90$ in A; $y = -1.01x + 103.43$ in B), 18:0 (▲, $y = 0.17x + 10.69$ in A) and 16:0 (●, $y = 0.60x + 31.42$ in A; $y = 0.95x - 2.53$ in B) at the sn-1 position of glycerol backbone in thylakoid membranes of squash (A) and spinach (B) cotyledons. Data points represent the means of two measures for squash (from 48 and 96 h illuminated seedlings) and those of four measures for spinach (from 9, 24, 48 and 96 h illuminated seedlings).

Figure 3. The relationship between growth temperatures and the relative content of the phosphatidylglycerol molecular species containing 18:3 (■, $y = -0.81x + 47.30$ in A; $y = -1.38x + 80.11$ in B), 18:2 (●, $y = -0.05x + 7.30$ in A; $y = 0.26x + 22.33$ in B) and 18:1 (▲, $y = 0.10x + 3.28$ in A; $y = 0.11x + 0.99$ in B) in thylakoid membranes of squash (A) and spinach (B) cotyledons. Data points represent the means of two measures for squash (from 48 and 96 h illuminated seedlings) and those of four measures for spinach (from 9, 24, 48 and 96 h illuminated seedlings).

incorporate rate of 18:1. When spinach cotyledons were grown at the temperatures in the range of 15 to 30°C (Figure 2B), decreasing 1°C induced a decrease of 0.95 mol%

in molecular species containing 16:0 at the *sn*-1 position of glycerol backbone and an increase rate (1.01 mol%) of other molecular species containing 18:1, 18:2 and 18:3 which represent the incorporate rate of 18:1.

Molecular species 18:1/16:0 is one of the initial products of PG biosynthesis. Fatty acid 18:1 in PG is desaturated by desaturases to produce 18:2 and 18:3 containing PG molecular species (Browse and Somerville, 1991). Therefore, we can measure the effects of temperatures on the desaturation of the fatty acids at the *sn*-1 position of glycerol backbone by comparing the molecular species change rate induced by temperature. The PG molecular species change rate induced by lowering 1°C in squash cotyledons are 0.81, 0.05 and –0.10 mol% for 18:3, 18:2 and 18:1 containing molecular species, respectively (Figure 3A). This indicates that about 0.86 mol% of 18:2 containing molecular species can be produced from 18:1 containing ones, and then about 0.81 mol% of them are further desaturated to 18:3 containing molecular species by decreasing 1°C. The PG desaturation induced by lowering temperature in spinach cotyledons displays the similar pattern to that in squash cotyledons (Figure 3B). In brief, about 1.38 mol% of 18:3 containing molecular species can be produced from 18:2 containing ones, and 1.12 mol% (-1.38 + 0.26) of 18:2 containing molecular species can be produced from 18:1 containing ones by decreasing 1°C.

Our results indicate that, of all the thylakoid PG in both spinach and squash cotyledons, only the levels of molecular species containing 16:0 and 18:3 at the *sn*-1 position of glycerol were greatly affected by temperature. Lowering temperature induced a decrease in the former and an increase in the latter. Several steps of PG biosynthesis, such as the fatty acid incorporation and desaturation, can be affected by temperature and each step of PG biosynthesis affected by changing temperature can be quantitatively detected.

4. References

Browse, J. and Somerville, C. (1991) Glycerolipid Synthesis: Biochemistry And Regulation. Annu Rev Plant Physiol Plant Mol Biol. 42, 467-506.

Kito, M., Takamura, H., Narita, H., and Urade, R. (1985) A sensitive method for quantitative analysis of phospholipid molecular species by high-performance liquid chromatography. J. Biochem. 98, 327-331.

Murata, N. (1983) Molecular species composition of phosphatidylglycerols from chilling-sensitive and chilling-resistant plants. Plant and Cell Physiol. 24, 81-86.

Murata, N., Ishizaki-Nishizawa, O., Higashi, S., Hayashi, H., Tasaka, Y., and Nishida, I. (1992) Genetically engineered alteration in the chilling sensitivity of plants. Nature. 356, 710-713.

Xu, Y.N. and Siegenthaler, P.A. (1996a) Effect of non-chilling temperature and light intensity during growth of squash cotyledons on the composition of thylakoid membrane lipids and fatty acids. Plant and Cell Physiology. 37, 471-479.

Xu, Y.N. and Siegenthaler, P.A. (1996b) Phosphatidylglycerol molecular species of photosynthetic membranes analyzed by high-performance liquid chromatography: theoretical considerations. Lipids. 31, 223-229.

MODULATION OF LIPID BODY SIZE AND PROTEIN PROFILES IN THE OLEAGINOUS FUNGUS BY CHANGING NITROGEN CONCENTRATION IN CULTURE MEDIUM

Y. KAMISAKA, N. NODA, and M. YAMAOKA
Research Institute of Biological Resources, AIST
Tsukuba, Ibaraki 305-8566, Japan

1. Introduction

Mortierella ramanniana var. *angulispora*, an oleaginous fungus, accumulates large amounts of triacylglycerol (TG) and lipid bodies [1,2]. Lipid bodies in this fungus have a diameter of about 1 μm at the initial culture phase and then enlarge into a diameter of about 2-3 μm during lipid accumulation [2]. To form matured lipid bodies, TG and/or its precursors have to transport to lipid bodies. We have characterized the lipid transport into lipid bodies using fluorescent phospholipid analogues to find several transport pathways for the lipid body formation [2,3]. The question arises how the lipid transport into lipid bodies is regulated. To address the question, we looked for culture conditions to change the size of lipid bodies and characterized the protein profile of lipid bodies with different size. We found that changing nitrogen concentration represented as carbon to nitrogen ratio (C/N ratio) in culture medium affected the lipid body size in addition to lipid content. Comparing the lipid body fraction obtained from fungal cells with different lipid body size revealed that tyrosine phosphorylation was enhanced in the lipid body fraction when fungal cells were cultured with a lower C/N ratio (higher concentration of nitrogen).

2. Materials and Methods

M. ramanniana var. *angulispora* (IFO 8187) was obtained from the culture collection of the Institute for Fermentation (Osaka, Japan). The fungi were cultured as described elsewhere [2]. Different C/N ratios were attained by varying the amount of $(NH_4)_2SO_4$ added to the medium. Lipid bodies stained by Nile red were observed with a laser scanning confocal microscope (LSM410, Zeiss) as described [2]. In some experiments, 42-hr fungal culture was centrifuged and added by the culture supernatant of 42-hr fungal culture at different C/N ratio. The fungal culture that replaced the culture supernatant was further cultured to investigate modulation of

N. Murata et al. (eds.), Advanced Research on Plant Lipids, 175–178.
© 2003 *Kluwer Academic Publishers. Printed in the Netherlands.*

lipid bodies.

Homogenization and subcellular fractionation of fungal cells were as described [1]. Polypeptide profiles in SDS-PAGE were revealed with a silver staining kit and tyrosine phosphorylation was measured by immunoblotting using a monoclonal antibody to phosphotyrosine antibody. Horseradish peroxidase conjugated anti-mouse IgG and ECL plus (Amersham Biosciences) were used to visualize the profiles.

3. Results

To investigate regulatory factors for lipid body biogenesis, we looked for culture conditions to modulate the lipid body size. When the C/N ratio in the culture medium was 38 as we usually used, lipid bodies were initially formed with a diameter of about 1 μm, then matured into a diameter of about 2-3 μm [2]. Decreasing the C/N ratio to 10 by adding $(NH_4)_2SO_4$ inhibited lipid body enlargement to about 2-3 μm, but kept most of lipid bodies with a diameter of around 1 μm throughout the incubation until the stationary phase. Lower C/N ratio or higher nitrogen concentration decreased the lipid content. Although it is already known that the C/N ratio affected the lipid content in fungi [4-6], lipid body morphology under the conditions has not been reported.

The lipid body fraction from fungal cells cultured at different C/N ratios had a similar polypeptide profile after 2-day culture, but had several polypeptide bands specific to each condition after 4-day culture. Since the 2-day fungal culture in the C/N 38 medium had lipid bodies already enlarged to a diameter of 2-3 μm, it was not likely that the existence of some major polypeptides affected the lipid body size. The specific polypeptides detected after 4-day culture are probably not regulatory factors of the lipid body size, but may be specifically transported to lipid bodies as a result of their size difference. To further characterize the protein profiles of the lipid body fraction, we investigated tyrosine phosphorylation that is known to play important roles in signal transduction. We found that several tyrosine phosphorylated polypeptides appeared in the lipid body fraction from fungal cells cultured in the C/N 10 medium, but much less tyrosine phosphorylated polypeptides appeared in the lipid body fraction from fungal cells cultured in the C/N 38 medium. The difference of tyrosine phosphorylation was evident after both 2-day and 4-day cultures. The results indicated that the modulation of the lipid body size induced by the C/N ratio change accompanied different protein profiles in the lipid body fraction.

To ensure the relationship between lipid body size and tyrosine phosphorylation of lipid body proteins, we replaced the culture supernatant of fungal culture by that of fungal culture at different C/N ratio. When the culture supernatant of 42-hr fungal culture in the C/N 10 medium was replaced by that in the C/N 38 medium, enlarged lipid bodies appeared after 15-hr incubation rather than smaller lipid bodies due to the initial C/N 10 medium. On the contrary, when the culture supernatant of 42-hr fungal culture in the C/N 38 medium was replaced by that in the C/N 10 medium, smaller lipid bodies were more prevalent after 15-hr incubation rather than larger lipid bodies due to the initial C/N 38 medium. In accordance with the change of lipid bodies, tyrosine phosphorylation of the lipid body fraction changed; appearance of smaller lipid bodies due to lower C/N ratios correlated with increase in tyrosine phosphorylation.

4. Discussion

Although more attention has been paid to the lipid body biogenesis lately, its molecular mechanisms are still poorly understood. Since the size of lipid bodies is unique depending on organisms and lipid bodies do not coalesce with adjacent ones, regulatory mechanisms for the lipid body size have been assumed [7]. The present study was aimed to investigate regulatory factors for the lipid body size in *M. ramanniana* var. *angulispora*. We found the C/N ratio or nitrogen concentration in culture medium as one of these factors. Furthermore, the modulation of the lipid body size induced by changing the C/N ratio accompanied alteration of tyrosine phosphorylation in the lipid body fraction. Although the link between the lipid body size and tyrosine phosphorylation in the lipid body fraction is not clear yet, tyrosine phosphorylation may play some roles in signal transduction to regulate the lipid body size.

Recently, signal transduction pathways after nutrient sensing have been vigorously studied in several fungi and yeasts, and protein phosphorylation such as the MAP kinase cascade is known to play important roles in these pathways [8]. Similar mechanisms may sense and transfer difference in the C/N ratio for cell growth in this fungus. Since lipid bodies are the intracellular sites of energy reservoir, it is possible that the lipid body size is regulated as one of cell responses to nutrient conditions. Furthermore, recent studies indicated that lipid bodies in mammalian cells form a raft-like membrane domain containing signaling molecules such as caveolin [9], suggesting that lipid bodies interact with surround organelles by signal transduction systems. Future studies are directed to elucidate molecular links among nutrient sensing, tyrosine phosphorylation in lipid bodies, and lipid body size in this fungus.

5. References

[1] Kamisaka, Y. and Nakahara, T. (1994) Characterization of the diacylglycerol acyltransferase activity in the lipid body fraction from an oleaginous fungus. J. Biochem. 116, 1295-1301.

[2] Kamisaka, Y., Noda, N., Sakai, T. and Kawasaki, K. (1999) Lipid bodies and lipid body formation in an oleaginous fungus, Mortierella ramanniana var. angulispora. Biochim. Biophys. Acta 1438, 185-198.

[3] Kamisaka, Y. and Noda, N. (2001) Intracellular transport of phosphatidic acid and phosphatidylcholine into lipid bodies in an oleaginous fungus, Mortierella ramanniana var. angulispora. J. Biochem. 129, 19-26.

[4]   Suzuki, O., Yokochi, T. and Yamashina, T. (1982) Studies on production of lipids in fungi. VIII. Influence of cultural conditions on lipid compositions of two strains of Mortierella isabellina. J. Jpn. Oil Chem. Soc. (Yukagaku) 31, 921-931.

[5]   Kamisaka, Y., Yokochi, T., Nakahara, T. and Suzuki, O. (1993) Characterization of the diacylglycerol acyltransferase activity in the membrane fraction from a fungus. Lipids 28, 583-587.

[6]   Ratledge, C. (1989) Biotechnology of oils and fats. in C. Ratledge and S.G. Wilkinson (eds.), Microbial Lipids, Vol. 2, Academic Press, London, pp. 567-668.

[7]   Murphy, D.J. (2001) The biogenesis and functions of lipid bodies in animals, plants and microorganisms. Prog. Lipid Res. 40, 325-438.

[8]   Rohde, J., Heitman, J. and Cardenas, M.E. (2001) The TOR kinases link nutrient sensing to cell growth. J. Biol. Chem. 276, 9583-9586.

[9]   Fujimoto, T., Kogo, H., Ishiguro, K., Tauchi, K. and Nomura, R. (2001) Caveolin-2 is targeted to lipid droplets, a new "membrane domain" in the cell. J. Cell Biol. 152, 1079-1085.

CLONING AND CHARACTERISATION OF A PHOSPHOLIPID: DIACYLGLYCEROL ACYLTRANSFERASE FROM *ARABIDOPSIS THALIANA*

A. BANAS[1], M. LENMAN[1], A. DAHLQVIST [1], S. STYMNE [2]

1-Scandinavian Biotechnology Research AB, PO Box 116, S-230 53 Alnarp, Sweden
2-Department of Crop Science, SLU, PO Box 44, S-230 53 Alnarp, Sweden

1. Introduction

In our earlier work (Dahlqvist et al., 2000) we cloned and characterised a gene encoding an enzyme, which in yeast catalyse an acyl-CoA-independent synthesis of TAG. This enzyme utilises phospholipids as acyl donors and diacylglycerol as an acyl acceptor and was named phospholipid: diacylglycerol acyltransferase or PDAT. We have also demonstrated that a similar enzyme activity is present in microsomal preparations of developing seeds of several plants (Dahlqvist et al., 2000; Banaś et al., 2000). Based on sequence homologies with the yeast PDAT, several candidate genes were identified in the *Arabidopsis thaliana* genome. In the presented study, one of these candidates: "AB006704" was cloned and characterised.

2. Material and methods

An AB006704 cDNA was ligated behind the 35S promoter into the shuttlevector pART27 and the constructs transformed into Arabidopsis cv. Columbia *via* vacuum infiltration (Clough and Bent, 1998). T2 seeds of the transformants were germinated on agar containing Kanamycin and then transferred into liquid ½ MS-medium supplemented with 1 % sucrose. The plants were grown for 27 days at 23°C under light. For evaluation of expression level of the inserted gene, total RNA was extracted from leaves and roots (according to modified "phenol extraction method"; Sambrook et al., 1989). 10 µg of extracted RNA was than run on 1,5% agarose gels with 5% formaldehyde and transferred to nylon (Hybond-N$^+$) membranes. Blots were hybridised with the AB006704 cDNA in 0,5 M NaHPO$_4$ (pH 7,2), 7% SDS and 1 mM EDTA at

179

N. Murata et al. (eds.), Advanced Research on Plant Lipids, 179–182.
© 2003 *Kluwer Academic Publishers. Printed in the Netherlands.*

64°C over night. After washing, hybridisation was visualised by electronic autoradiography.

Phospholipid: diacylglycerol acyltransferase activity towards different substrates, were evaluated in microsomes prepared (according to procedure of Stobart and Stymne, 1985) from roots and leaves of transformed and control plants grown as indicated above. Aliquots of microsomal fractions (corresponding to 12 nmol of microsomal PC) were lyophilised over night and used for enzyme assays. Substrates (2,5 nmol of PC, PE or PA with ^{14}C-labelled acyl groups in the sn-2 position and 1,5 nmol of di-oleoyl- or sn-1-oleoyl-sn-2-epoxy-DAG) dissolved in 14 µl of benzene were added to dried microsomes. The solvent was immediately evaporated under a stream of nitrogen (leaving the added lipids in direct contact with the membranes) and 0,1 ml of 50 mM potassium phosphate (pH 7,2) was added. The suspension was thoroughly mixed and incubated at 30°C for 60 min. Lipids were extracted from the reaction mixture into chloroform (Blight & Dyer, 1959) and separated by thin layer chromatography. The radioactive lipids were visualised and quantified on the plates by electronic autoradiography.

3. Results and discussion

After a preliminary study of the PDAT activity in microsomal fractions prepared from leaves of several transformants, three transformed plants with different level of enzyme activity were selected for further study. In all three plants the expression of "AB006704" gene was clearly detected in both leaves and roots. The transformed gene was most highly expressed in plant number "1-1-6". Transformant "1-3b-44" had a middle level of expression and "1-2-13" the lowest (approximately 70 and 25 % intensity of the "1-1-6", respectively). The expression of this gene in control plants was hardly detectable.

PDAT activity in microsomal fractions prepared from leaves and roots of the three transformants was compared using sn-1-oleoyl-sn-2- [^{14}C]-ricinoleoyl-PC as an acyl donor and di-oleoyl-DAG as an acyl acceptor. The amount of de novo synthesised 1-OH-TAG generally reflected the expression pattern of the "AB006704" gene in plant material used for microsomal preparation. The TAG synthesis activity was highest in microsomes of transformant "1-1-6", medium in microsomes of transformant "1-3b-44" and the lowest (but clearly higher than in the control) in microsomes of transformant "1-2-13" (Table 1.).

To verify that formation of ^{14}C-labelled TAG occurred via a transfer of an acyl group from the radiolabelled phospholipid to DAG (i.e. catalysed by PDAT), sn-1-oleoyl-sn-2- [^{14}C]-ricinoleoyl-PC was used as an acyl donor and sn-1-oleoyl-sn-2-epoxy-DAG was used as an acyl acceptor. It was clearly shown in this experiment that the synthesised TAG contained both [^{14}C]-ricinoleoyl and vernoloyl groups. The de novo TAG synthesis also reflected well the expression pattern of the "AB006704" gene (Table 2). Thus these data strongly suggests that the transgene can use fatty acids from the sn-2 position of PC to acylate sn-1-oleoyl-sn-2-epoxy-DAG in the formation of TAG. Therefore we conclude that the "AB006704" gene encode a PDAT enzyme which is capable of using PC as the immediate acyl donor and DAG as the acyl acceptor in acyl-CoA-independent formation of TAG.

Substrate specificity of the cloned PDAT was tested with microsomal preparation of leaves of transformant "1-1-6". The enzyme activity was dependent on the degree of unsaturation of the acyl group at position *sn*-2 of PC. Among 18-C fatty acids, the stearic acid (18:0) was most poorly transferred, whereas the linolenic acid (18:3) was transferred most efficiently. The Arabidopsis PDAT also showed higher specificity towards short acyl chains; "10:0" and "18:1" was more efficiently transferred as compared to "18:0" and "22:1", respectively. Furthermore, the enzyme has high specificity for the transfer of acyl chain with hydroxy and epoxy groups whereas the replacement of a double bond with a triple bond decreased the activity (Table 3). The activity was also dependent on the phospholipid head group. Phosphatidylethanolamine was utilised by the enzyme better than phosphatidylcholine and phosphatidic acid worse (Table 3). Acyl composition of the acceptor had also some influence on the enzyme activity. TAG was more efficiently synthesised with *sn*-1-oleoyl-*sn*-2-epoxy-DAG than with di-oleoyl-DAG as the acyl acceptor (Table 1 & 2).

TABLE 1. *In vitro* synthesis of TAG carrying [^{14}C]-ricinoleoyl group by microsomes prepared from leaves and roots of three different Arabidopsis plants transformed with the "AB006704" gene and by microsomes prepared from an untransformed Arabidopsis plants (control).

Plant material	Radioactivity in TAG (% of added)	
	Leaves	Roots
Control	1,3	2,2
"1-1-6"	10,9	12,2
"1-3b-44"	6,8	6,6
"1-2-13"	3,2	4,8

TABLE 2. *In vitro* synthesis of TAG carrying one vernoloyl and one [^{14}C]-ricinoleoyl group by microsomes prepared from leaves of three different Arabidopsis plants transformed with the "AB006704" gene and from an untransformed control.

Plant material	Radioactivity in TAG (% of added)	
	TAG carrying [^{14}C]ricinoleoyl and vernoloyl groups	TAG carrying [^{14}C]ricinoleoyl group (endogenous DAG acceptor)
Control	0,7	0,7
"1-1-6"	10,6	4,1
"1-3b-44"	5,9	2,8
"1-2-13"	1,4	2,0

TABLE 3. Substrate specificity of the cloned Arabidopsis PDAT (activity towards 18:1-PC was calculated as 1)

Acyl group in position sn-2 of a donor	Relative TAG synthesis (different donors of acyl group)		
	PC	**PE**	**PA**
$[^{14}C]$-10:0	1,4	2,0	-
$[^{14}C]$-18:0	0,3	-	-
$[^{14}C]$-18:1	1,0	1,7	0,3
$[^{14}C]$-18:2	2,2	-	-
$[^{14}C]$-18:3	3,8	-	-
$[^{14}C]$-22:1	0.3	-	-
$[^{14}C]$-crep	1,3	-	-
$[^{14}C]$- ver	4,2	-	-
$[^{14}C]$-ric	6,9	-	1,5

Dioleoyl-DAG together with sn-1-oleoyl-sn-2-$[^{14}C]$-FA-phospholipid (PC, PE or PA) were used as a substrates (10:0-PC, 18:0-PC, 22:1-PC and 10:0-PE in position sn-1 had 16:0); crep – crepenynic acid, ver –vernolic acid, ric –ricinoleic acid

Acknowledgement:
This work was supported by financial contributions from the BASF Plant Science Ltd

4. References

Banaś, A., Dahlqvist, A., Ståhl, U., Lenman, M., Stymne, S. (2000) The involvement of phospholipid: diacylglycerol acyltransferases in triacylglycerol production. Biochemical Society Transaction, 28 (6), 703-705.

Blight, E.G. and Dyer, A. (1959) A rapid method of total lipid extraction and purification. Can. J. Biochem. Physiol., 37, 911-917.

Clough, S.J. and Bent, A.F. (1998) Floral dip: simplified method for *Agrobacterium*-mediated transformation of *Arabidopsis thaliana*. The Plant Journal. 16 (6), 735-743.

Dahlqvist, A., Ståhl, U., Lenman, M. Banaś, A., Lee, M., Sandager, L., Ronne, H., Stymne, S. (2000) Phospholipid:diacylglycerol acyltransferase: An enzyme that catalyzes the acyl-CoA-independent formation of triacylglycerol in yeast and plants. PNAS, 97 (12), 6487-6492.

Sambrook, J., Fritsch, E.F., Maniatis, T. (1989) Molecular Cloning. Cold Spring Harbor Laboratory Press.

Stobart, A.K. and Stymne, S. (1985) The regulation of fatty acid composition of the triacylglycerols in microsomal preparations from avocado mesocarp and the developing cotyledons of safflower. Planta, 163, 119-125.

A MULTIGENE FAMILY OF LYSOPHOSPHATIDIC ACID ACYLTRANSFERASES OF ARABIDOPSIS THALIANA

S. MAISONNEUVE, R. GUYOT, M. DELSENY AND T. ROSCOE*

Laboratoire de Génome et Développement des Plantes, C.N.R.S. UMR 5096, Université de Perpignan, 52 Avenue de Villeneuve, 66860 Perpignan, France.

Abstract

A genomics based approach has been used to identify members of the lysophosphatidic acid acyltransferase (LPAAT) multigene family in Arabidopsis thaliana. Ten putative members of this family containing conserved motifs known to be present in LPAAT and in glycerol-3-phosphate acyltransferase (GPAT) proteins have been identified in the Arabidopsis thaliana genome. These acyltransferases were classified according to their sequence similarity. A member of the first class of genes encodes a LPAAT implicated in glycerolipid biosynthesis in the eukaryotic pathway. The second class contains a gene that encodes an LPAAT of the prokaryotic pathway of lipid biosynthesis in plastids. An additional class containing three genes encodes proteins of unknown function previously undescribed as acyltransferases. The sequence divergence, compartmentalisation and differential expression of members of the gene family are consistent with a specific role for each LPAAT isoform in the production of phosphatidic acid.

1. Introduction

Phosphatidic acid is an important intermediate in the synthesis of glycerolipids and is synthesed by sequential acylation of glycerol phosphate, catalysed by glycerol-3-phosphate acyltransferase (GPAT) and lysophosphatidic acid acyltransferase (LPAAT) to produce lysophosphatidic acid and phosphatidic acid respectively using acyl thioester donors. Isozymes of these acyltransferases that differ in their substrate specificity, in particular LPAAT, are found in organellar and endomembrane compartments and thus play an important role in determining the acyl composition of membrane lipids, (Frentzen, 1998). In addition, the seeds of species that accumulate unusual fatty acids contain microsomal LPAATs that exhibit a strong variation in substrate preference.

N. Murata et al. (eds.), Advanced Research on Plant Lipids, 183–186.

We are interested in the structure and function relationships of plant LPAATs and in their role in determining lipid acyl composition in developing seeds. The availability of the complete Arabidopsis genomic sequence allowed us to identify several additional genes encoding proteins possessing structural features present in all LPAAT proteins.

2. Methods

Screening for Arabidopsis LPAAT genes

Putative LPAAT genes were identified from annotated and anomymous genomic sequences of Arabidopsis present in the databases using the NCBI server. The nucleotide sequences encoding the Brassica napus chloroplastic LPAAT (Accession no. A111161) and the cytoplasmic LPAAT (accession no. Z95673) were used as search sequences database using BLAST algorithms. Where possible the predicted LPAAT encoding genomic sequences were verified by comparison to available cDNA sequences. The deduced amino acid sequence corresponding to a central core of approximately 120 residues were aligned using CLUSTAL and Phylip (EBI server) and a denrogram produced using Treeview.

3. Results

3.1 *A multigene family of Arabidopsis LPAATs*

A total of ten distinct genes encoding putative LPAATs were identified in the Arabidopsis genomic sequence. These genes were distributed among the five chromosomes and did not show any clustering, (Figure 1). Four of these genes were localised in chromosomal regions that were duplicated however the LPAAT genes were not duplicated. Dot plot analysis of BACs corresponding to duplicated regions where the LPAAT genes were located verified the lack of LPAAT gene duplication.

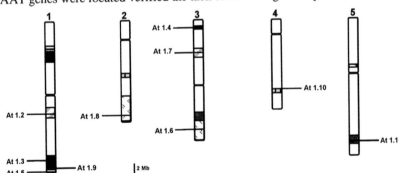

Figure 1. Chromosomal localisation of acyltransferase genes in Arabidopsis thaliana. The positions of the acyltransferase genes were deduced from the Arabidopsis physical map at the TAIR site. The shaded regions indicated duplicated Arabidopsis chromosomal segments

3.2 *Structural relationships of plant LPAATs*

The deduced amino acid sequence corresponding to the central core of each of the ten putative LPAATs was used to produce a dendrogram illustrating the structural relationships between the LPAAT proteins, (Figure 2). Two conserved motifs were present in each of these deduced protein sequences, the first motif contained the residues NH(X)4(D/E) and the second contained the residues F(P/V)EG. The number of residues and the amino acid sequences were highly divergent among the ten proteins. All deduced protein sequences were predicted to contain at least one transmembrane spanning domain. Three main classes of proteins were evident and share only 16-19% amino acid identity over the full length of the sequences although the central core of the proteins share upto 27% identity. The first class (Class A) of genes encodes proteins that are found only in eukaryotic species and share between 33-62% amino acid identity among the four members. A second class (Class B) shares contains a gene that encodes an plastidial LPAAT. The members of this group were divergent in sequence and the amino acid identity varied between 24-30%. The third class (Class C) represents a novel, previously unrecognised group of acyltransferases and contains three genes encoding proteins of unknown function. Members of this class were also divergent in sequence and share 28-30% amino acid identity in their central core region.

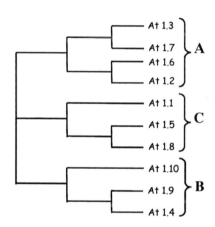

Figure 2. Phylogram of Arabidopsis acyltransferases derived from amino acid sequences alignments

4. Discussion

We have identified a minium of ten Arabidopsis genes coding for proteins that contain conserved motifs present in glycerolipid acyltransferases. The deduced proteins share sequence homology only with LPAATs. The sequence divergence among the putative LPAAT proteins, the variation in the pattern of expression of members of each class of genes (data not shown) together with their different subcellular localisations argues against redundancy and suggests distinct roles for each LPAAT isoform.

Alignment of the ten putative LPAATs according to their sequence homology and revealed the existence of three classes of LPAAT genes. Previously, LPAAT isozymes have been associated with organellar and cytoplasmic fractions and cloning of cDNAs confirmed the existence of two distinct classes of plant LPAAT genes that encode microsomal proteins (Frentzen 1998) that were divergent in sequence, their patterns of expression and their substrate preferences. Thus, upto four distinct LPAATs are known in plants. What then are the functions of the isoforms encoded by the additional LPAAT like genes present in the Arabidopsis genome?

The Arabidopsis sequence At1.2 of class A is the homologue of the maize LPAAT (Brown et al 1994) demonstrated to encode a ubiquitously expressed microsomal enzyme associated with the eukaryotic pathway. The relative sequence similarity of the three other class A isoforms to the gene At1.2 together with a predicted localisation in the endoplasmic reticulum may indicate a role the production of phosphatidic acid for membrane lipid biosynthesis. The Arabidopsis sequence AT1.10 of class B is the homologue of the rapeseed plastidial LPAAT (Bourgis et al 1999) and suggests that class B genes may encode proteins localised in organelles. The homology of the AT1.9 sequence to the taffazin acyltransferase proteins possibly located in mitochrondria seems to confirm the prokaryotic nature of proteins in class B. We did not detect an Arabidopsis homologue of the seed specific microsomal LPAATs identified in Limnanthes (Hanke et al 1995) and other species that incorporate unusual fatty acids at sn-2 of triacylglycerol. The role of the class C proteins is obscure, since the members of this class show limited homology only with acyltransferase proteins. Two of the deduced proteins are predicted to be localised at the plasmalemma and one of which may contain a protein dimerisation domain and the second a EF hand calcium binding motif. It is therefore possible that members of this group may play a role in signalling pathways. We anticipate that this work will lead to an understanding of role played by LPAAT isoforms in the production of phosphatidic acid required for diverse cellular functions such as biosynthesis of membrane and storage lipids, vesicle formation and signalling pathways.

References

Barth PG, Wanders RJ, Vreken P, Janssen EA, Lam J, Baas F. (1999) X-linked cardioskeletal myopathy and neutropenia (Barth syndrome) (MIM 302060). J Inherit Metab Dis 22:555-67

Bourgis F, Kader J-C, Barret P, Renard M, Robinson D, Robinson C, Delseny M, Roscoe TJ (1999). A plastidial 1-acyl-Glycerol-3-phosphate acyltransferase of Brassica napus. Plant Physiol 120: 913-921

Brown A.P, Coleman J, Tommey AM, Watson MD, Slabas AR. (1994) Isolation and characterisation of a maize cDNA that complements a 1-acyl sn-glycerol-3 phosphate acyltransferase mutant of Escherichia coli and encodes a protein which has similarities to other acyltransferases. Plant Mol Biol. 26:211-23.

Frentzen M, (1998) Acyltransferases from basic science to modified seed oils. Fett/Lipid 100:161-166

Hanke C, Wolter FP, Coleman J, Peterek G, Frentzen M. (1994) A plant acyltransferase involved in triacylglycerol biosynthesis complements an Escherichia coli sn-1-acylglycerol-3-phosphate acyltransferase mutant. Eur J Biochem. 232:806-10.

KINETIC ANALYSIS OF MICROSOMAL *SN*-GLYCEROL-3-PHOSPHATE ACYLTRANSFERASE FROM THE DEVELOPING SEEDS OF SUNFLOWER AND *MORTIERELLA ALPINA*

T.C.M. FRASER, S. CHATRATTANAKUNCHAI, A.D. Waters and A.K. STOBART
School of Biological Sciences, University of Bristol, Woodland Road, Bristol BS8 1UG, UK

1. Abstract

Glycerol-3-phosphate acyltransferase (GPAT) is the first committed step in glycerolipid synthesis and presumed to be rate-limiting. It catalyses the esterification of the *sn*-1 hydroxyl of glycerol-3-phosphate (G3P) with an activated fatty acid forming 1-acyl-*sn*-glycerol-3-phosphate (LPA). Kinetic data reported here are consistent with the GPAT catalysed acylation of G3P occurring by single and double displacement reaction mechanisms in sunflower and *Mortierella alpina*, respectively, and therefore suggest that there may be no evolutionary link between fungal and higher plant microsomal GPAT.

2. Materials and Methods

2.1 *Microsomal membrane preparation from sunflower developing seed cotyledons.*
Developing seeds were harvested 10-12 days post anthesis and suspended in ice cold 0.1 M potassium phosphate buffer, pH 7.2, containing 0.33 M sucrose and 0.1% bovine serum albumin (BSA) (fatty acid free). After grinding in a glass homogenizer, the homogenate was centrifuged at 20,000 g for 30 minutes. The supernatant was filtered through a layer of Miracloth and centrifuged at 100,000 g for 90 minutes. The pelleted microsomal membranes were resuspended in a small volume of 0.1 M potassium phosphate buffer, pH 7.2 using a glass homogeniser. The microsomal membrane preparations were either used immediately or stored at –80 °C until required.

2.2 *Fungal microsomal membrane preparation*
Cultures were harvested by filtration after 5 days growth. The mycelia were suspended in cold 0.1 M potassium phosphate buffer, pH 7.2, containing 0.33 M sucrose, 0.1%

N. Murata et al. (eds.), Advanced Research on Plant Lipids, 187–190.

BSA (fatty acid free), 1000 units of catalase•ml^{-1} and 1 mM Pefabloc. Subsequent procedures were as described above.

2.3 *Enzyme assay*

Microsomal membranes equivalent to 50 μg protein were incubated for 3 minutes with various concentrations of [^{14}C]G3P and oleoyl-CoA and 0.5 mg of BSA in a total volume of 0.1 ml of 0.1 M phosphate buffer, pH 7.2 at 20 ^{0}C. The reaction was initiated by the addition of microsomal membranes. The labelled lipid products were extracted in 1 ml of butan-1-ol/water/acetic acid (20:19:1 v/v) and the resulting butanol phase was washed three times with butanol saturated acidified aqueous phase (1 ml). Radioactive incorporation was quantified by liquid scintillation counting. Incorporation was proportional to incubation time and protein concentration.

3. Results

3.1 *Effect of acyl-CoA on GPAT activity*

Microsomal membranes were incubated for 3 minutes with various concentrations of [^{14}C]G3P and oleoyl-CoA and the radioactive incorporation into lipid was determined. In sunflower GPAT initial rate (referred to here as activity) increased with oleoyl-CoA concentration and reached a maximum at 0.2 mM and then decreased (Fig. 1a). In *M. alpina*, an optimum oleoyl-CoA concentration was also observed, unlike sunflower however it increased with G3P concentration (Fig. 1b). Inhibition of activity at high acyl-CoA concentration has been reported in numerous studies (Green et al., 1981; Mishra & Kamisaka, 2001).

3.2 *Effect of G3P on GPAT activity*

In sunflower (Fig. 2a) GPAT activity increased with G3P concentration up to 10 mM irrespective of the oleoyl-CoA concentration. In *M. alpina*, however, the activity became saturated at G3P concentrations which increased with the acyl-CoA concentration (Fig. 2b).

Double reciprocal plots of initial rate and G3P concentration were made at distinct acyl-CoA concentrations for sunflower (Fig. 3a) and *M. alpina* (Fig. 3b). Kinetic data reported here are consistent with the GPAT catalysed acylation of G3P occurring by single and double displacement reaction mechanisms in sunflower and *Mortierella alpina*, respectively.

4. Discussion

Preparations from prokaryote, yeast, plant and animal cells all have measurable GPAT activity. In *E.coli* GPAT has been purified (Green et al., 1981) and the GPAT gene identified (Lightner et al., 1980). In eukaryotic cells, multiple GPAT isoforms are present (Dirks & Sul, 1997; Athenstaedt & Daum, 1999). Mammalian mitochondrial and plant plastidial GPAT genes have now been isolated and characterised (Dirks & Sul, 1997; Murata & Tasaka, 1997). Two yeast putative GPAT genes have also been identified in *S. cerevisiae* (Zeng & Zou, 2001) and one of these genes has been implicated in encoding a microsomal GPAT. A putative microsomal GPAT has now been purified from *M. ramanniana* (Mishra & Kamisaka, 2001). In contrast plant and animal microsomal counterparts have remained elusive. The data presented here are consistent with the sunflower microsomal GPAT catalysing the esterification of G3P by a single displacement reaction mechanism involving the formation of a ternary complex between the enzyme and both substrates. In *M. Alpina,* however, the data suggest that a double displacement 'ping-pong' type reaction mechanism occurs in which the first substrate combines with the enzyme forming a substituted enzyme and releasing the first product before the second substrate binds to the substituted enzyme. The difference in reaction mechanism between higher plant and fungal GPAT therefore suggest that there may be no evolutionary link between them. The kinetic data presented here do not demonstrate the presence of multiple GPAT isoforms in the microsomal membranes of sunflower or *M. alpina*, although this possibility cannot be excluded.

5. References

Athenstaedt, K. Daum, G. (1999) Phosphatidic acid, a key intermediate in lipid metabolism. Eur. J. Biochem. 266, 1-16.

Dirks, L.K., and Sul. (1997) Mammalian mitochondrial glycerol-3-phosphate acyltransferase. Biochim. Biophys. Acta 1348, 17-26.

Green, P.R. Merril Jr, A.H. and Bell, R.M. (1981) Membrane phospholipid synthesis in *Escherichia coli.* Purification, reconstitution, and characterisation of *sn*-glycerol-3-phosphate acyltransferase. J. Biol. Chem. 256, 11151-11159.

Lightner, V.A. Larson, T.J., Tailleur, P., Kantor, G.D. Raetz, C.R.H., Bell, R.M. and Modrich, P. (1980) Membrane phospholipid synthesis in *Escherichia coli.* Cloning of a structural gene (plsB) of the *sn*-glycerol-3-phosphate acyltransferase. J. Biol. Chem. 255, 9413-9420.

Mishra, S. and Kamisaka, Y. (2001) Purification and characterisation of a thiol-reagent-sensitive glycerol-3-phosphate acyltransferase from the membrane fraction of an oleaginous fungus. Biochem. J. 355, 315-322.

Murata, N. and Tasaka, Y. (1997) Glycerol-3-phosphate acyltransferase in plants. Biochim. Biophys. Acta 1384, 10-16.

Zeng, Z. and Zou, J. (2001) and Bell, R.M. (1981) The Initial Step of the Glycerolipid Pathway. Identification of glycerol 3-phosphate/dihydroxyacetone phosphate dual substrate acyltransferases in *Saccharomyces. Cerevisiae.* J. Biol. Chem. 276, 41710-41716.

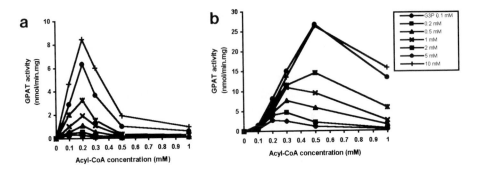

Figure 1. Effect of acyl-CoA on GPAT activity in sunflower and and *M. alpina*
Microsomal membranes (equivalent to 50 μg protein) from sunflower (a) and *M. alpina* (b) were incubated for 3 minutes with various concentrations of [^{14}C]G3P and oleoyl-CoA and the radioactivity incorporated into glycerolipid was determined.

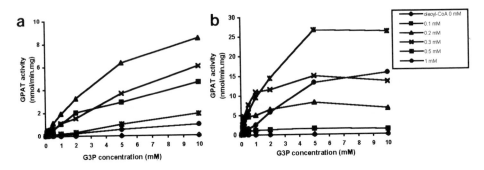

Figure 2. Effect of G3P on GPAT activity in sunflower and *M. alpina*
Microsomal membranes (equivalent to 50 μg protein) from sunflower (a) and *M. alpina* (b) were incubated for 3 minutes with various concentrations of [^{14}C]G3P and oleoyl-CoA and the radioactivity incorporated into glycerolipid was determined.

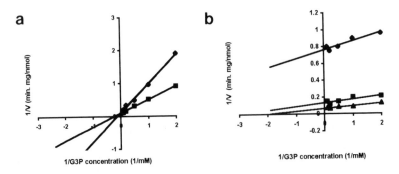

Figure 3. Double reciprocal plots
Double reciprocal plots of sunflower (a) and *M. alpina* (b) GPAT activity and G3P substrate concentration in the presence of 0.1 (♦), 0.2 (■) or 0.3 mM (▲) oleoyl-CoA.

ISOLATION AND CHARACTERIZATION OF DISRUPTANTS FOR CTP:PHOSPHORYLCHOLINE CYTIDYLYLTRANS- FERASE ISOGENES IN *ARABIDOPSIS*

R. INATSUGI, M. NAKAMURA AND I. NISHIDA

Department of Biological Sciences, Graduate School of Science
The University of Tokyo
7-3-1 Hongo, Bunkyo-ku, Tokyo, 113-0033 Japan

1. Introduction

Phosphatidylcholine (PC) is a major class of bilayer-forming lipids in eukaryotes and is widely distributed in almost all membranes in plant cells except thylakoid membranes. Besides its structural role as a membrane component, PC serves as metabolic precursors to the glycolipids of plastid membranes, monogalactosyldiacylglycerol (MGDG), digalactosyldiacylglycerol (DGDG) and sulfoquinovosyldiacylglycerol (SQDG) and to triacylglycerol in oil-accumulating tissues (for review, see Ohlrogge and Browse, 1995). PC also serves as a major substrate for fatty acid desaturation and, hence, is potentially important for the regulation of membrane fluidity in temperature adaptation of plant cells. Thus, the efficient synthesis of PC is potentially important for the growth and development of plant cells.

In plant cells, PC is largely synthesized via the nucleotide pathway, which includes three steps of sequential reactions catalyzed by choline kinase (CKI; EC 2.7.1.32), CTP:phosphorylcholine cytidylyltransferase (CCT; EC 2.7.7.15) and CDP-choline:diacylglycerol cholinephosphotransferase (CPT; EC 2.7.8.2). CCT is a rate-limiting enzyme in this pathway. *Arabidopsis thaliana* contained two genes for CCT in chromosome 2 and chromosome 4, which we designated *AtCCT1* and *AtCCT2*, respectively (Nishida et al., in this volume).

We recently found that the expression of *AtCCT2* but not that of *AtCCT1* was enhanced in the rosette leaves of *Arabidopsis* after cold treatment (Nishida et al., in this volume). Enhanced levels of AtCCT2 protein were associated with the increased levels of the CCT activity and the increase in PC content in the rosette leaves of *Arabidopsis* after cold treatment. We therefore conclude that the expression of *AtCCT1* and *AtCCT2* is differentially regulated in the rosette leaves of *Arabidopsis* for the biosynthesis and/or turnover of PC at low temperatures.

N. Murata et al. (eds.), Advanced Research on Plant Lipids, 191–194.
© 2003 *Kluwer Academic Publishers. Printed in the Netherlands.*

To investigate the physiological roles of PC in plant cells, especially in relation to cold acclimation, it is quite important to isolate mutants defective in the biosynthesis of PC. In this communication, we report isolation and characterization of T-DNA-tagged mutants for *AtCCT1* and *AtCCT2*. The result not only confirms our recent finding that the expression of AtCCT1 and AtCCT2 is differentially regulated at low temperatures but also provides us with a preliminary but first opportunity to test the functional cooperation and differentiation between AtCCT1 and AtCCT2 in the biosynthesis of PC in *Arabidopsis* at low temperatures.

2. Experimental Procedures

T-DNA-tagged lines of *Arabidopsis thaliana* were obtained from the Arabidopsis Biological Resource Center, and the screening for gene disruptants for *AtCCT1* and *AtCCT2* was performed with the aid of the Arabidopsis Knockout Facility at the University of Wisconsin Biotechnology Center.

3. Results and Discussion

3. 1. *Isolation of Gene Disruptants for* AtCCT1 *and* AtCCT2
We screened for disruptants for *AtCCT1* and *AtCCT2* in T-DNA-tagged lines of *Arabidopsis* by PCR, using primers shown in Fig.1. The PCR was designed to amplify DNA fragments including the left border sequence of T-DNA and its flanking genomic sequence for either *AtCCT1* or *AtCCT2*. Amplified DNA fragments were sequenced, and T-DNA-tagged lines that contained the insertion of T-DNA into the 6th intron of *AtCCT1* or that into the 2nd exon of *AtCCT2* were isolated (Fig. 1). Homozygous disruptants were selected from the descendents of isolated T-DNA-tagged lines with respect to T-DNA insertion into each of *AtCCT1* and *AtCCT2*, and the resultant homozygous disruptants for *AtCCT1* and *AtCCT2* were tentatively termed *cct1* and *cct2*, respectively. However, it should be noted that genome blot analysis showed that the T-DNA-tagged line for *AtCCT1* still contained several T-DNA insertion sites other than *AtCCT1*, whereas *AtCCT2* is the sole site for T-DNA insertion in *cct2* plants.

3. 2. *Expression of* AtCCT1 and AtCCT2 *in the Rosette Leaves of* cct1 *and* cct2 *Plants*
Expression of *AtCCT1* in *cct1* plants was examined by RNA gel blot and immunoblot analyses (Fig. 2). Using a probe specific to a 3' untranslated region of *AtCCT1*, the transcripts of *AtCCT1* were detected as a 1.3-kb band in the wild type (Fig. 2A). By contrast, no band was detected on RNA gel blots prepared from RNA from *cct1* plants grown at 23 °C. Immunoblot analysis revealed that AtCCT1 protein was detected as a 44.0-kDa band in the wild type, whereas no band for AtCCT1 was detected on immunoblots prepared from the protein extracts from *cct1* plants grown at 23 °C (Fig.

2B). These results suggest that insertion of T-DNA in the 6th intron of *AtCCT1* decreased the expression of *AtCCT1* in *cct1* plants grown at 23 °C below detectable levels.

Expression of *AtCCT2* in *cct2* plants was similarly examined (Fig. 2). RNA gel blot analysis using a probe specific to a 3' untranslated region of *AtCCT2* revealed that the transcripts of *AtCCT2* were detected as a 1.2-kb band in the wild type (Fig. 2A), and immunoblot analysis revealed that AtCCT2 protein was detected as a 38.5 kDa band in the wild type (Fig. 2B). By contrast, RNA gel blot analysis revealed that two bands (1.2 kb and 1.4 kb) were unexpectedly detected in *cct2* plants (asterisks in Fig. 2A). However, we think these bands could be artifacts of hybridization, because no significant band for AtCCT2 protein was detected by the immunoblot analysis of the protein extracts from the rosette leaves of *cct2* plants. These results suggest that insertion of T-DNA in the 2nd exon of *AtCCT2* disrupted the expression of *AtCCT2* in *cct2* plants grown at 23 °C.

3. 3. *Contribution of AtCCT2 in the Biosynthesis of PC at Low Temperatures*

To evaluate the degree of contribution of AtCCT2 in the biosynthesis of PC at low temperatures, we measured the total CCT activity and the PC content in the rosette leaves of the wild type and *cct2* plants that were grown at 23 °C under continuous light for 15 d and for another 7 d at 5 °C under 8-h light/16-h dark cycles. Interestingly, the level of total CCT activity in the homogenates of rosette leaves from *cct2* plants was almost equal to that in the wild type, suggesting that the effect of disruption of *AtCCT2* was complemented by some mechanism in *cct2* plants. In the wild type, the total CCT activity in the homogenates of rosettes increased twofold after 48 h at 5 °C. By contrast, the CCT activity in the homogenates of rosette leaves from *cct2* plants did not increase after cold treatment. However, the PC content in the rosette leaves was similar between the wild type and *cct2* plants after 7 d at 5 °C. These results suggest that although the increase of AtCCT2 protein is likely to increase the CCT activity in the rosette leaves of the wild type after cold treatment, the increase in PC content in the cold-treated rosette leaves is not necessarily dependent on the activity of AtCCT2.

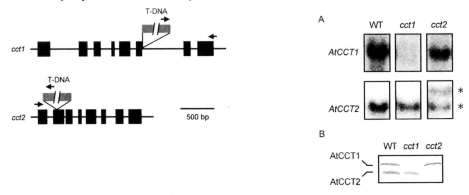

Figure 1 (left). Structure of *cct1* and *cct2*
Figure 2 (right). Expression of *AtCCT1* and *AtCCT2* in *cct1 and cct2* plants

194

3.4. *Future Perspectives*

In this work, T-DNA-tagged gene disruptants for *AtCCT1* and *AtCCT2* were isolated. Although *AtCCT2* is likely to contribute to the enhancement of the total CCT activity in the rosette leaves of the wild type at low temperatures, the increase in PC content at low temperatures seemed to be supported by multiple mechanisms. In *cct2* plants, the total CCT activity was maintained at similar levels to that in the wild type. Therefore, it is interesting to clarify whether the level of expression of *AtCCT1* is enhanced or the activity of AtCCT1 was up-regulated, or both, in *cct2* plants. To obtain a clear view over the regulation of PC metabolism, isolation of *cct1 cct2* double mutants is essential.

In rat cells, it is suggested that the CCT is localized in the nuclear membranes (DeLong et al., 2000). It is also suggested that rat CCT binds to a protein associated with vesicles involved in transcytosis (Feldman and Weinhold, 1998). These studies suggest that subcellular localization of CCT protein is important for the function of CCT, although CDP-choline is a diffusible substrate for PC biosynthesis in the cytoplasm. In this context, it is interesting to determine the subcellular localization of AtCCT1 and AtCCT2 in *Arabidopsis* cells. For this purpose, cct1 and cct2 plants should provide us with a suitable experimental system.

At moment, we do not find any changes in the growth features between the wild type and *cct2* plants at low temperatures. However, isolation of *cct1*, *cct2* and *cct1 cct2* plants will provide us with the first opportunity to evaluate the physiological importance of PC biosynthesis in plants under various stress conditions under which rapid or local biosynthesis of PC might be important for adaptation or survival.

4. References

Ohlrogge, J. and Browse, J. (1995) Lipid biosynthesis. Plant Cell 7, 957-970.

DeLong, C. J., Qin, L., and Cui, Z. (2000) Nuclear localization of enzymatically active green fluorescent protein-CTP:phosphocholine cytidylyltransferase α fusion protein is independent of cell cycle conditions and cell types. J. Biol. Chem. 275, 32325-32330.

Feldman, D. A. and Weinhold P. A. (1998) Cytidylyltransferase-binding protein is identical to transcytosis-associated protein (TAP/p115) and enhances the lipid activation of cytidylyltransferase. J. Biol. Chem. 273, 102-109.

Acknowledgement: This work was supported in parts by the Program for Promotion of Basic Research Activities for Innovative Biosciences (PROBRAIN) and by Grants-in-Aid for Scientific Research from the Japanese Society for the Promotion of Science.

THE PLASTIDIAL LYSO-PC ACYLTRANSFERASE FROM *ALLIUM PORRUM* LEAVES: INTERACTION WITH METAL SALTS AND PARTIAL PURIFICATION

AKERMOUN MALIKA, TESTET ERIC, SANTARELLI XAVIER,
CASSAGNE CLAUDE, and BESSOULE JEAN-JACQUES
*Laboratoire de Biogenèse Membranaire UMR 5544/ESTBB, CNRS-
Université Victor Segalen*
Bordeaux 2, 146 rue Léo Saignat, 33076 Bordeaux cedex, France

Introduction

It has been hypothesized that the import of extraplastidial lipids involves a PC deacylation in the ER, leading to the formation of lysophosphatidylcholine (lyso-PC). These molecules might partition between ER, the cytosol and the plastidial membranes. In the latter, lyso-PC might be acylated by a plastidial lyso-PC acyltransferase located in the envelope (Bessoule et al., 1995, Testet et al., 1999).

Several data obtained from *in vitro* (Bessoule et al., 1995; Testet et al., 1999)and *in vivo* (Mongrand et al., 1997; Mongrand et al., 2000) studies justified thorough investigation of the role of lyso-PC acyltransferase in the import of extraplastidial precursors of plastid lipids. Ongoing work aimed at obtaining plants overexpressing or devoid of lyso-PC acyltransferase activity. Because no sequence of lyso-PC acyltransferase has so far been reported, the purification of this enzyme has been undertaken. As a prerequisite for the purification of membrane-bound proteins, the lyso-PC acyltransferase was solubilized from the membrane fraction by treatment with CHAPS (CHAPS/protein (w/w)= 3) (Akermoun et al., 2000). We report in the present paper the effect of metal salts on the membrane-bound and on the plastidial solubilized- lyso-PC acyltransferase activity. We also provide results showing that Immobilized Metal Affinity Chromatography (IMAC) can be used to purify the enzyme under study.

Experimental

Materials
High-performance thin layer chromatography (HP-TLC) plates were Silicagel 60 F254 (Merck, Rahway, NJ). [1-14 C] lyso-phosphatidylcholine (2 GBq/mmol) was obtained

195

N. Murata et al. (eds.), Advanced Research on Plant Lipids, 195–198.
© 2003 *Kluwer Academic Publishers. Printed in the Netherlands.*

from Dupont-NEN (Les Ulis, France). Other reagents were from Sigma Chemical (St. Louis, MO, U.S.A.).

Methods
Purified plastids from leek leaves were obtained as described in (Akermoun et al., 2002). The protein content was determined according to (Bradford at al., 1976), using BSA as a standard.

Plastidial membranes were obtained by homogenizing the « green » part of leek leaves (250g) in the homogenization buffer (10 mM Tris, pH 7.5, 2 mM EDTA, 0.33 M Sorbitol and 0.1% BSA). The homogenate was strained through two layers of Miracloth and centrifuged for 5 min at 100g. The supernatant was filtered through Miracloth and centrifuged for 6 min at 3000g. The pellet was suspended in the homogenization buffer and recentrifuged for 6 min at 3000g. The resulting pellet was suspended in 10 mM Tris, pH 7.5 and centrifuged at 150 000g for 15 min using an Hitachi CS100 micro-ultracentrifuge. The pellet (plastidial membranes) was resuspended in 10 mM Tris, pH 7.5. The protein content was determined using BSA as a standard. The plastidial membranes were incubated with CHAPS at a CHAPS/protein ratio (w/w) equal to 3 for 5 min at 4°C. The mixture was then centrifuged three times at 150 000g for 15 min. The resulting supernatant was used as the solubilized enzyme source.

Purified chloroplasts, plastidial membranes or solubilized fraction were incubated with various amounts of labeled lyso-PC, unlabelled oleoyl-CoA and metal salts, as indicated in the figure legends. After incubation, lipids were extracted and purified by high-performance TLC, as indicated in (Bessoule et al., 1995). After autoradiography, PC was scraped from the plates and its label was determined by liquid scintillation counting. Alternatively, the PC was quantified by using a PhosphorImager (Pharmacia-Amersham).

Results and discussion

Effect of silver on lyso-PC acyltransferase activity
Among 12 heavy metals studied, silver, copper, mercury and lead inactivated the lyso-PC acyltransferase activity (Akermoun et al., 2002). Figure 1 shows the results obtained after the addition of various amounts of $AgNO_3$ to different chloroplast amounts. The lyso-PC acyltransferase was inactivated by $AgNO_3$ and the percent of inhibition of this enzyme depended on the amount of proteins in the assays : 5nmoles of $AgNO_3$ were sufficient to induce a complete "inhibition" when 8µg of plastids were used, whereas when the same amount of salt was incubated with 50µg of plastids, only 50% of inactivation was observed. When 200µg of plastids were used, 10nmoles of $AgNO_3$ were necessary to obtain a 50% "inhibition".

After solubilization from the membrane fraction (Akermoun et al., 2000), the effect of $AgNO_3$ on the solubilized lyso-PC acyltransferase activity was studied. The results reported in figure 2 show the effect of increasing $AgNO_3$ concentrations on the solubilized lyso-PC acyltransferase activity. Whereas at 3µM AgNO₃ there is some residual activity, inhibition at 6µM and 15µM is (almost) complete.

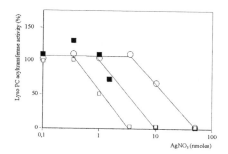

Figure 1: Eight micrograms (open squares) 50 micrograms (closed squares) or 200 micrograms (open circles) of purified plastids were incubated one hour at 30°C in the presence of oleoyl-CoA (5nmoles), labeled lyso-PC (0.5nmoles) and various amounts of AgNO₃ (final volume 0.1ml). For a given amount of proteins, results are expressed as percent of the activity determined in the abscence of AgNO₃.

Reversal of silver-inhibition

We intended to find out whether reducing-thiol reagents (i.e. ß-mercaptoethanol), metal chelator (EDTA), or imidazole reactivated lyso-PC acyltransferase activity. Figure 2 shows that the inhibition of the solubilized enzyme by AgNO₃ was not reversed by the addition of EDTA (50mM) or of imidazole (50mM). By contrast in the presence of ß-mercaptoethanol (10mM), 100% of the solubilized lyso-PC acyltransferase activity was recovered.

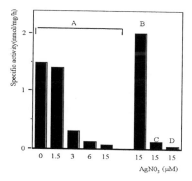

Figure 2 : Inhibition of the solubilized fraction by silver salts ; effect of B-mercaptoethanol, EDTA and imidazole. Solubilized proteins were incubated with substrates and various amount of AgNO3 (A) or with AgNO3 (15µM) plus ß-mercaptoethanol 10mM (B), or plus EDTA 100mM (C) or plus Imidazole 50mM (D).

Partial purification

The results reported above allowed us to undertake the purification of the lyso-PC acyltransferase by IMAC (Immobilized Metal Affinity Chromatography). The metal ions predominantly used with this chromatography are Cu^{2+} and Zn^{2+}. Cu^{2+} affords strong binding whereas Zn^{2+} gives a weaker binding.

Since $AgNO_3$ could not be used with this column chromatography, the effect of $(CH_3CO_2)_2Cu$ was studied to elucidate the potential interaction of the solubilized lyso-PC acyltransferase with this metal. Like the membrane-bound lyso-PC acyltransferase activity (Akermoun et al., 2002), the solubilized-enzyme was inhibited by copper salts (data not shown). Moreover, as observed with silver salts, the solubilized-lyso-PC acyltransferase activity inhibited by copper salts was reactivated in the presence of ß-mercaptoethanol. However, in contrast with what was observed after inhibition by silver salts, when the solubilized lyso-PC acyltransferase was inhibited by copper salts, a full reversal of the lyso-PC acyltransferase activity was observed in the presence of EDTA (100mM). These results allowed us to use an IMAC-Cu2+ to purify the lyso-PC acyltransferase, and it was found that most of the solubilized-lyso-PC acyltransferase was retained on IMAC-Cu^{2+}. In the fraction eluted by imidazole (100mM), an acylation of lyso-PC was detected, but this acylation was catalyzed not by the enzyme under study but by imidazole itself when used at 100mM (Testet et al., 2002). The lyso-PC acyltransferase activity could be eluted by using EDTA 100mM (data not shown).

References

Akermoun M., Testet E, Cassagne C.and Bessoule J.J., (2000) Solubilization of the plastidial lysophosphatidylcholine acyltransferase from *Allium porrum*: towards plants devoid of eukaryotic plastid lipids? Biochem.Soc.Trans. 28, 713-715.

Akermoun M., Testet E, Cassagne C. and Bessoule J.J. (2002) Inhibition of the plastidial phosphatidylcholine synthesis by silver, copper, lead and mercury induced by formation of mercaptides with the lyso-PC acyltransferase.BBA, Molecular and cell Biology of Lipids 1581, 21-28.

Bradford M.M. (1976) A rapid and sensitive method for the quantitation of microgram quantities of protein utilizing the principle of protein-dye binding. *Anal. Biochem.* 72, 248-254.

Bessoule J-J., Testet E. and Cassagne C. (1995) Synthesis of phosphatidylcholine in the chloroplast envelope after import of lysophosphatidylcholine from endoplasmic reticulum membranes. *Eur. J. Biochem.* 227,490-497.

Mongrand S., Bessoule J-J. and Cassagne C. (1997) A re-examination *in vivo* of the phosphatidylcholine galactolipid metabolic relationship during plant lipid biosynthesis. *Biochem.J.* 327, 853-858.

Mongrand S., Cassagne C. and Bessoule J-J. (2000) Import of lyso-phosphatidylcholine into chloroplasts likely at the origin of eukaryotic plastidial lipids. *Plant Physiol.* 122, 845-852.

Testet E., Verdoni N., Cassagne C. and BessouleJ- J. (1999) Transfer and subsequent metabolism of lysolipids studied by immobilizing subcellular compartments in alginate beads. *Biochim Biophys. Acta* 1440,73-80.

Testet E., Akermoun M., Shimoji M., CassagneC. and J.J. Bessoule (2002) Non-enzymatic synthesis of glycerolipids catalysed by imidazole. J. lipid Research., In press.

HISTOCHEMICAL ANALYSIS OF THREE MGD GENES USING PROMOTER-GUS FUSION SYSTEM IN ARABIDOPSIS

K. KOBAYASHI, K. AWAI, H. SHIMADA, T. MASUDA, K. TAKAMIYA and H. OHTA

Graduate School of Bioscience and Biotechnology, Tokyo Institute of Technology, 4259 Nagatsuta-cho, Midori-ku, Yokohama 226-8501, Japan.

Abstract

Monogalactosyldiacylglycerol (MGDG) is a membrane lipid unique to photosynthetic organisms, and a major structural component of plastid membrane. Its formation is catalyzed in plastid envelope membrane by MGDG synthase. To date, three types of MGDG synthase gene have been identified in Arabidopsis and designated as *atMGD1, 2* and *3*. Our previous analyses revealed that expression of each gene was detected in various organs and developmental stages. In this study, we present the histochemical analysis of these genes in major organs and several conditions using β-glucuronidase (GUS) assay in Arabidopsis. GUS staining of whole transformant revealed that the expression of *atMGD1-GUS* was detected highly in green tissues, such as cotyledons, stems and sepals, while the expression of *atMGD2-GUS* and *atMGD3-GUS* was observed only a few parts, such as tips of leaves. In floral tissues, GUS stainings of all atMGD-GUS constructs was detected in pollen grains. Moreover, only the pollen tube of atMGD2-GUS transformant was stained by cross-pollination of the transformant pollen and wild type stigma. Thus, atMGD2 might have special roles on the event of pollen germination. Under phosphate-deprived condition, the expression of *atMGD2-GUS* and *atMGD3-GUS* was detected intensely, especially in root, while *atMGD1-GUS* did not show significant difference. Considering that DGDG is accumulated under phosphate-deprived condition, particularly in root, type B MGDG synthase might supply MGDG to DGDG synthesis as a substrate.

Introduction

Monogalactosyldiacylglycerol (MGDG) is a typical glycolipid in some green bacteria and oxygenic photosynthetic organisms such as plants and cyanobacteria, and is a major membrane lipid constituting up to 50% of the total lipids of thylakoid membrane of

N. Murata et al. (eds.), Advanced Research on Plant Lipids, 199–202.

chloroplast. Recently, it is reported that MGDG is tightly bound to photosystem I reaction center, and therefore MGDG suggested to be required for photosynthetic reaction itself (Jordan et al., 2001). Another galactolipid, digalactosyldiacylglycerol (DGDG) is second abundant lipid in thylakoid membrane, and these galactolipids occupy 60% of the total membrane lipid even in nonphotosynthetic plastid (Alban et al., 1988). However, the significance of these galactolipids in nonphotosynthetic plastid is not known well.

The final step in MGDG biosynthesis occurs in the plastid envelope and is catalyzed by MGDG synthase, which transfers D-galactose from UDP-galactose to *sn*-1,2-diacylglycerol. DGDG is synthesized by galactosylation of MGDG, and therefore MGDG synthase is the key enzyme for synthesis of galactolipids, namely construction of the plastid membrane. We have cloned three MGDG synthase genes from Arabidopsis and designated these genes as *atMGD1*, *atMGD2* and *atMGD3* (Miége et al., 1999, Awai et al., 2001). From amino acid identity, these genes are classified into two types (type A and type B). Analysis for organ specificity of the expression of atMGD genes revealed that the expression of type B MGDG synthases (*atMGD2* and *atMGD3*) was detected specifically in flower and root, respectively, while type A (*atMGD1*) expression was dominantly observed in all organs analyzed. During the development of Arabidopsis, the expression of *atMGD1* and *atMGD2* were detected constantly. On the other hand, *atMGD3* showed intense expression in the early stage, and as the plant grew, the expression decreased to undetectable level (Awai et al., 2001).

To investigate the localization of the expression of these genes in more detail, we constructed three *atMGD* promoter-*GUS* fusion genes and produced their transgenic lines. Here, we report histochemical analyses of these transformants in several organs, conditions and stages of development. From these results presented here, we suggested the possible function of type B MGDG synthases, which might have unknown role particularly in nonphotosynthetic organs.

Materials and methods

Plant Material and Growth Condition
Arabidopsis thaliana (Columbia) were grown on Murashige and Skoog (MS) medium containing 1% sucrose, or on soil, at 23°C in continuous white light. For phosphate limitation experiments, plants were prepared as described by Härtel et al. (2000).

Construction of GUS fusion vectors
5'-upstream region of *atMGD1*, which was amplified by PCR from −1359 bp to +442 bp, was translationally ligated to a GUS gene in a pBI101 plasmid. By the same way, 5'-upstream region of *atMGD2* (from −1409 bp to +316 bp), and *atMGD3* (from −1409 bp to +74 bp) were inserted into pBI101. All plants were transformed by modified versions of the vacuum-infiltration method.

GUS Assay

Plant samples were soaked at 37°C for 1 day in the GUS assay solution which included 1mM 5-bromo-4chloro-3-indolylglucronide (X-Gluc), 0.5mM $K_3Fe(CN)_6$, 0.5mM $K_4Fe(CN)_6$, 0.3% TritonX-100, 20% methanol, and 50mM PBS. Then the samples were soaked in 70% ethanol for 1 day to stop reaction and remove chlorophyll.

In vivo Pollen Germination Experiment

Wild type Arabidopsis stigma was prepared by removing other tissues from buds. Then the stigma was brushed with anthers of transformants until it was completely covered with pollen. Pollinated stigma was incubated on solidified MS medium for 12h. Samples were then applied for GUS assay.

Results and discussion

GUS staining of whole transformant of *atMGD1-GUS* revealed that the expression of *atMGD1-GUS* occurred mostly in green tissues. This observation agreed with a hypothesis that type A MGDG synthase contributes to bulk of MGDG synthesis in chloroplast (Jarvis et al., 2000, Awai et al., 2001). In atMGD2-GUS transformants, in contrast, GUS staining was not detected until 14-day old plants, and after 14 days, only tips of the leaves were stained. Staining of atMGD3-GUS transformants was most intense in 3-day old plants, and as the plants grew, the staining gradually became weak. After 8 days, the staining was detected only in the tips of the leaves. These observations clearly matched with the results from RT-PCR analysis (Awai et al., 2001). Our construct of GUS-fusion system, therefore, was enough to analyze the expression sites of atMGD genes in more detail.

In floral tissue of atMGD1-GUS transformants, in addition to green tissues such as sepal and a part of immature petal, intense staining was observed in anthers. Anthers were also stained specifically in atMGD2-GUS and atMGD3-GUS transformants. When we observed the thin sections of these anthers with microscope, pollen was mostly stained. To investigate the expression of MGDG synthase in pollen, pollen of each transformant were cross-pollinated to stigmas of wild type, and the pistils were applied for GUS assay. Only the pistils which were pollinated with atMGD2-GUS transformants showed GUS staining, and the staining seemed to occur in pollen tubes. These results indicate that atMGD2 might have special roles in the process of fertilization.

Etiolated seedlings of each atMGD-GUS transformant were also applied for GUS assay. In atMGD1-GUS transformants, cotyledon was stained intensely, while atMGD2-GUS and atMGD3-GUS transformants were not stained. This result suggests the possibility that atMGD1 have a main role in MGDG synthesis for construction of prolamellar body in addition to illuminated plants.

Recently, Härtel et al. (2000) described the accumulation of DGDG in extra-chloroplastic membranes in response to phosphate deprivation. In atMGD1-GUS transformants, differences of staining pattern between phosphate-deprived and normal condition could not be distinguished. On the contrary, atMGD2-GUS and atMGD3-GUS transformants were stained very intensely, especially in roots, under phosphate-deprived condition, while the staining was scarcely detected in normal condition. Härtel et al. (2000) described that under phosphate-deprived condition, Arabidopsis accumulated DGDG, particularly in root, while MGDG was not accumulated specially. Considering that DGDG synthesis occurs in outer envelope of plastid (Dorne et al., 1982), and type B MGDG synthases are thought to be located outer envelope (Awai et al., 2001), type B MGDG synthases might supply MGDG to DGDG synthesis.

In the past, galactolipids were thought to be important only in photosynthetic organs particularly for constructing thylakoid membrane. In this report, we suggest the possibility of other roles of MGDG or MGDG synthases which might function in non-photosynthetic organs. Roles of galactolipid and the synthases are not still understood in non-photosynthetic organs. Analysis of transgenic plant such as T-DNA tagged mutant(s) will provide further information to solve these complicated problems.

References

Alban, C., Joyard, J. and Douce, R. (1988) Preparation and characterization of envelope membranes from nongreen plastids. Plant Pysiol. 88: 709-717.

Awai, K., Maréchal, E., Block, M. A., Brun, D., Masuda, T., Shimada, H., Takamiya, K., Ohta, H. and Joyard, J. (2001) Two types of MGDG synthase genes, found widely in both 16:3 and 18:3 plants, differentially mediate galactolipid syntheses in photosynthetic and nonphotosynthetic tissues in Arabidopsis thaliana. Proc. Natl. Acad. Sci. USA. 98: 10960-10965.

Dorne A. J., Block, M. A., Joyard, J. and Douce R. (1982) The galactolipid:galactolipid galactosyltransferase is located on the outer surface of the outer membrane of the chloroplast envelope. FEBS Lett. 145: 30-34

Härtel, H., Dörmann, P. and Benning, C. (2000) DGD1-independent biosynthesis of extraplastidic galactolipids after phosphate deprivation in Arabidopsis. Proc. Natl. Acad. Sci. USA. 97: 10649-10654

Jarvis, P., Dörmann, P., Peto, C.A., Lutes, J., Benning, C. and Chory, J. (2000) Galactolipid deficiency and abnormal chloroplast development in the Arabidopsis MGD synthase 1 mutant. Proc. Natl. Acad. Sci. USA. 97: 8175-8179.

Jordan, P., Fromme, P., Witt, HT., Klukas, O., Saenger, W. and Krauß, N. (2001) Three-dimensional structure of cyanobacterial photosystem I at 2.5 A resolution. Nature. 411: 909-917.

Miége, C., Maréchal, E., Shimojima, M., Awai, K., Block, M. A., Ohta, H., Takamiya, K., Douce, R. and Joyard, J. (1999) Biochemical and topological properties of type A MGDG synthase, a spinach chloroplast envelope enzyme catalyzing the synthesis of both prokaryotic and eukaryotic MGDG. Eur J Biochem. 265: 990-1001.

OIL BIOSYNTHESIS IN *MORTIERELLA ALPINA* MICROSOMAL MEMBRANES

S. CHATRATTANAKUNCHAI, T.C.M. FRASER, A.D. WATERS, and A.K. STOBART
School of Biological Sciences, University of Bristol, Woodland Road, Bristol BS8 1UG, UK

1. Abstract

Microsomal membrane preparations from *Mortierella alpina* incubated with *sn*-glycerol-3-phosphate (G3P) and [^{14}C]oleoyl-CoA catalysed the formation of radioactive phosphatidic acid (PA), diacylglycerol (DAG), triacylglycerol (TAG) and phosphatidylcholine (PC). A similar pattern of radiolabelling was observed when [^{14}C]G3P and non-labelled oleoyl-CoA were used except that no radioactivity was found in PC. This lipid was, however, labelled from [^{14}C]G3P and oleoyl-CoA in the presence of CDP-choline. Equilibration between [^{14}C]glycerol labelled DAG and PC in *M. alpina* microsomal membranes was dependent on the addition of cytidine diphosphocholine (CDP-choline). A concomitant decrease in label in DAG and TAG occurred in conjunction with the increase in radioactivity in PC. The results indicate that PC and TAG synthesis via diacylglycerol:cholinephosphotransferase (DAG:CPT) and diacylglycerol acyltransferase (DAGAT), respectively, share a common pool of DAG.

2. Materials and methods

2.1 Fungal growth and microsomal membrane preparation
M. alpina (CBS 210.32) was grown in liquid culture in potato dextrose medium at 28 °C. Microsomal membranes were prepared as described elsewhere (Chatrattanakunchai et al., 2000)

2.2 Microsomal Incubations
Incubations were carried out at 30 °C. Reaction mixtures contained BSA (5 mg), substrates and cofactors (as stated in the figure) and microsomal membranes made up to a final volume of 1 ml with potassium phosphate buffer (0.1 M, pH 7.2). Reactions were started by the addition of the microsomal membranes to the reaction mixture.

N. Murata et al. (eds.), Advanced Research on Plant Lipids, 203–206.

2.3 *Analytical procedures*

Reactions were terminated and lipids extracted and purified as given elsewhere (Chatrattanakunchai et al., 2000). Radioactivity in individual lipids was quantified using electronic autoradiography (Instant imager; Packard).

3. Results

3.1 *G3P acylation in the presence of [^{14}C]oleoyl-CoA*

Microsomal membrane preparations were incubated with G3P and [^{14}C]oleoyl-CoA and at regular intervals the radioactive incorporation into lipids was determined. The results (Fig 1a.) show that there was initially a rapid incorporation of radioactivity into PA, which reached a maximum after 40 minutes incubation and then declined. Compared with the radioactivity in PA there was a slight lag in the appearance of label in DAG and TAG, after which it increased almost linearly for 80 minutes. Radioactivity also appeared in PC and this continued to accumulate for 80 minutes. After this time little increase in radioactive product was observed due to limiting substrate(s). The increase in radiolabelled PA indicated active glycerol-3-phosphate acyltransferase and lysophosphatidic acid acyltransferase (LPAAT) enzymes in the microsomal membranes. No radioactivity was observed in lysophosphtidic acid indicating the relatively higher activity of the LPAAT.

3.2 *[^{14}C]G3P acylation in the presence of oleoyl-CoA*

The results indicated that microsomal membrane preparation of *M. alpina* were capable of synthesising TAG by the acylation of G3P. To confirm the presence of an active Kennedy pathway in *M. alpina* microsomal membranes, experiments were conducted using [^{14}C]G3P and non-radioactive oleoyl-CoA. The results (Fig 1b.) show that the labelled glycerol moiety was rapidly acylated to yield PA and this lipid was the most radioactive complex lipid throughout the time course. The radioactivity in PA declined somewhat after 40 min similarly to that observed in the previous studies (Fig 1a.). After a slight lag substantial glycerol was present in DAG and TAG and this continued to accumulate throughout the incubation period. No radioactivity was found in PC indicating little or no conversion of DAG to PC during the flow of the glycerol moiety through the Kennedy pathway in the microsomal membranes.

3.3. *Effect of CDP-Choline on G3P utilisation in the presence of oleoyl-CoA*

Evidence suggests that the enzyme responsible for the interconversion of DAG and PC is DAG:CPT (Slack *et al.*, 1985) which may be involved in catalysing the movement of DAG through PC and back into the DAG pool for TAG synthesis thus enriching the glycerol backbone with polyunsaturated fatty acids (Stobart & Stymne, 1985). The previous experiment (Fig 1b.) showed that the forward reaction of this enzyme i.e. DAG to PC was not occurring in the microsomal membranes possibly indicating that CDP-choline was limiting. To test this, microsomal membranes were incubated with [^{14}C]G3P, oleoyl-CoA and CDP-choline (20 mM) and the incorporation of radioactivity in the complex lipids determined. The results (Fig 1c.) show that in the presence of CDP-choline there was a relatively large amount of radioactivity incorporated into PC and this continue to increase throughout the incubation period. The radiolabelled DAG in the presence of CDP-choline was lower than that found in the absence of CDP-choline (Fig 1b.). A slight decrease in TAG formation was also observed. However, the radiolabelled PA produced in the presence of CDP-choline was similar to that observed in controls. These results indicate therefore that the formation of PC from DAG by DAG:CPT required exogenously supplied CDP-choline.

4. Discussion

Microsomal membranes preparations from *M. alpina* catalysed the efficient acylation of G3P and the accumulation of TAG. The microsomes, therefore, contained all the enzymes necessary for oil assembly. The *in vitro* conversion of DAG to PC, however, did not occur without exogenously supplied CDP-choline. In the presence of CDP-choline, there was a decrease in DAG and TAG formation, perhaps due to some of the DAG pool being diverted to PC. These results may indicate that DAGAT and DAG:CPT share a common pool of DAG. These data on the CDP-choline dependent PC synthesis and a common pool of DAG for these two enzymes (DAGAT and DAG:CPT) is similar to that reported for TAG biosynthesis in preparations of permeabilized rat hepatocytes where exogenous CDP-choline was aslo shown to stimulate PC formation (Stals et al., 1992). Whilst CDP-choline is reported to be limiting in rat hepatocytes (Stals et al., 1992), it is, however, not required for PC-DAG interconversion in microsomal membranes from developing oil seeds (Stobart & Stymne, 1985). The accumulation of label in PC from [^{14}C]oleoyl-CoA was not G3P dependent and hence is most probably due to LPCAT activity, an enzyme previously shown to be active in *Mortierella* microsomal preparations (Chatrattanakunchai et al., 2000)

5. References

Chatrattanakunchai, S.; Fraser, T. and Stobart, K. (2000) Oil biosynthesis in microsomal membrane preparations from *Mortierella alpina*. Trans. Biochem. Soc. 28, 707-709.

Slack, C.R., Roughan, P.G., Browse, J.A. and Gardiner, S.E. (1985) Some properties of cholinephosphotransferase from developing safflower cotyledons. Biochem. Biophys. Acta 833, 438-448.

Stals, H.K., Mannaerts,G.P. and Declercq, P.E. (1992) Factors influencing triacylglycerol synthesis in permeabilized rat hepatocytes. Biochem. J. 283, 719-725.

Stobart, A.K. and Stymne, S. (1985) The interconversion of diacylglycerol and phosphatidylcholine during triacylglycerol production in microsomal preparations of developing cotyledon of safflower (*Carthamus tintorius* L.) Biochem. J. 232, 217-221.

206

a. G3P acylation in the presence of [^{14}C]oleoyl-CoA

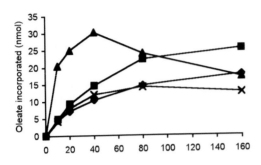

b. [^{14}C]G3P acylation in the presence of oleoyl-CoA

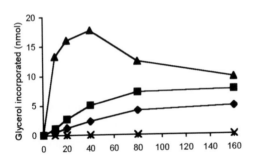

c. Effect of CDP-Choline on G3P utilisation in the presence of oleoyl-CoA

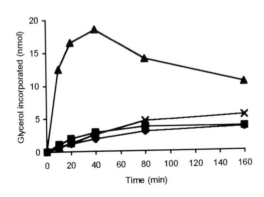

Figure 1. Microsomal membranes (equivalent to 0.3 mg protein) were incubated with glycerol-3-phosphate (400 nmol) and oleoyl-CoA (100 nmol) and in the presence of 20 mM cytidine diphosphocholine (c.). The radioactivity incorporated into various lipids was determined at regular intervals. ◆, Triacylglycerol; ■, diacylglycerol; ▲, phosphatidic acid; ✖, phosphatidylcholine. Results are expressed in nmol oleate or glycerol incorporated per mg protein.

GALACTOLIPID SYNTHESIS COORDINATES WITH THE EXPRESSION OF PHOTOSYNTHETIC ACTIVITY VIA REDOX CASCADE

Y.YAMARYO[1], K.MOTOHASHI[2], T.MASUDA[1], H.SHIMADA[1], K.TAKAMIYA[1], T.HISABORI[2] and H.OHTA[1].

1 Graduate school of Bioscience and Biotechnology, Tokyo Institute of Technology, Yokohama, 226-8501, Japan
2 Chemical Resources Laboratory, Tokyo Institute of Technology, Yokohama, 226-8503, Japan

Introduction

MGDG synthase (MGD, UDP-galactose: 1,2-diacylglycerol 3-β-D-galactosyltransferase; EC 2.4.1.46) catalyzes the final step of MGDG synthesis. This enzyme transfers a galactose from UDP-galactose to 1,2-diacylglycerol in chloroplast envelope (Douce, 1974). We have revealed that irradiation of white light on etiolated seedlings induced a transient increase in the level of MGD transcripts, followed by activation of the enzyme and a light-dependent accumulation of MGDG and DGDG (Yamaryo et al. 2000). Light-dependent expression of photosynthesis-related genes is generally mediated by the photoreceptor phytochrome. However, it is still unknown which photoreceptor is responsible for light-dependent changes in MGDG synthesis.

We first determined the wavelength specificity and red light $\leq$ 700 nm was necessary for the accumulation of galactolipids although far-red light $\geq$ 700 nm was sufficient for the light-dependent accumulation of MGD transcripts and increase in activity. The red light $\leq$ 700 nm implies the possibility of a link between the expression of photosynthetic activity and galactolipid synthesis. Since the key enzyme of fatty acid synthesis, acetyl-CoA carboxylase was shown to be regulated by redox cascade and the link between light and fatty acid synthesis was proposed (Sasaki et al., 1997), we examined the regulation of MGDG synthase by changes in redox potential of chloroplast.

Results

1.1 Effects of far-red light > 700 nm on galactolipid synthesis

Irradiation with red light caused cucumber cotyledons to green normally, and the expression of both *csMGD* mRNA and MGD enzyme activity was similar to that

N. Murata et al. (eds.), Advanced Research on Plant Lipids, 207–210.

observed during irradiation with white light (data not shown). Irradiation with far-red light caused *csMGD* mRNA levels to increase to almost the level induced with white light irradiation, as shown in Figure 1a. However, irradiation with far-red light caused no greening.

Enzyme activity was enhanced by far-red light and attained almost the same level as that induced by illumination with white light (data not shown). Consistent with this, both MGDG and DGDG levels increased in far-red light. However, the increase in MGDG was less than half that observed under white light (data not shown), although red light (600–700 nm) induced a similar accumulation of galactolipids as white light (data not shown). This strongly suggests that red light is required for the accumulation of galactolipids.

1.2 Expression of MGDG synthase in the lh mutant

In a complementary experiment, we analyzed the light-dependent accumulation of *csMGD* mRNA in the long hypocotyl (*lh*) mutant, which has a mutation in phytochrome B (PhyB) (López-Juez et al. 1992). We detected *csMGD* mRNA at levels identical to that in the wild type during both white-light and far-red-light illumination, indicating that PhyB is not required for the light-dependent induction of the *csMGD* gene (data not shown).

2.1 The effects of dithiol and monothiol on MGDG synthase activation

MGD needs some SH compounds to maintain its enzymatic activity (Covès et al. 1986) and also has Cys-residue(s) related to the expression of enzyme activity. We, therefore, figured out whether MGD needs the reduction of disulfide bond(s) by thiol compound.

We used partially purified recombinant protein of MGD-GST fusion as enzyme for the following experiments. As shown in Figure 1, MGD preferred to DTT, the dithiol compound, rather than monothiol compound such as β-mercaptoethanol or reduced-glutathione. DTT activated MGD in parallel to DTT concentration. This suggested that MGD has some disulfide bonds and the reduction of the disulfide bond is essential for the expression of enzyme activity.

2.2 The effects of thioredoxin on the activity of MGDG synthase

Thioredoxin (Trx) has disulfide bond and regulates many chloroplastidial proteins by the ferredoxin-thioredoxin system, redox cascade by photosynthesis (Dai et al. 2000). Therefore, Trx-regulated protein should be strongly related to the expression of photosynthetic activity. We examined the regulation of MGD by Trx in low DTT concentration.

Addition of 4 μM Trx-m from spinach, as shown in Figure 2, well activated MGD more than Trx-f or DTT only did. DTT reduced Trx to an active form, and subsequently it reduced MGD to the active form. This result indicated that MGD should be Trx-regulated protein.

Discussion

Using far-red light ≥ 700 nm, the accumulation of MGD mRNA was induced and its level was almost the same as that of white light as shown in Figure 1. PhyA can accept light with a broad range of wavelength from 280 to over 700 nm. In the case of far-red light over 700 nm, PhyA is a molecular switch for turning on the light-dependent gene expression, whereas PhyB is that for turning off the effect of red light (Shinomura et al. 1996). Because our result shows that the light over 700 nm sets off the expression of MGD gene, PhyA is most likely to be the photoreceptor responsible for its expression. The analysis of the *lh* mutant revealed that the PhyB is not necessary for mRNA expression of the *csMGD*.

However, only a slight accumulation of galactolipids was observed after 25 and 50 h of far-red light irradiation. This result indicated that far-red light is not sufficient to induce the full activity of MGDG synthesis in cucumber cotyledons. Obviously, other factors, such as the accumulation of substrates or further activation of the enzyme *in vivo*, are required to achieve the maximal rate of synthesis of galactolipids.

It is possible that galactolipid synthesis itself is regulated in cooperation with the expression of photosynthetic activity. The *in vitro* assay for MGD activity strictly requires reducing condition maintained by the addition of SH-compounds like DTT (Covès et al., 1986), suggesting that the MGD activity itself is also regulated by redox potential *in vivo* similar to acetyl-CoA carboxylase. In fact, as shown in Figure 1, MGD has a disulfide bond and its reduction is necessary for the enzyme activation. The addition of Trx enhanced MGD activation (Fig. 2), proving the requirement of a photosynthetic electron flow. Trx-regulated MGD activation suggested that the expression of photosynthetic activity linked to galactolipid synthesis, namely the development of chloroplast.

Reference:

Covès, J., Block, M.A., Joyard, J. and Douce, R. (1986) Solubilization and partial purification of UDP-galactose:diacylglycerol galactosyltransferase activity from spinach chloroplast envelope. *FEBS Lett.* 208: 401-406.

Dai, S., Schwendtmayer, C., Johansson, K., Ramaswamy, S., Schürmann, P. and Eklund, H. (2000) How does light regulate chloroplast enzymes? Structure–function studies of the ferredoxin/thioredoxin system. *Q. Rev. Biophys.* 33: 67-108.

Douce, R. (1974) Site of biosynthesis of galactolipids in spinach chloroplasts. *Science* 183: 852-853.

López-Juez, E., Nagatani, A., Tomizawa, K., Deak, M., Kern, R., Kendrick, R.E. and Furuya, M. (1992) The cucumber long hypocotyl mutant lacks a light-stable PHYB-like Phytochrome. *Plant Cell* 4: 241-251.

Sasaki, Y., Kozaki, A. and Hatano, M. (1997) Link between light and fatty acid synthesis: Thioredoxin-linked reductive activation of plastidic acetyl-CoA carboxylase. *Proc. Natl. Acad. Sci. USA* 94: 11096-11101.

Shinomura, T., Nagatani, A., Hanzawa, H., Kubota, M., Watanabe, M., and Furuya M. (1996) Action spectra for phytochrome A- and B-specific photoinduction of seed germination in *Arabidopsis thaliana*. *Proc. Natl. Acad. Sci. USA* 93: 8129-8133.

Yamaryo, Y., Kanai, D., Awai, K., Masuda, T., Shimada, H., Takamiya., K and Ohta., H. (2000) Transcriptional regulation by light and phytohormones of *MGD* gene in cucumber. *Recent Advances in the Biochemistry of Plant Lipids* (Harwood L. J., ed.) 738-740, Portland Press Ltd. London

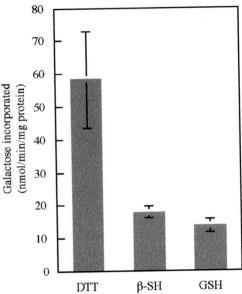

Figure 1. Effect of thiol compounds on MGD activity.
1 mM DTT, 2-mercaptoethanol (β-SH) or reduced
glutathione (GSH).

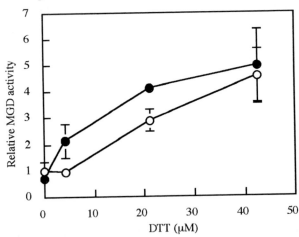

Figure 2 Effect of thioredoxin on MGD activity. 4 µM spinach Trx-m was
added with DTT. ●, +Trx; ○, -Trx.

CHANGES IN OIL COMPOSITION OF TWO SUNFLOWER VARIETIES INDUCED BY SALT-STRESS

H. JABRI[1], A. SMAOUI [1], M. ZARROUK[1] and A. CHERIF[2].
[1]L.A.A.P.- I.N.R.S.T., BP 95 -2050, Hammam-Lif, TUNISIA.
[2]E.S.I.A.T., 1002, El Khadra Tunis, TUNISIA.
Email : Abderrazak.Smaoui@inrst.rnrt.tn

1. Introduction

Few studies have concerned salt stress on lipid seeds (Gharsalli et al., 1982, Smaoui et al., 1992). However, effects of some environmental factors on developing seeds, such as light or temperature were studied. Trémolières (1984) indicated that temperature affect fatty acid composition of sunflower oil and rape during maturation. Cautison et al. (1998) investigated maturation changes and temperature effects on fatty acid composition of developing high-saturated sunflower seeds.

.In some areas of our country, sunflower plants are irrigated with brackish water containing important levels of chloride and sodium ions. A preliminary study was carried out in order to select two contrasting varieties: a salt tolerant variety and a salt sensitive one among 23 tested genotypes of sunflower. Moreover, we investigated sodium chloride effect on some agronomic features and quality of oil extracted from full-ripened seeds of the two selected varieties.

2. Experimental

Two types of cultures have been carried out. In the first, sunflower plants from 23 varieties were grown in a culture solution. One set of seedlings (two weeks old) was cultivated on hydroponic solution containing 150 mM NaCl during 15 days, the second set used as control, was grown on salt free hydroponic medium. At the end of experiment, plants were harvested for determination of biomass production. In the second, plants were grown in green house on pots filled with inert sand. They were irrigated with either a nutritive solution for control or with solution added with 150 mM NaCl for stressed plants. Seeds were harvested at complete maturity when whole plants were desiccated, indicating end of plant development cycle.

N. Murata et al. (eds.), Advanced Research on Plant Lipids, 211–214.
© 2003 Kluwer Academic Publishers. Printed in the Netherlands.

Seeds total lipids were extracted according to a previous report (Smaoui 1992). TAG fractions were obtained by the method described earlier (Smaoui 2000).

3. Results and discussion

3.1. *Agronomic parameters*

Salt susceptibility index (SI) determined at juvenile stage is expressed by the ratio: dry matter from NaCl treated plants minus dry matter of control ones x 100 on dry matter of control. Calculation of SI allowed us to establish a salt tolerance scale for 23 varieties (data none shown). Makil is the most salt tolerant genotype (SI = -31%) and Rigasol is the most salt sensitive one (SI=-85%). When exposed to high salinity all their life cycle along, plants of the two genotypes expressed different behaviour, which correlate well with that of juvenile stage.

Table 1 shows that salt affect yield of sunflower plants. So harvest index expressed as ratio: dry matter of seeds / dry matter of plant aerial parts, was severely dropped by salt in Rigasol variety. In the same way, oil content of seeds was drastically lowered by salt (-70%) in Rigasol but it was slightly affected in Makil. However, sodium chloride treatment decreased seeds dry matter in both varieties. All observed changes might be related to decrease in photosynthesis efficiency due to deleterious effect of chloride and sodium ions in the cells, and consequently lower carbohydrates production which is the main source of lipid biosynthesis.

TABLE 1. : Changes in harvest index, dry matter of seeds and oil content of Makil (salt tolerant) and Rigasol (salt susceptible) exposed to high NaCl concentration.

	NaCl	Harvest index	Dry matter of 1000 seeds	Oil content
Cv Makil	control	0.06	54.6	36.1
	150 mM	0.02	20.2	33.6
Cv Rigasol	control	0.09	48.9	48.9
	150 mM	0.01	24.9	33.6

3.2. *Total Lipid composition*

Total fatty acid composition of (TFA) of Makil control seeds is different from Rigasol ones. In Makil, the linoleic and oleic acids represent about 45% each one. In Rigasol, oleic acid is the major fatty acid (60% of TFA) and linoleic acid represent about 30% (figures 1, A and B). Rigasol genotype is known as "oleic sunflower".

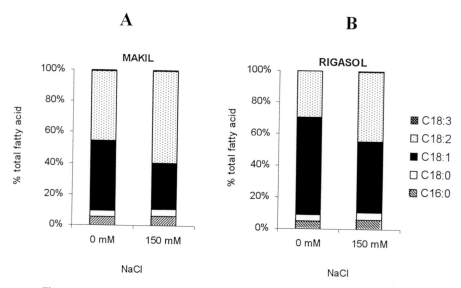

Figures 1, A and B. Changes in total fatty acid composition of seeds harvested from control (0 mM) and treated plants (150 mM) in Makil (A) and Rigasol (B).

In both varieties, palmitic and stearic acids do not exceed 10% of TFA. Linolenic acid, a minor fatty acid, represents less than 1%. Salt stress decreased oleic acid percentage and increased linoleic acid level in the two varieties studied. Level reduction varied between about 26 and 34 % of control. Palmitic, stearic acids rates remained practically unchanged (figure 1, A and B).

3.3. Changes in molecular TAG species

Triacylglycerols (TAGs), principal lipid fraction, represent about 90% of total fatty acids in full-ripened seeds. The fatty acids fixed on TAG molecule bone are: palmitic acid (P), stearic acid (S), oleic acid (O) and linoleic acid (L). HPLC analysis of control TAGs categories separated from the two varieties showed that oleodilinolein (OLL), dioleolinolein (OOL), and triolein (OOO) form the major molecular species and reach 65 to 71 % of TAGs. The other eight molecular species are minor and do not exceed 10 % (figure 2, A and B). High salt concentration decreased rich oleic acid fractions of TAGs (e.g. OOl, OOO) and increased rich linoleic acid species.(e.g. LLL , PLL). Changes induced by salt stress changes are similar for the two sunflower varieties. These results confirm previous ones relative to decreased oleic acid percentage observed in total lipid composition of seeds (figures 1, A and B). These findings are in agreement with experiment of Zarrouk (2000) on salt stressed olive, who indicated a decrease of oleic acid and triolein.

Irrigation of sunflower plants with water containing sodium chloride depressed crop yield, oil content and affected oil quality essentially by decreasing oleic acid percentage.

214

We concluded that Makil variety is more salt tolerant and its productivity in saline environment is better than Rigasol variety.

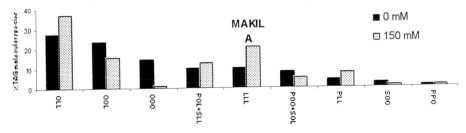

Figures 2, A and B. Changes in molecular species of TAGs sinduced by 150
mM NaCl in Makil (A) and Rigasol(B) varieties.

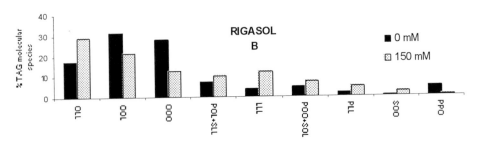

4. References

Cautisan, S., Martinez-Force, E., Alvarez-Ortega, R. and Garces, R. (1998) Maturation changes and temperature effect on fatty acid composition in developing high saturated sunflower (*Helianthus annuus*) seeds. In J. Shanchez, E. Cerdà-Olmedo, and E. Martínez-Force (eds.), Advances in Plant Lipid Resarch. Universidad de Sevilla, pp. 60-62.

Gharsalli, M., Djemal-Daouadi, F. and Chérif, A. (1982) The effect of sodium chloride on the lipid composition of sunflower oil. In 10th Intern. Sunflower Conference. Surfers Paradise, Australia, pp.71-72, ,

Smaoui, A. and Chérif, A. (1992) Glycerolipids in salt stressed cotton seeds. In A. Chérif, D. Ben Miled-Daoud, , B. Marzouk, A. Smaoui, and M. Zarrouk (eds.), Metabolism, Structure and Utilization of Plant Lipids, C. N. P., Tunis, pp. 341-343.

Smaoui, A., and Chérif, A., (2000) Changes in molecular species of triacylglycerols in developing cotton seeds under salt stress. Biochemical Society Transactions.28, 6 , 902-905.

Trémolières, A. (1984) Régulation de la synthèse des acides gras en fonction de la température dans quelques tissus végétaux. Oléagineux 39, 227-231.

Zarrouk, M. (1999) in Thèse de Doctorat d'Etat es Sciences, Faculté des Sciences de Tunis, 220 p.

TAPETOSOMES AND ELAIOPLASTS IN *BRASSICA* AND *ARABIDOPSIS* FLORAL TAPETUM

KAI HSIEH, SHERRY S. H. WU, CHANDRA RATNAYAKE,
AND ANTHONY H. C. HUANG
Center for Plant Cell Biology
Department of Botany and Plant Sciences
University of California, Riverside, CA 92521, USA

1. Introduction

Insect/self-pollinating species such as *Brassica, Arabidopsis,* tobacco and lily have abundant neutral lipids on the pollen surface. The lipids form a continuous layer to waterproof the pollen and provide a medium in which other important molecules (e.g., flavonoids, cell wall digestive enzymes, and self incompatibility molecules) are dissolved.

The pollen-surface lipids originate from the floral tapetum cells enclosing the locule, in which the pollen mature. In the tapetum cells, they are housed in two predominant organelles, the tapetosomes and the elaioplasts.

2. The tapetosomes and elaioplasts in *Brassica*

The tapetosomes and elaioplasts have been isolated from the florets (tapetum) of *Brassica* (Wu, et al., 1997). The tapetosomes contain triacylglycerol (TAG) droplets situated among densely packed vesicles and do not have an enclosing membrane. They possess osmotic properties, presumably exerted by the individual vesicles. They contain oleosins, which have been previously known to be associated with seed storage oil bodies. Presumably, the oleosins stabilize the TAG droplets inside the organelles. The *in vitro* morphologic features, as described, are clearly visible only after the organelles have been isolated and artificially swollen. When the isolated organelles are suspended in a high osmotica solution, the internal structures of the organelles cannot be resolved, and the organelles resemble those *in situ*. Thus, although the organelles apparently are present in diverse insect/self-pollinating dicots and monocots, as observed by electron microscopy, the distinct features indicative of their being a unique organelle in the tapetum have not been previously revealed. The

215

N. Murata et al. (eds.), Advanced Research on Plant Lipids, 215–218.
© 2003 *Kluwer Academic Publishers. Printed in the Netherlands.*

synthesis of the tapetosomes is intimately related to the rough endoplasmic reticulum (ER). Electron microscopy has shown that different nascent tapetosomes in a tapetum cell are interconnected via the ER, which may produce the tapetosome vesicles by a budding mechanism.

The elaioplasts have few internal membranes and are packed with globuli (plastoglobuli) consisting of steryl esters. The steryl esters in the globuli are stabilized by surface glycolipids and structural proteins (Wu et al., 1999). Monogalactose diacylglycerols account for 16% of the total lipids in the elaioplasts and are likely to be the predominant amphipathic lipid on the globuli. The structural protein is plastid-lipid-associated protein (PAP), of ~33 kD. The elaioplasts are derived from proplastids during development. The proplastids have limited internal membranes and a large stroma. As development proceeds, droplets of lipids (globuli) accumulate in the organelles.

At the late stage of anther development, the tapetum cells undergo programmed cell death. Of the tapetosomes, the oleosins but not the TAGs are preserved and transferred to the pollen coat. Of the elaioplasts, only the steryl esters are preserved and transferred to the pollen surface, forming the bulk of the pollen coat lipids.

3. The tapetosomes and elaioplasts in *Arabidopsis*

Electron microscopy of *Arabidopsis* tapetum cells carried out by several laboratories has indicated the presence of organelles with morphologic features similar to those of the tapetosomes and elaioplasts in *Brassica*. It would be beneficial to isolate these *Arabidopsis* organelles for biochemical studies; then, further investigations of these organelles could be facilitated with the use of the available genomic database and mutants. Nevertheless, the minute size of *Arabidopsis* and its flowers has imposed a technical difficulty for the initial biochemical studies.

We labored to obtain sufficient *Arabidopsis* florets and were able to isolate the tapetosomes and elaioplasts using a procedure modified from that for the *Brassica* organelles. The isolated *Arabidopsis* organelles were subjected to SDS-PAGE for separation of the protein constituents (Fig. 1A). As expected, the tapetosomes and the elaioplasts had different complements of proteins, and their proteins represented minute amounts of the total floret proteins. Some proteins appeared to be present in both organelle fractions; they could represent authentic, identical proteins present in both organelles, or cross contaminants, or different organelle-specific proteins of very similar molecular weight. Antibodies were raised against a synthetic oleosin peptide, which represents a segment of a putative major *Arabidopsis* tapetosome oleosin of 55 kD; similar segments with different degrees of residue variation are present in the other putative tapetum oleosins. The antibodies were used for immunoblotting after SDS-PAGE of the isolated tapetosomes (Fig. 1B). They recognized the expected 55-kD oleosin (the major oleosin, encoded by CAB87942 or *Atgrp*-7) and a 15-kD oleosin (by CAB87945 or *Atgrp*-14) (Kim et al., 2002).

With the isolated tapetosomes and elaioplasts, we were able to identify more of the organelle-specific proteins with the use of the *Arabidopsis* genomic database. The isolated proteins were subjected to direct N-terminal sequence without or with trypsin treatment and MALDI-TOF MS analyses. For the tapetosomes, the identified proteins include several more oleosins and two proteins known to be associated with the ER. For the elaioplasts, the identified proteins include a PAP and the small subunit of Rubisco.

The successful identification shows that we may be able to further identify other proteins in the two organelles with the use of the genomic database. These proteins could include enzymes for the synthesis of specific lipids, especially TAGs and steryl esters, and some metabolites known to be synthesized in the ER or plastids. Such information will allow us to identify specific mutants of the organelles via the *Arabidopsis* mutant stock.

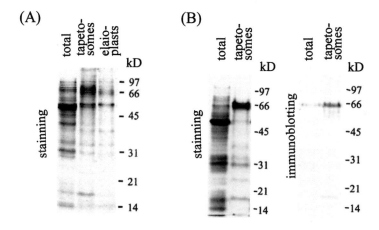

Figure 1. SDS-PAGE of proteins of samples from *Arabidopsis* florets (tapetum). The total floret extract, isolated tapetosomes and isolated elaioplasts were subjected to SDS-PAGE. Panel A shows a comparison among the total floret extract, isolated tapetosomes and isolated elaioplasts. Panel B shows two identical gels of the total floret extract and the isolated tapetosomes, one stained for proteins and the other treated for immunoblotting with the use of antibodies against a synthetic peptide representing a segment of the *Arabidopsis* tapetum 55-kD oleosin. Positions of molecular weight markers are shown on the right.

References

Wu, S. S. H., Moreau, R. A., Whitaker, B. D., Huang, A. H. C. (1999) Steryl esters in the elaioplasts of the tapetum in developing *Brassica* anthers and their recovery on the pollen surface. Lipids 34, 517-523.

Wu, S. S. H., Platt, K. A., Ratnayake, C., Wang, T. W., Ting, J. T. L., Huang, A. H. C. (1997) Isolation and characterization of novel neutral-lipid-containing organelles and globuli-filled plastids from *Brassica napus* tapetum. Proc. Natl. Acad. Sci. (US) 94, 12711-12716.

Kim, H. U., Hsieh, K., Ratnayake, C., and Huang, A. H. C. (2002) A novel group of oleosins is present inside the pollen of *Arabidopsis*. J. Biol. Chem. (in press)

Chapter 6:

Waxes, Sphingolipids and Isoprenoids

CLONING AND GENE EXPRESSION OF THE 3-KETOACYL-CoA REDUCTASE IN B. napus SEEDS

PUYAUBERT J.[a], DIERYCK W. [a,b], CHEVALIER S. [a], COSTAGIOLI P. [a,b], GARBAY B. [a,b], CASSAGNE C. [a,b] AND LESSIRE R. [a]

a-Laboratoire de Biogenèse Membranaire, b. ESTBB Université V. Segalen Bx 2, 146 rue Léo Saignat 33076 Bordeaux Cédex, France

Introduction

The membrane-bound acyl-CoA elongases are responsible for the synthesis of Very Long Chain Fatty Acids (VLCFA) which are the primer of wax constituents and are the major fatty acid is some seeds. VLCFA result from elongation of oleoyl-CoA by malonyl-CoA through four successive reactions: condensation (CE), keto-acyl-CoA reduction (KCR), dehydratation (DH) and enoyl-CoA reduction (ER) (Domergue *et al.*, 1998). The elongation mechanism has been extensively studied but the organization of the acyl-CoA elongase in the membrane remains unclear although genes encoding 3-ketoacyl-CoA synthases have been cloned. Recent advances in the knowledge of the other proteins of the elongation complex from the leaf have been made. Firstly, the characterization of maize gl8 cDNA as a ketoacyl-CoA redutase of the acyl-CoA elongase involved in wax biosynthesis was reported (Xu *et al.*, 1997, 2002) and secondly an *A.thaliana* clone was also characterized as a ketoacyl-CoA redutase by the complementation of a yeast mutant (Beaudoin *et al.*, 2002). In high erucic acid rapeseed (HEAR), the knowledge of the elongation complex which synsizes erucic acid is incomplete despite its biotechnological importance. In this paper, we describe the isolation of two cDNAs encoding a 3-ketoacyl-CoA synthase using a *B.napus* probe made from *A.thaliana* Glossy 8 cDNA sequence. The analysis of Bn KCR 1 and Bn KCR 18 amino acids sequences and the expression of the corresponding mRNA during seed development are also reported.

Material and Methods

The library was prepared using the vector lambda Zap II (Stratagene) and was a gift from T. Roscoe (Université de Perpignan). One million plaques were screened with a cDNA *B.napus* probe. This probe was synthesized by RT-PCR with primers

221

N. Murata et al. (eds.), Advanced Research on Plant Lipids, 221–224.
© 2003 *Kluwer Academic Publishers. Printed in the Netherlands.*

designed from the sequence of *A.thaliana* glossy 8 cDNA . Other methods were
described in Puyaubert *et al.*, 2001.

Results and Discussion

Cloning of B.napus cDNAs derived from Arabidopsis gl8 cDNA

One million plaques were screened from a *B.napus* embryos cDNA library with a
637 b *B.napus* cDNA probe. Five cDNAs were obtained and sequenced. Among
them, four had the same open reading frame, so only two different clones were
obtained. These clones Bn KCR 1 and Bn KCR 18 encoded a 319 amino acid
protein with a predicted molecular mass of 35 kDA. The comparison of the amino
acid sequences deduced from Bn KCR 1 and Bn KCR 18 cDNAs showed 96.8% of
identity, suggesting that Bn KCR 1 and Bn KCR 18 are two isoforms (Fig 1).

```
BN1.AMI     1  MEICTYFKSQPTULLVLFSLGSISILRFTFTLLTSLYIYFLRPGKNLRRY   50
BN18.AMI    1  MEICTYFKSQPTULLVLFSLGSISILRFTLTLLTSLYIYFLRPGKNLRRY   50

                  GXXXGXGXXXAXXXAXXG
BN1.AMI    51  GSWAIITGPTDGIGKAFAFQLAQKGLULVLVARNPDKLKDYSDSIQAKYS  100
BN18.AMI   51  GSWAIITGPTDGIGKAFAFQLAQKGLHLVLVARNPDKLKAYSDSIQAKGS  100

BN1.AMI   101  NTQIKTVWMDFSGDIDGGVRRIKEAIEGLEVGILINNAGVSYPYAKYFHE  150
BN18.AMI  101  TTQIKTVLMDFSGDIDAGVRRIKEAIEGLEVGILINNAGVSYPYAKYFHE  150

                                                          *
BN1.AMI   151  VDEBMLGNLIKINVEGTTKVTQAVLVNMLKRKRGAIVNNGSGAAALIPST  200
BN18.AMI  151  VDEBLLGNLIKINVEGTTKVTQAVLVNMLKRKRGAIVNNGSGAAALIPST  200

                 *   *
BN1.AMI   201  PFYSVYAGAKTYVDQFSRCLHVEYKKSGIDVQCQVPLYVATKMTKIRRAS  250
BN18.AMI  201  PFYSVYAGAKTYVDQFSRCLHVEYKKSGIDVQCQVPLYVATKMTKIRRAS  250

BN1.AMI   251  FLVASPEGYAKAALRFVGYEPRCTPYUPHALNGYVVSALPESVFESFNIK  300
BN18.AMI  251  FLVASPEGYAKAALRFVGYEPRCTPYUPHALNGYVVSALPESVFESFNIK  300

BN1.AMI   301  RCLQIRKKGMLKDSTSKKE...........................  350
BN18.AMI  301  RCLQIRKKGMLKDSSSKKE...........................  350
```

Figure 1 : Comparison of amino acids sequences of Bn KCR 1 and Bn KCR 18
The amino acids of the reductase catalytic triad are indicated with stars. The consensus sequence of
the NADP binding domain is indicated. The reticucum endoplasmic targeting signal is boxed.

Structure of Bn KCR 1 and Bn KCR 18 proteins

The amino acid sequences of the KCR proteins were analyzed using different
softwares. The PSORT algorithm predicted that the KCR proteins are membrane
proteins with a single putative transmembrane helix in N-terminal. The hydrophobic
characteristics of the KCR proteins were confirmed by using Kyte and Doolittle
hydrophobicity plot (Fig 2). The profiles of Bn KCR 1, Bn KCR 18 and maize
glossy 8 proteins were nearly the same thus indicating that these proteins could be
membrane components. This hypothesis is also supported by the prediction of a
peptide signal in N-terminal (1-30) using PSORT algorithm analysis. This is also in

agreement with Signal IP algorithm analysis which predicted a peptide signal or a membrane anchor signal in N-terminal with a greater probability.

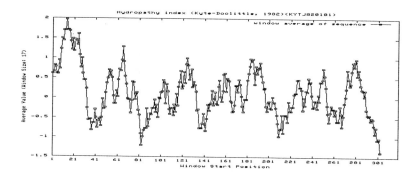

Figure 2 : Kyte and Doolittle hydrophobicity plot of the deduced amino acid sequence of the Bn KCR 1 protein.

The Bn KCR 1 and Bn KCR 18 proteins contain three conserved amino acids which constitue a catalytic triad (Ser-Xaa_{10}-Tyr-Xaa_4-Lys) and a signature pattern of the Short-Chain-Alcohol Dehydrogenases Reductases family (Fig1). The deduced amino acid sequences of Bn KCR 1 and Bn KCR 18 contain an NADP-binding consensus sequence which is also a signature pattern for the reductases (Fig1). All these results strongly suggest that the Bn KCR 1 and Bn KCR 18 cDNAs encode NAPH membrane-bound keto reductases.

Finally, the presence in the C-terminal of the KKX motif strongly suggests that the KCR proteins could be targeted to the endoplasmic reticulum (Fig 1). Since in *B.napus* seeds the acyl-CoA elongase is mainly located in microsomes, this finding is very consistent with that KCR protein could be one of the acyl-CoA elongase complex enzyme (Domergue *et al.*, 1998).

Identification and Analysis of Bn KCR 1 and Bn KCR 18 proteins

The Bn KCR 1 and Bn KCR 18 amino acid sequences were compared with maize and *A.thaliana* glossy 8 proteins. The proteins encode by *A.thaliana* and maize were respectively 86.1% and 51.1% identical, to the Bn KCR 1 and Bn KCR 18 proteins. Unsurprisingly, the Bn KCR 1, Bn KCR 18 clones are more identical to *A.thaliana* than maize since *B.napus* and *A.thaliana* are *Brassicacae*. The maize and *A.thaliana* Glossy 8 clones have recently been identified as ketoacyl-CoA reductase involved in the VLCFA biosynthesis in cuticular waxes (Xu *et al.*, 2002; Beaudoin *et al.*, 2002). The large percentage of identity between *B.napus* and *A.thaliana* clones strongly suggest that we have cloned a ketoacyl-CoA reductase from *B.napus* seed.

The deduced amino acid sequences of Bn KCR 1 and Bn KCR 18 were compared with several reductases: ketoacyl-CoA reductases of maize and *A.thaliana*; ketoacyl-ACP reductase of *E.coli*, *A.thaliana* and *B.napus*; enoyl-ACP reductase of *E.coli* and *B.napus*. and a human tropinone reductase. Three groups could be distinguished from this alignment. The first was constituted by ketoacyl-ACP reductases (*E.coli*, *A.thaliana*, *B.napus*), the second group by enoyl-ACP reductases (*E.coli* and *B.napus*) and the third by reductases which catalyse the reduction of a ketone group into alcohol. The latter was constituted by Bn KCR 1 and Bn KCR 18, human tropinone reductase, maize Glossy 8, and *A.thaliana* Glossy 8. The ketoacyl-ACP reductases which are involved in the *de novo* synthesis pathway are more distantly. This is not surprising as they are hydrophyllic globular proteins.

Expression of ketoacyl-CoA reductase in different tissues from Hear variety and during the development of the seed
A cDNA probe corresponding to a highly homologous region of Bn KCR 1 and Bn KCR 18 was used to determine ketoacyl-CoA reductase expression by northern-blot. A 1000 b transcript was detected in all tissues, and the highest accumulation of ketoacyl-reductase mRNA was recovered in the seed. The ketoacyl-CoA reductase was expressed throughout the seed maturation and the maximal expression occurred at the 8 week after pollination. A similar pattern of expression was observed for the ketoacyl-CoA reductase and the condensing enzyme (Puyaubert *et al.*, 2001). These enzymes seem to be coordinately regulated In conclusion, two genes Bn KCR 1 and Bn KCR 18 have been cloned. These genes presumably encode a ketoacyl-CoA reductase, a membrane-bound protein certainly involved in the erucic acid synthesis. This enzyme could be a constituent of the elongase complex. These experiments are underway to clarify this latter point.

Acknowledgments : This work was supported by MESR, ONIDOL, CETIOM, RUSTICA and SERASEM. J. Puyaubert is a recipient of a grant from l'ADEME. The help of Conseil Régional d'Aquitaine is also gratefully acknowledged. We also thank T. Roscoe for providing the cDNA library and, M.Renard and H. Picault for harversting the seed.

References

Beaudoin F, Gable K, Sayanova O, DunnT, Napier J (2002) A saccharomyces cerevisiae gene required for heterologous fatty acid elongase activity encodes a microsomal keto-reductase. *J.Biol.Chem* **277** (13): 11481-8.

Domergue F, J-J Bessoule, Moreau P, Lessire R and Cassagne C (1998) Recent advances in plant fatty acid elongation. *Plant Lipid Biosynthesis: Fundamentals and Agricultural Applications,* 185-222.

Puyaubert J, Gaebay B, Costaglioli P, Dieryck W, Roscoe T, Renard M, Cassagne C, Lessire R (2001) Acyl-CoA elongase expression during seed development in Brassica napus. *Biochimica et Biophysica Acta*, 141-152.

Xu X, Dietrich C, Delledone M, Xia Y, Wen T, Robertson D, Nikolau B, Schnable P (1997) Sequence Analysis of the Cloned glossy 8 gene of maize suggests thet it may code for a ketoacyl reductase required for the biosynthesis of cuticular waxes. *Plant Physiol* **115**: 501-510.

 Xu X, Dietrich C, Delledone M, Lessire R, Nikolau B, Schnable P (2002) The endoplasmic reticulum-associated maize GL8 protein is a component of the acyl-CoA elongase involved in the production of cuticular waxes. *Plant Physiol* **128** (3): 924-934.

DISSECTING THE MAIZE EPICUTICULAR WAX BIOSYNTHETIC PATHWAY VIA THE CHARACTERIZATION OF AN EXTENSIVE COLLECTION OF *GLOSSY* MUTANTS

M. A. D. N. PERERA, C. R. DIETRICH, R. MEELEY, P. S. SCHNABLE, B. J. NIKOLAU
Iowa State University, Ames, Iowa, U.S.A., and Pioneer Hi-Bred International, Inc. Johnston, Iowa, U.S.A.

1. Introduction

Being a major component of the cuticle, cuticular waxes serve as the outer boundary between plants and their environment and thus provide multiple protective functions (Kolattukudy, 1987; Jenks et al., 1994). Cuticular waxes are complex mixtures of very long chain fatty acids (VLCFA; > C18), and VLCFA-derivatives such as alcohols, aldehydes, esters, alkanes, and ketones, (Walton, 1990). Thus, VLCFAs serve as the precursors for the cuticular wax biosynthesis. Elucidating the complex processes by which cuticular waxes are biosynthesized is the long-term objective of our genetic studies in maize. We have characterized a collection of 186 *glossy* (*gl*) mutants of maize, which are affected in the normal accumulation of epicuticular waxes. The focus of this report is the molecular characterization of the *gl8* gene, which encodes one of the components of the acyl-CoA elongase system that is involved in the biosynthesis of VLCFA for cuticle deposition.

2. Results & Discussion

2.1 *Genetics of Epicuticular Waxes in Maize*

As a result of mutagenesis experiments to identify maize genes that are involved in the production of epicuticular wax, 186 *gl* mutants have been identified. These *gl* mutants were selected based on the water-beading phenotype that is expressed on juvenile leaves (Schnable et al. 1994). These 186 independently isolated *gl* mutants have been genetically characterized and have been placed into 26 complementation groups, nine of which define previously undefined *gl* loci. For each mutant locus, the reference mutant allele has been backcrossed into a constant genetic background (the maize inbred B73). In addition, each new *gl* locus has been placed on the maize genetic map.

2.2 *Molecular Cloning of gl8 Paralogs*

The *gl8* mutant was originally isolated in 1935 (Emerson et al., 1935). Based upon the phenotype associated with this mutant, we cloned this gene using a *Mutator*-tagged *gl8* mutant allele (Xu et al., 1997). Subsequent biochemical characterizations using antibodies against the GL8 protein, established that this gene product accumulates in the

225

N. Murata et al. (eds.), Advanced Research on Plant Lipids, 225–228.

226

endoplasmic reticulum membranes, and is the 3-ketoacyl reductase component of the acyl-CoA elongase that produces VLCFA for epicuticular wax deposition (Xu et al., 2002).

Molecular analyses indicate that the maize genome contains two *gl8* paralogs. One of these, defined by the Emerson mutant allele is located on chromosome 5L, we now term this locus *gl8a*. The paralogous locus is located on chromosome 4L, and we call it *gl8b*. The *gl8b* gene has been cloned and characterized. The proteins encoded by the *gl8a* and *gl8b* genes share over 97% sequence identity. The identification of these two near identical paralogs begs the questions of their individual functions.

2.2 *Reverse Genetics of* gl8b

To address this question we used the Trait Utility System for Corn (TUSC) (Bensen et al., 1995) to isolate a *Mutator*-induced mutation in the *gl8b* locus. This allele contains a *Mutator1* insertion 140-bp downstream of the start codon in exon 1 of the *gl8b* gene.

The *gl8b* mutant does not present a visible glossy phenotype. However, upon comparing the chemical composition of the cuticular waxes of *gl8a* and *gl8b* mutants to wild type seedlings, it is apparent that the former mutation eliminates 95% of the wax present on maize seedling leaves, whereas the latter mutation eliminates only 45% of the wax (Figure 1). Furthermore, whereas the *gl8a* mutation nearly eliminates the

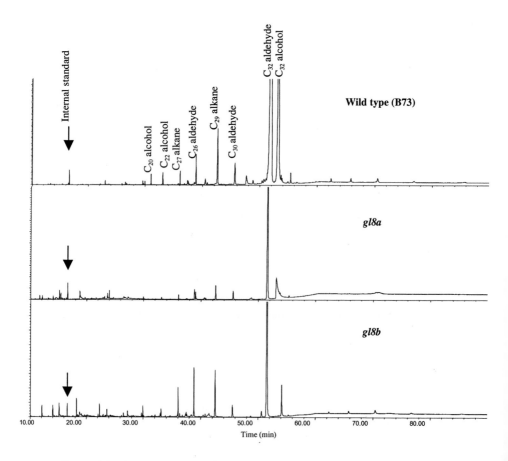

Figure1. GC-MS analysis of the cuticular waxes from wild type, *gl8a* and *gl8b* maize seedlings

accumulation of the C_{32} alcohols and aldehydes, the *gl8b* mutation reduces these compounds to about 50% of wild-type levels. In addition, unlike the *gl8a* mutation, the *gl8b* mutation causes a nearly 3-fold increase in the accumulation of the esters and a near doubling of the free fatty acid fraction.

2.3 *Expression of* gl8a *and* gl8b *Paralogs*

Expression patterns of the *gl8a* and *gl8b* genes have been studied by RNA-gel bolt analysis, *in situ* RNA hybridization and by the analysis of *gl8a* and *gl8b* promoter-mediated GUS expression in transgenic maize lines. These studies indicate that the two genes have highly similar expression patterns. However, overall we find that *gl8a* is expressed at 3- to 5-fold higher levels than *gl8b*. Furthermore, we find that these two genes are not only expressed in the epidermis, which is expected for cuticular wax biosynthesis, but also in the internal tissues of a number of organs (e.g., young unfurled leaves, roots, silks, embryos and scutellum).

2.3 *The* gl8a *and* gl8b *Functions are Essential for Normal Kernel Development*

With the exception of the effect on cuticular wax deposition, the *gl8a* and *gl8b* mutations have no apparent effects on the normal growth and development. To generate plants that carry both mutations, we selfed plants that were either homozygous mutant at the *gl8a* locus and heterozygous at the *gl8b* locus (i.e., the self of *gl8a/gl8a; Gl8b/gl8b*), or homozygous mutant at the *gl8b* locus and heterozygous at the *gl8a* locus (i.e., the self of *Gl8a/gl8a; gl8b/gl8b*). One quarter of the progeny kernels were expected to be double mutants (i.e., *gl8a/gl8a; gl8b/gl8b*). However, we found that approximately 25% of these kernels did not produce viable seedlings. These kernels had normal endosperm but the scutellum and embryos had in most instances degenerated. DNA isolated from endosperms of such kernels confirmed that these kernels are double mutant for the *gl8a* and *gl8b* mutations. Hence, the *gl8a gl8b* double mutant conditions a lethal phenotype that is expressed during kernel development.

3. Discussion

We have generated a large collection of maize mutants that are affected in the normal deposition of cuticular waxes. These *glossy* plants are a unique resource for elucidating the biological function of the cuticle and for dissecting the molecular mechanisms of cuticular wax biosynthesis. From this mutant collection we have cloned the *gl1* (Hansen et al., 1997) and *gl8* (Xu et al., 1997) loci. Biochemical and molecular characterizations of the cloned *gl8* gene have established that it encodes one of the four enzyme components of the acyl-CoA elongase; specifically the 3-ketoacyl reductase component (Xu et al., 2002).

Additional molecular and genetic characterizations have lead to the identification and cloning of two paralogous genes (*gl8a* and *gl8b*) that encode the 3-ketoacyl reductase component of the acyl-CoA elongase. These two genes encode very similar proteins and are expressed in near identical patterns. Yet mutations at each locus condition different morphological and chemical phenotypes on the cuticle. Furthermore, plants that carry mutations in both loci fail to develop normal kernels. These characteristics indicate that these two genes have overlapping but non-identical functions. Moreover, these data indicate that VLCFAs may be essential to the normal development of maize kernels. Whether this is due to the inability to generate a cuticle during kernel development or an inability to generate another VLCFA-containing lipid (e.g., sphingolipids) is yet to be determined.

4. References

Bensen R.J., Johal G.S., Crane V.C., Tossberg T.J., Schnable P.S., Meeley R.B., Briggs S.P. (1995) Cloning and characterization of the maize An1 gene. Plant Cell 7, 75-84.

Emerson R.A., Beadle G.W., Fraser A.C. (1935) A summary of linkage studies in maize. Cornell University Agric. Exp. Stn. Memoir 180: 1-83.

Hanson J.D., Pyee J., Xia Y., Wen T.-J., Robertson D.S., Kolattukudy P.E., Nikolau B.J., Schnable P.S. (1997) The glossy 1 locus of maize and an epidermis-specific cDNA from Kleinia odora define a class of receptor-like proteins required for the normal accumulation of cuticular waxes1. Plant Physiol 113, 1091-1100.

Jenks M.A., Joly R.J., Peters P.J., Rich P.J., Axtell J.D., Ashworth E.N. (1994) Chemically induced cuticle mutation affecting epidermal conductance to water vapor and disease susceptibility in Sorghum bicolor (L.) Moench. Plant Physiol 105, 1239-1245.

Kolattukudy P.E. (1987) Lipid-derived defense polymers and waxes and their role in plant-microbe interaction. In: Stumpf PK ed, The Biochemistry of Plants. Academic Press, New York, pp 291-314.

Walton T.J. (1990) Waxes, cutin and suberin In JL Harwood JR Bowyer, eds. Methods in Plant Biochemistry:Lipids

Schnable P.S., Stinard P.S., Wen T.-J., Heinen S., Weber D., Schneerman M., Zhang L., Hansen J., Nikolau B.J. (1994) The genetics of cuticular wax biosynthesis. Maydica 39, 279-287.

Xu X., Dietrich C.R., Delledonne M., Xia Y., Wen T.J., Robertson D.S., Nikolau B.J., and Schnable P.S. (1997) Sequence analysis of the cloned glossy8 gene of maize suggests that it may code for a beta-ketoacyl reductase required for the biosynthesis of cuticular waxes. Plant Physiol. 115, 501-510.

Xu X., Dietrich C.R., Lessire R., Nikolau B.J., and Schnable P.S. (2002) The endoplasmic reticulum-associated maize GL8 protein is a component of the acyl-CoA elongase involved in the production of cuticular waxes. Plant Physiol. 128, 924-934.

DISTRIBUTION OF CEREBROSIDE IN GENUS *SACCHAROMYCES* AND ITS CLOSELY RELATED YEASTS, AND CLONING OF CEREBROSIDE METABOLISM-RELATED GENES

N. TAKAKUWA[1], K. YAMANE[1], M. KINOSHITA[1], Y. ODA[2], and M. OHNISHI[1]

[1]*Obihiro University of Agriculture and Veterinary Medicine, Obihiro, Hokkaido 080-8555, Japan;* [2]*National Agricultural Research Center for Hokkaido Region, Memuro, Kasai, Hokkaido 082-0071, Japan*

1. Abstract

As sphingolipids, cerebroside (glucosylceramide) as well as free ceramide and inositolphosphate-containing ceramide, are widely distributed in fungi and plants. However, there has been no evidence of cerebroside or unsaturated sphingoid bases found in *Saccharomyces* yeasts. When thirty-one strains accepted in the genus *Saccharomyces* and its closely related yeasts were analyzed for sphingolipids, cerebrosides were found in *Saccharomyces kluyveri*, *Zygosaccharomyces cidri*, *Z. fermentati*, *Kluyveromyces lactis*, *K. thermotolerans*, and *K. waltii*. The cerebrosides of *S. kluyveri* and *K. lactis* included 9-methyl-*trans*-4, *trans*-8-sphingadienine and its putative metabolic intermediates; 4-unsaturated and 4,8-diunsaturated sphingoid bases. We further cloned genes encoding glucosylceramide synthase and Δ^8-sphingolipid desaturase from *S. kluyveri* and *K. lactis*.

2. Introduction

Glucosylceramide, namely cerebroside, is a typical membrane lipid of plants, fungi, and animals. Among yeasts, it is found in the *Candida albicans*, *Pichia pastoris*, *Pichia, anomala*, and *Cryptococcus* species but is lacking in *Saccharomyces cerevisiae*, which instead synthesizes inositol phosphorylceramides as essential components for viable growth (Dickson and Lester, 1999; Levery et al., 2000; Sakaki et al., 2001). In this study, we surveyed the *Saccharomyces* species and its related genera extensively and carried out phylogenetic analysis. The putative genes encoding glucosylceramide synthase (GCS) and Δ^8-sphingolipid desaturase (SLD) were then

229

N. Murata et al. (eds.), Advanced Research on Plant Lipids, 229–232.

isolated from the yeast strains to compare to those from other organisms.

3. Experimental

Extraction, purification and analysis of cerebrosides

Total lipids were extracted from the lyophilized cells with chloroform/methanol (2:1 and 1:2, v/v) and hydrolyzed in 0.4N KOH-methanol. Cerebroside was isolated on preparative thin layer chromatography (TLC) developed by chloroform/methanol (95:12, v/v) from the lipids. After hydrolysis of cerebroside to sphingoid bases and fatty acids, their compositions were determined by gas-liquid chromatography according to the methods previously described (Fujino and Ohnishi, 1977; Ohnishi et al., 1996).

PCR-based cloning of GCS and SLD genes

To determine the internal sequence of GCS and SLD genes, PCRs were performed by using the genomic DNA of *S. kluyveri* and *K. lactis* as the template and degenerate primers: 5'-AAYCCNAARRTNMRNAAYHT-3' and 5'-ADRAACATYTCNTCNAR-3' for the GCS and 5'-AARDVNCAYCCNGGNGGNGA-3' and 5'-AVRTGYTGNAVRTCNGGRTC-3' for the SLD. Based on the PCR fragments, the entire regions spanning the ORF were sequenced by thermal asymmetric interlaced (TAIL)-PCR.

Expression in Saccharomyces cerevisiae

PCR fragments corresponding to the GCS and SLD genes were inserted into pYES2 as the expression vector. The cells of *S. cerevisiae* strain INVSc1 (Invitrogen) transformed with the plasmid were grown in a uracil-deficient medium. The medium containing cells in the mid-exponential growth was supplemented with 2.0% galactose as the final concentration and incubated for up to 24 h.

4. Results and discussion

When alkali-stable lipids extracted from the cells of thirty-one strains were separated on TLC and visualized by heating after spraying with an anthrone reagent, six strains had a component with an Rf value identical to the authentic cerebroside preparation. These were *Saccharomyces kluyveri*, *Zygosaccharomyces cidri*, *Z. fermentati*, *Kluyveromyces lactis*, *K. thermotolerans*, and *K. waltii*. In the phylogenetic tree depicted on the basis of the 18S rDNA of the species examined in these experiments, most species of *Saccharomyces*, *Torulapora*, and *Zygosaccharomyces* exhibited a genetic association within each genus (Figure 1). *S. kluyveri*, *Z. cidri*, and *Z. fermentati*, which were shown to contain cerebrosides, deviated from their accepted genera and rather related to the *Kluyveromyces* species tested. The evolution of *S.*

cerevisiae and close relatives has recently been studied in detail after sequencing the whole genomic DNA of *S. cerevisiae*. The large duplicated chromosomal regions suggested that *S. cerevisiae* is a degenerate tetraploid generated by genome duplication approximately 10^8 years ago. *S. cerevisiae* and *S. kluyveri* are assumed to diverge from the lineage leading to *K. lactis* before genome duplication in *S. cerevisiae* (Keoch et al., 1998). Since an ancestral organism for these yeasts selected inositol phosphorylceramide as an essential constituent for growth, the biosynthetic pathway of cerebroside may narrow in *K. lactis* and *S. kluyveri* and finally be lost in *S. cerevisiae* during the process of divergence.

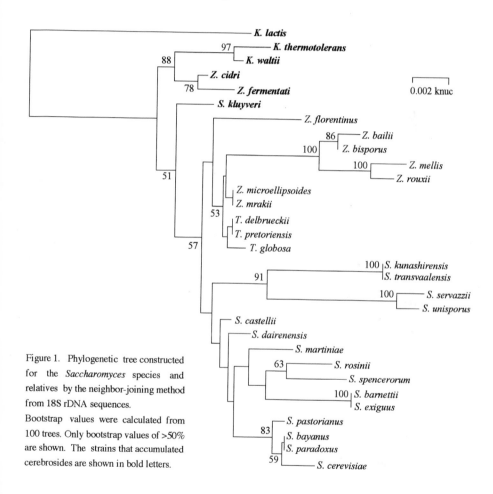

Figure 1. Phylogenetic tree constructed for the *Saccharomyces* species and relatives by the neighbor-joining method from 18S rDNA sequences.
Bootstrap values were calculated from 100 trees. Only bootstrap values of >50% are shown. The strains that accumulated cerebrosides are shown in bold letters.

Cerebrosides were extracted and purified from the cells of *S. kluyveri* and *K. lactis* and analyzed after hydrolysis. Cerebrosides of both strains included 9-methyl-*trans*-4, *trans*-8-sphingadienine (9-Me d18:2^{4t,8t}), a major and characteristic component in fungi, and its possible metabolic intermediates, 4-unsaturated and 4,8-diunsaturated sphingoid bases. Interestingly, the trihydroxy sphingoid base, which occurs exclusively in sphingolipids in *S. cerevisiae*, covered 30% of the component sphingoid bases of the cerebroside in *S. kluyveri*. In the fatty acid components, 2-hydroxystearic acid was usually predominant. Thus, in cerebrosides of *S. kluyveri* and *K. lactis*, the proportion of 9-Me d18:$2^{4t,8t}$ (39 and 49%, respectively) was much lower than that in other fungi, suggesting that C9-methylation of the sphingoid base in ceramide is limited to form the usual acceptor for glucosyl transfer. A trihydroxy sphingoid base, which is rarely found in fungal cerebrosides, may be produced from sphinganine and used as the substrate for GCS. These observations mean that the capacity for cerebroside synthesis before GCS in *S. kluyveri* is smaller than that in *K. lactis*.

PCR fragments containing putative GCS genes of *S. kluyveri* (AB080772) and *K. lactis* (AB080773) were introduced to *S. cerevisiae* after insertion into the expression vector. Cerebroside appeared in expression of the gene from *S. kluyveri* but not from *K. lactis*. Similar reported findings show that GCS from *Candida albicans* and not from *Pichia pastoris* was expressed in *S. cerevisiae* (Leipelt et al., 2001). In the putative protein sequences of SLD genes from yeasts, HPGG and three histidine boxes were conserved, similar to plant SLD (Sperling et al., 1998). The heterologous expression of *K. lactis* SLD gene in *S. cerevisiae* resulted in the formation of 4-hydroxy-*trans*-8-sphingenine. These results represent the first identification of gene encoding for a SLD from fungi.

5. References

Dickson, R. C. and Lester, R. L. (1999) Yeast sphingolipids. Biochim. Biophys. Acta 1426, 347-357.

Fujino, Y. and Ohnishi, M. (1977) Structure of cerebroside in *Aspergillus oryzae*. Biochim. Biophys. Acta 486, 161-171.

Keoch, R. S., Seoighe, C. and Wolfe, K. H. (1998) Evolution of gene order and chromosome number in *Saccharomyces*, *Kluyveromyces*, and related fungi. Yeast 14, 443-457.

Leipelt, M., Warnecke, D., Zähringer, U., Ott, C., Müller, F., Hube, B. and Heinz, E. (2001) Glucosylceramide synthases, a gene family responsible for the biosynthesis of glucosphingolipids in animals, plants, and fungi. J. Biol. Chem. 276, 33621-33629.

Levery, S. B., Toledo, M. S., Doong, R. L., Straus, A. H. and Takahashi, H. K. (2000) Comparative analysis of ceramide structural modification found in fungal cerebrosides by electrospray tandem mass spectrometry with low energy collision-induced dissociation of Li$^+$ adduct ions. Rapid Commun. Mass Spectrom. 14, 551-563.

Ohnishi, M., Kawase, S., Kondo, Y., Fujino, Y. and Ito, S. (1996) Identification of major cerebroside species in seven edible mushrooms. J. Jpn. Oil Chem. Soc. 45, 51-56.

Sakaki, T., Zähringer, U., Warnecke, D. C., Fahl, A., Knogge, W. and Heinz, E. (2001) Sterol glycosides and cerebrosides accumulate in *Pichia pastoris*, *Rhynchosporium secalis*, and other fungi under normal conditions or under heat shock and ethanol stress. Yeast 18, 679-695.

Sperling, P., Zähringer, U. and Heinz, E. (1998) A sphingolipid desaturase from higher plants. J. Biol. Chem. 273, 28590-28596.

PROPERTIES AND PHYSIOLOGICAL EFFECTS OF PLANT CEREBROSIDE SPECIES AS FUNCTIONAL LIPIDS

K. AIDA [1,2], N. TAKAKUWA[1], M. KINOSHITA[1], T. SUGAWARA[3], H. IMAI[4], J. ONO[2], and M. OHNISHI[1]

[1]*Obihiro University of Agriculture and Veterinary Medicine, Obihiro 080-8555, Japan;* [2]*Nippon Flour Mills Co. Ltd., Atsugi 243-0041, Japan;* [3]*National Institute of Health and Nutrition, Tokyo 162-8636, Japan;* [4]*Faculty of Science and Technology, Konan University, Kobe 658-8501, Japan*

1. Abstract

The properties and physiological effects of plant cerebrosides were investigated. The major sphingoid bases of cerebrosides from soybean, maize, and rice bran were *trans*-4, *cis* or *trans*-8-sphingadienine. However, the major sphingoid bases in wheat and rye grains were 8-sphingenine mainly with the *cis*-configuration (more than 60%). The phase transition temperature of plant cerebrosides measured by differential scan calorimetric method was significant lower than that of bovine brain cerebroside. Plant cerebroside-derived compounds such as sphingoid bases induced apoptosis in human colon cancer cell lines. This result suggested that dietary plant cerebrosides have a potent physiological function similar to that of animal sphingolipids.

2. Introduction

Sphingolipids appear to be ubiquitous among eukaryotic organisms. The most prevalent sphingoid base of mammalian sphingolipids is *trans*-4-sphingenine (sphingosine), and smaller amounts of others are frequently encountered, such as sphinganine (dihydrosphingosine) and 4-hydroxysphinganine (phytosphingosine). Plant sphingolipids have more diverse sphingoid base structures, such as 8-sphingenine, 4,8-sphingadienine and 4-hydroxy-8-sphingenine (Ohnishi et al., 1982; Ohnishi et al.,1988). The average daily intake of plant-origin sphingolipids has been estimated as 50 mg in man (Sugawara and Miyazawa, 1999). Recent reports indicated that the intake of sphingomyelin, a major animal-origin sphingophospholipid has potent physiological effects (Vesper et al., 1999). Sphingolipids and their metabolites are now well recognized as playing important roles as an intracellular mediator of cell differentiation and apoptosis (Vesper et al., 1999; Hannun, 1996). However, there is no evidence that

233

N. Murata et al. (eds.), Advanced Research on Plant Lipids, 233–236.
© 2003 *Kluwer Academic Publishers. Printed in the Netherlands.*

dietary plant sphingolipids contribute to human health. Here, we report the properties and physiological effects of plant cerebroside species as functional lipids.

3. Experimental

Separation and analysis of plant cerebrosides

The plant cerebrosides used in this study were extracted and purified from rice bran, maize seeds, soybeans, and rye and wheat grains. The purification and component analysis of plant cerebrosides were conducted as reported previously (Ohnishi et al., 1982; Ohnishi et al., 1988). Differential scan calorimetric (DSC) study and fluorescence depolarization measurement were performed as described previously (Ohnishi et al., 1988; Kasamo et al., 2000).

Quantitative analysis of apoptotic cells

Human colon cancer cells (Caco-2) were cultured in DMEM supplemented with penicillin (100 units/ml), streptomycin (100 μg/ml) and 10% FCS. Apoptotic cells were quantified by counting cells with characteristic fragmented-nuclei under a fluorescent microscope (Harada-Shiba et al., 1998). The percentage of apoptotic cells was calculated as the number of apoptotic cells / the number of total cells x 100.

4. Results and discussion

The ceramide moiety constituents of plant cerebrosides are shown in Table 1. The major

Table 1. 2-Hydoroxy fatty acid and sphingoid base compositions of plant cerebrosides (%)

	Wheat grain	Rye grain	Rice bran	Maize seed	Soybean seed
2-Hydroxy fatty acid					
16h:0	39	26	1	6	82
18h:0	7	5	6	16	<1
20h:0	38	36	50	39	<1
22h:0	4	8	14	12	7
24h:0	4	7	21	21	8
24h: 1	1	8	ND	ND	ND
26h:0	2	2	3	3	1
26h: 1	1	1	ND	ND	ND
Others	4	7	5	3	2
Sphingoid base					
Sphinganine	5	4	1	1	<1
trans-4-Sphingenine	1	<1	4	3	<1
trans-8-Sphingenine	21	21	1	<1	<1
cis-8-Sphingenine	50	50	<1	<1	5
trans-4, trans-8-Sphingadienine	4	3	13	17	49
trans-4, cis-8-Sphingadienine	9	5	45	54	24
4-Hydroxysphinganine	1	2	7	2	<1
4-Hydroxy-trans-8-Sphingenine	1	2	3	2	9
4-Hydroxy-cis-8-Sphingenine	8	12	26	21	13

sphingoid bases of cerebrosides from soybean, maize, and rice bran were *trans*-4-*cis* or *trans*-8-sphingadienine, and those from wheat and rye grains were mainly 8-sphingenine with the *cis*-configuration. Major 2-hydroxy fatty acids also showed species differences. The major 2-hydroxy fatty acids from rice bran and maize cerebrosides were 20h:0, 22h:0 and 24h:0; 16h:0 in soybean and 16h:0 and 20h:0 in wheat and rye grains, respectively. The unsaturated 2-hydroxy fatty acids were only found in wheat and rye grain cerebrosides.

DSC profiles of plant cerebrosides are shown in Figure 1. The sharp phase transition temperature's peak of bovine brain cerebroside (major components are 24h:0 and *trans*-4-sphingenine) was observed at near 60°C. On the other hand, plant cerebrosides showed broad peaks with significantly lower temperature compared with bovine brain cerebroside. The reason for this would be due to the diversity of molecular species of cerebroside especially in terms of sphingoid base and the existence of 8-*cis* configuration in the ceramide moiety. The fluorescence depolarization values of liposome composed of asolecthin were increased by the addition of plant cerebrosides. However, the existence of the *cis*-8 double bond in the component sphingoid base suppressed increases in the depolarization value (data not shown).

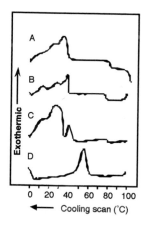

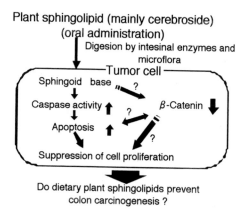

Figure 1. DSC profiles of plant cerebrosides.
A: soybean, B: maize, C: rice bran, D: bovine brain.

Scheme 1. Speculated physiological effects of dietary plant sphingolipids.

The physiological effects of dietary sphingolipids are shown in Scheme 1(Vesper et al., 1999). Sphingolipids such as sphingomyelin, cerebroside and complex sphingolipids would be hydrolyzed in all subsequent regions of the small intestine and colon of several animals. Sphingoid base and ceramide affect cell growth, differentiation and apoptosis in most types of cells that have been studied in culture cells. This raises the possibility that release of these compounds during digestion of dietary sphingolipids may alter the behavior of normal or transformed cells, especially of the intestine. In this study, we focused on the apoptosis inducing activity of plant sphingolipid metabolites to human colon cancer cell lines. As shown in Figure 2, plant

236

sphingoid bases derived from wheat grain cerebroside were found to induce apoptosis in a dose-dependent manner. This result suggested that dietary plant cerebroside has a potent physiological function similar to that of animal sphingolipids.

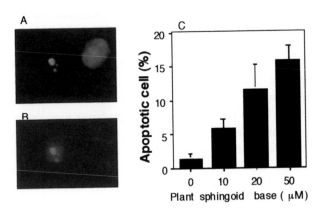

Figure 2. Apoptosis induced by plant sphingoid bases in human colon cancer cell line (Caco-2). A and B, morphology of apoptotic cell (A, DAPI stain; B, Tunel stain.); C, dose-dependent induction of apoptosis in Caco-2 by plant sphingoid bases derived from wheat grains.

5. References

Hannun, Y. A. (1996) Functions of ceramide in coordinating cellular responses to stress. Science 274, 1855-1859.

Harada-Shiba, M., Kinoshita, M., Kamido, H. and Shimokado, K. (1998) Oxidized low density lipoprotein induces apoptosis in cultured human umbilical vein endothelial cells by common and unique mechanisms. J. Biol. Chem. 273, 9681-9687.

Kasamo, K., Yamaguchi, M. and Nakamura, Y. (2000) Mechanism of the chilling –induced decrease in proton pumping across the tonoplast of rice cells. Plant Cell Physiol. 41, 840-849.

Ohnishi, M., Ito, S. and Fujino, Y. (1982) Characterization of sphingolipids in spinach leaves. Biochim. Biophys. Acta 752, 416–422.

Ohnishi, M., Imai, H., Kojima, M., Yoshida, S., Murata, N., Fujino, Y. and Ito, S. (1988) Separation of cerebroside species in plant by reversed-phase HPLC and their phase transition temperature. Proc. ISF-JOCS World Congress, II, pp. 930-935, The Japan Oil Chemists' Society, Tokyo.

Sugawara, T. and Miyazawa, T. (1999) Separation and determination of glycolipids from edible plant sources by high-performance liquid chromatography and evaporative light-scattering detection. Lipids 34, 1231-1237.

Vesper, H., Schmelz, E. M., Nikolova-Karakashian, M. N., Dillehay, D. L., Lynch, D. V. and Merrill, A. H. (1999) Sphingolipids in food and the emerging importance of sphingolipids to nutrition. J. Nutr. 129, 1239-1250.

METABOLIC ENGINEERING OF THE MEVALONATE AND NON-MEVALONATE PATHWAYS IN TOMATO

E. M. ENFISSI[1], P. D. FRASER[1], L. M. LOIS[2], A. BORONAT[2] , W. SCHUCH[3] AND P. M. BRAMLEY[1].
[1]School of Biological Sciences, Royal Holloway, University of London, Egham, Surrey TW20 0EX, U.K.
[2]Departament de Bioquimica I Biologia Molecular, Facultat de Quimica, Universitat de Barcelona, Marti I Franques 1-7, 08028 Barcelona, Spain.
[3] CellFor Inc, 355 Burrard Street, Vancouver, BC Canada V6C 2G8

1. Introduction

In higher plants isoprenoids have many essential roles e.g. in membrane structure (phytosterols), redox chemistry (plastoquinone) and as antioxidants (carotenoids). Despite their functional and chemical diversity all isoprenoids are biosynthetically related from a common precursor, isopentenyl pyrophosphate (IPP). In higher plants two pathways exist for the formation of IPP, the mevalonate pathway where IPP is formed from acetyl-CoA via mevalonic acid (MVA) and a second, recently proven, pathway where IPP is formed from D-glyceraldehyde-3-phosphate and pyruvate via 1-deoxy-D-xylulose 5-phosphate (DXP) (Rohmer, 1999). In the cytosol/endoplasmic reticulum (ER) and mitochondria the mevalonate pathway is responsible for the synthesis of phytosterols and ubiquinone. 3-Hydroxy-3-methylglutaryl Co-enzyme A reductase (HMGR) is a key regulatory enzyme in the mevalonate pathway and catalyses the formation of MVA from 3-hydroxy-3-methylglutaryl Co-A (HMG-CoA). Plastid isoprenoids such as carotenoids and tocopherols are formed via the non-MVA pathway. The formation of DXP, catalysed by 1-deoxy-D-xylulose 5-phosphate synthase (DXS), has been shown to be a regulator of flux through this pathway to carotenoids (Lois et al., 2000).

Many isoprenoids are of commercial interest as flavoring agents, pigments and drugs. For example, carotenoids and tocopherols reduce the onset of chronic diseases such as cancer and cardiovascular disease. Dietary phytosterols reduce intestinal absorption of cholesterol and hence have favorable effects on plasma lipid profiles. These properties of isoprenoids have led to intensive efforts to elevate their levels in food crops. Here we have attempted to elevate the levels of phytosterols and carotenoids by the

N. Murata et al. (eds.), Advanced Research on Plant Lipids, 237–240.

manipulation of HMGR and DXS, respectively in tomato fruit. HMGR1 from *A. thaliana* was constitutively expressed in the cytosol/ER, whilst DXS was used in the antisense orientation and also expressed in the plastids as an *E. coli* DXS. The resultant transgenic plants were analyzed for changes in isoprenoid levels.

2. Experimental

2.1 Generation of Primary Plants

The wild type tomato line was *Lycopersicon esculentum* Mill cv. Ailsa Craig. *Agrobacterium tumefaciens* strain LBA4404 was used to transform tomato stem explants as described in Bird et al. (1988). The constructs transformed were a) a full length *A. thaliana hmg1* cDNA under the control of the constitutive CaMV 35S promoter, b) full length tomato *dxs* cDNA in the antisense orientation under the control of the fruit ripening enhanced promoter, fibrillin, c) full length *E. coli dxs* cDNA fused to the putative tomato *dxs* transit sequence to facilitate plastid targeting under the fibrillin or CaMV 35S promoters. Kanamycin was used as a selection marker.

2.2 Molecular Analyses

Standard molecular biology techniques were used (Sambrook et al., 1989). Kanamycin resistant transformants were tested by PCR for the presence of the transgene. PCR-positive plants were then subjected to Southern blot analysis to determine the insertion number and RT-PCR was carried out on selected lines using primers specific to the transgene to confirm expression.

2.3 Biochemical Analyses

Proteins were separated by SDS/PAGE (Laemmli, 1970) then electroblotted onto PVDF membranes. Immunodetection of DXS was performed as in Fraser et al. (1994). HPLC analysis was according to Fraser et al. (2000). Chlorophylls and total pigments were determined spectroscopically (Wellburn, 1994; Schiedt and Liaaen-Jensen, 1995). Total phytosterols were extracted and analysed by GC-FID according to Fraser et al. (1995). HMGR and DXS activities were determined by monitoring the incorporation of [3-^{14}C]HMG-CoA into MVA and [2-^{14}C]pyruvate into DXP. IPP isomerase and phytoene synthase activities were determined as detailed in Fraser et al. (2002). *In vivo* labeling studies were carried out on uniform pericarp discs. Discs were incubated in 25mM Tris HCL pH7.5 buffer containing sorbitol 200mM, sucrose 10mM, KCl 50mM, succinic acid 5mM, DTT 1mM, EGTA 1mM, MgCl$_2$ 5mM, thiamine 0.1mM, D-glyceraldehyde 3-phosphate 4mM and 0.5µCi [2-^{14}C]pyruvate as substrate, total volume 250µl. Incubations were shaken at 28°C for 40h in dim light. Reaction products were separated and quantified by TLC.

3.   Results and Discussion

From 20 plants transformed with HMGR, 2 contained single insertions of the transgene (H3-1 and H3-6). Two multiple insert plants were also selected (H3-10, 2 inserts and H7-1, 4 inserts). RT-PCR showed that the transgene was expressed in all cases. Isoprenoids showed no change in leaf tissue or ripe fruit from single insert plants. Multiple insert plants had a 50% reduction in total fruit carotenoids, mostly due to a reduction in lycopene. GC-FID analysis of the phytosterols identified increases in all samples analyzed (Table1). The total level in single insert plants was increased by >2-fold compared to the wild type. Campesterol increased 5-fold, β-sitosterol 2.7-fold and stigmasterol 1.7-fold. The levels of cycloartenol were increased by less than 1.5-fold. H3-10 had a phytosterol content similar to the wild type. Total phytosterols were elevated 1.6-fold in line H7-10 but the sterol profile was altered. 20 progeny plants were generated from both single insert lines. Although RT-PCR analysis showed the transgene to be expressed, none of the lines showed significant increases in phytosterols and neither HMGR nor IPI activities were increased. However, this study has shown that elevation of phytosterols is possible in a crop plant without detrimental pleiotrophic effects or the accumulation of cycloartenol.

TABLE 1. Phytosterol content of ripe fruit from tomatoes transformed with an additional HMGR

Plant ID	Insert # or zygosity	Campesterol μg/g DW	Stigmasterol μg/g DW	β-Sitosterol μg/g DW	Cycloartenol μg/g DW
Wild Type	-----------------	8.78 ± 0.52	65.84 ± 0.21	29.60 ± 0.54	5.65 ± 0.23
H3-1	1	50.51 ± 2.92	126.43 ± 9.19	75.91 ± 9.02	7.74 ± 0.25
H3-6	1	44.04 ± 4.04	100.70 ± 7.67	87.17 ± 6.56	8.38 ± 1.08
H3-10	2	6.95 ± 3.54	75.15 ± 13.04	32.48 ± 3.91	5.75 ± 0.00
H7-1	4	18.52 ± 2.39	113.70 ± 5.81	32.99 ± 3.24	12.36 ± 0.00
H3-1/11	Homozygous	9.79 ± 0.17	51.08 ± 0.29	43.31 ± 3.85	13.55 ± 3.54
H3-1/12	Homozygous	15.21 ± 0.09	61.12 ± 2.24	80.03 ± 4.63	23.33 ± 0.86
H3-6/3	Homozygous	16.38 ± 0.19	57.88 ± 1.13	84.55 ± 1.51	17.82 ± 5.49
H3-6/13	Azygous	9.60 ± 0.62	41.02 ± 0.89	71.61 ± 3.53	16.07 ± 3.71

Transformation of tomato with the antisense *dxs* produced calli with limited leaf tissue but no shoots. Growth under different reduced light regimes and supplementation with erythrose or erythritol (1g/l) did not overcome this problem. Carotenoids were reduced by an order of magnitude. Two single insert plants from the fibrillin-*dxs* transformation were identified, Fdx-3 and Fdx-4. No single inserts of the CaMV 35S-*dxs* transformation were produced and so a plant containing 2 inserts was analyzed, Sdx-1. However, carotenoid analysis showed it to have a wild type phenotype. The carotenoid content of ripe fruit from Fdx-3 and Fdx-4 was increased 1.5-fold. RT-PCR showed expression of the transgene and Western blotting indicated that the protein was expressed and the transit sequence cleaved. The increase in carotenoids was inherited and phytoene accumulated to 2-fold that of the control (Table 2). Western blotting confirmed expression of the protein and DXS activity was increased 2-fold. PSY activity was also elevated, although IPI remained at the wild type level. *In vivo* labeling studies showed that incorporation from pyruvate into DXP and carotenoids was increased. This suggests that although DXS activity is essential for plant viability it is not the rate-limiting enzyme in carotenoid biosynthesis. Rather, as indicated by the labeling studies, availability of substrate for DXS is a limiting factor.

TABLE 2. Carotenoid content of ripe fruit from the T1 progeny of plants carrying a single insertion of the Fibrillin-*E. coli dxs*

Plant ID	Zygosity	Phytoene µg/g DW	Lycopene µg/g DW	β-Carotene µg/g DW	Total mg/g DW
Fdx-3/7	Azygous	1536.67 ±32.83	9770.00 ±641.33	790.00 ±49.33	6.60 ±0.07
Fdx-3/6	Homozygous	3185.00 ±45.00	10170.00 ±440.00	1450.00 ±20.00	9.29 ±0.07
Fdx-3/15	Homozygous	3516.67 ±56.08	13236.67 ±1125.13	1090.00 ±15.28	8.22 ±0.08
Fdx-3/19	Homozygous	2806.67 ±263.40	11936.67 ±1587.37	1240.00 ±34.29	7.37 ±0.71

4. References

Bird, C.R., Smith, C.J.S., Ray, J.A., Moreau, P., Bevan, M.W., Bird, A.S., Hughes, S., Morris, P.C., Grierson, D. and Schuch, W. (1988) Plant Mol. Biol. 11,651-662.

Fraser, P.D., Truesdale, M.R., Bird, C.R., Schuch, W. and Bramley, P.M. (1994) Plant Physiol. 105,405-413.

Fraser, P.D., Hedden, P., Cooke, D.T., Bird, C.R., Schuch, W. and Bramley, P.M. (1995) Planta. 196,321-326.

Fraser, P.D., Pinto, M.E.S., Holloway, D.E. and Bramley, P.M. (2000) Plant J. 24,1-10

Laemmli, U.K. (1970) Nature. 227,680-685.

Lois, L.M., Rodriguez-Concepcion, M., Gallego, F., Campos, N. and Boronat, A. (2000) Plant. J. 22,503-513.

Rohmer, M. (1999) Nat. Prod. Report. 16:565-574.

Sambrook, J., Fritsch, E.F. and Maniatis, T. (1989) Ed. Nolan, C. Cold Harbor Laboratory Press.

Schiedt, K and Liaaen-Jensen. (1995) Ed. Britton, G., Liaaen-Jensen, S. and Pfander, H. Birkhäuser.

Wellburn, A.R. (1994) J. Plant Physiol. 144,307-313.

5. Acknowledgements

We wish to thank Syngenta and the BBSRC for financial support.

CHARACTERIZATION OF THE MOLECULAR SPECIES OF PHYTOSTEROL FATTY ACYL ESTERS IN CORN (MAIZE)

ROBERT A. MOREAU[1], VIJAY SINGH[2], KAREN M. KOHOUT[1], AND KEVIN B. HICKS[1]

[1]Eastern Regional Research Center, ARS, USDA, 600 East Mermaid Lane, Wyndmoor, PA, 19038, USA

[2]Department of Agricultural Engineering, University of Illinois, Champaign, IL 61801, USA

Introduction

Previous attempts at separating nonpolar lipid esters (including wax esters, sterol esters, and methyl esters) have only achieved limited success (Moreau & Christie, 1999). We recently found that wax esters co-chromatograph with phytosteryl esters in our previously reported normal-phase HPLC method (Moreau et al, 1996). Several normal phase methods have attempted to improve this separation (Inger-Elfman-Börjesson & Härröd, 1997, El-Hamdy and Christie, 1993, and Foglia and Jones, 1997). Recently a report of a method employing an alumina column at 30°C, described promising results at separating these nonpolar lipid classes (Nordbäck and Lundberg, 1999). In the current study, modification of the alumina method by increasing the column

241

N. Murata et al. (eds.), Advanced Research on Plant Lipids, 241–244.

temperature to 75°C improved the separation of standards of wax esters and sterol esters.

Materials and Methods

Lipid standards were purchased from Sigma Chemical Co. Corn fiber oil (unrefined) was extracted as previously described (Moreau et al, 1996). All HPLC analyses were performed with a Hewlett Packard Model 1100 HPLC, with autosampler, column heater, and detection by both an HP Model 1100 diode-array UV-visible detector (Agilent Technologies, Collegeville, PA), and a Sedex Model 55 Evaporative Light Scattering Detector (Richard Scientific, Modesto, CA), operated at 30°C and a nitrogen gas pressure of 2 bars. For alumina column separations, the column was an Aluspher AL 100, 5 micron column (125 H 4 mm) packed in a LiChroCART cartridge (Merck KgaA, Darmstadt, Germany). The binary gradient had a constant flow rate of 0.6 ml/min, with Solvent A = hexane/tetrahydrofuran, 1000/1, Solvent B = isopropanol. Gradient timetable: at 0 min, 100% A/O% B; at 10 min, 100%A/ 0% B; at 20 min, 95%A/ 5% B, at 21 min, 100% A/ 0% B, at 60 min, 100 %A/ 0 %B.

Results and Discussion

Standards of hydrocarbon (squalene), wax ester (stearyl stearate), and sterol ester (cholesterol stearate) were separated at a column temperature of 25°C using an Aluspher column and a gradient method very similar to that reported by Nordbäck and Lundburg (1999). However, a fourth standard, methyl stearate, was also included in this mixture, and it appeared as a shoulder that preceded the peak of steryl ester. Increasing the column temperature to 50°C enhanced the separation of all four components, and further increasing to 75°C

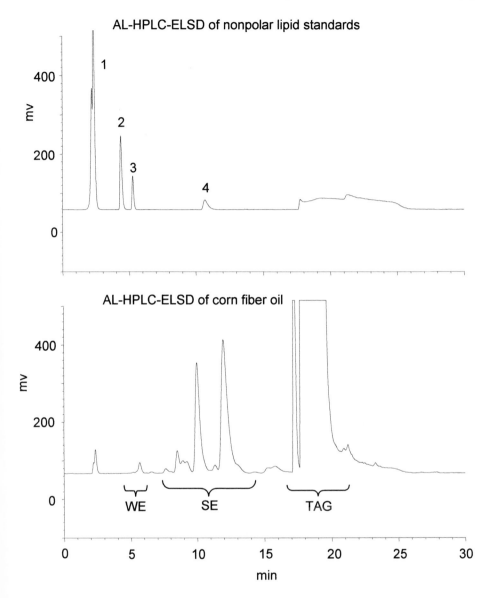

Figure 1. HPLC-ELSD Chromatogram obtained using the alumina gradient method with a column temperature of 75 °C. A. Injection of a standard mixture of 1 μg each of hydrocarbon (squalene, 1), wax ester (stearyl stearate, 2), fatty acid methyl ester (methyl stearate, 30, and phytosterol fatty acyl ester (cholesterol stearate, 4). B. Injection of μg 165 unrefined corn fiber oil. Abbreviations: WE, wax esters; SE, phytosteryl esters; and TAG, triacylglycerols.

further enhanced the separation, and selectively increased the retention time of cholesterol-stearate from about 6 to about 9 minutes (Figure 1A).

Injection of corn fiber oil in this same AL-HPLC system with a column temperature of 75°C (Figure 1B), resulted in major peaks at 10 and 12 minutes, and several minor peaks at 7-9 minutes. The peaks in this 7-12 minute region appear to be molecular species of phytosteryl esters. A small peak at about 5 minutes co-chromatographed with the wax ester standard.

With the increased interest in phytosterol- and phytostanol-ester enriched functional foods (Hicks & Moreau, 2001), this method should provide an additional valuable technique to characterize and compare these products.

References

El-Hamdy, A.H., and. Christie, W.W. (1993) Separation of Non-polar Lipids by High Performance Liquid Chromatography on a Cyanopropyl Column, J. High Resol. Chromatogr, 16,55-57.

Foglia, T.S., and Jones, K.C. (1997) Quantitation of Neutral Lipid Mixtures using High Performance Liquid Chromatography with Light Scattering Detection. J. Liq. Chrom & Rel. Technol. 20,1829-1838.

Hicks, K.B., Moreau, and R.A. (2001) Phytosterols and Phytostanols: Functional Food Cholesterol Busters. Food Technology 55,63-67.

Inger-Elfman-Börjesson, I., and M. Härröd, Analysis of Non-polar Lipids by HPLC on a Diol Column, J. High Resol. Chromatogr, 20:516-518 (1997).

Moreau, R.A., and Christie, W.W. (1999) The Impact of the Evaporative Light Scattering Detector on Lipid Research, International News on Fats, Oils and Related Materials 10,471-478.

Moreau, R.A., Powell, M.J., and Hicks, K.B. (1996) Extraction and Quantitative Analysis of Oil from Commercial Corn Fiber, J. Agric. Food Chem. 44,2149-2154.

Nordbäck, J., and Lundburg, E. (1999) High Resolution Separation of Non-polar Lipid Classes by HPLC-ELSD using Alumina as Stationary Phase, J. High Resol. Chromatogr. 22,483-486.

ACTION OF MEVALONIC ACID-DERIVED ISOPRENOIDS IN ARABIDOPSIS

N. NAGATA, M. SUZUKI, S. YOSHIDA and T. MURANAKA
Laboratory for Biochemical Resources, Plant Science Center, RIKEN,
2-1 Hirosawa, Wako, Saitama 351-0198, Japan

1. Abstract

Mevalonic acid (MVA) is a key intermediate leading to the synthesis of a large number of isoprenoid compounds including plant hormones such as brassinolide (BL) and zeatin (ZA)(Fig. 1). However, Isopentenyl diphosphate (IPP) is the product of two independent synthetic pathways in higher plants: the mevalonate (MVA) pathway in the cytoplasm, and the non-mevalonate 2-C-methyl-D-erythritol-4-phosphate (MEP) pathway in plastids. Whether IPP or IPP-derived products undergo exchange between the cytoplasm and plastids has previously been unresolved. Here, we demonstrate that IPP or IPP-derived products flow from the MVA pathway to the MEP pathway, and that MVA-derived products contribute to formation of functional products used in plastid development, in both light-grown and dark-grown seedlings of Arabidopsis. Furthermore, we showed a possibility that MVA flows to gibberellins (GA) in addition to BL, but not to ZA, by examinations of cell elongation or endoreduplication.

2. Introduction

The classical MVA pathway has been the subject of extensive research for many years, whereas the MEP pathway was more recently discovered. The remaining gaps in the sequence of enzymatic steps in the *E. coli* MEP pathway were filled quite recently (Rohdich et al. 2002). In higher plant cells, the compartmentalization of IPP synthesis such that cytosolic and plastidic IPP are synthesized *via* different pathways leads to compartmentalization of isoprenoid synthesis: sterols and brassinosteroids are formed in the cytoplasm, whereas carotenoids, phytol, abscisic acid and GA intermediates, and chlorophyll side chains are synthesized inside the plastid compartment (Fig. 1)(Eisenreich et al. 1998; Lichtenthaler 1999). On the other hand, Cytokinins are thought to be synthesized de novo from 5'-AMP and IPP's isomer dimethylallyl diphosphate (DMAPP), but their biosynthesis has not been established.

N. Murata et al. (eds.), Advanced Research on Plant Lipids, 245–248.
© 2003 *Kluwer Academic Publishers. Printed in the Netherlands.*

The question of whether IPP pools can be exchanged between the cytoplasm and plastids has proved difficult to answer conclusively. Feeding experiments with labeled ^{14}C- or ^{13}C-MVA have been resulted in labeling only of the cytosolic sterols, saponins, and brassinosteroids. On the other hand, sparsely labeled plastidic isoprenoids were reported to appear in some plants following feeding of labeled MVA. Furthermore, the feeding experiments with labeled compounds, although a basic and powerful technique in metabolic studies, are limited in their ability to clarify whether the presence of a labeled product reflects efficient synthesis of functional isoprenoids that are relevant in plant development.

The rate-limiting step in the MVA pathway is generally regarded to be the reduction of 3-hydroxy-3-methylglutaryl-coenzyme A (HMG-CoA) by HMG-CoA reductase. Lovastatin, also referred to as mevinolin, and its natural analogues are highly potent competitive inhibitors of HMG-CoA reductase, not only in animals and microbes but also in plants. On the other hand, the first enzyme of the MEP pathway is 1-deoxy-D-xylulose-5-phosphate (DX) synthase (Estévez et al. 2001). To examine the possible exchange of products between the two pathways by using an approach that differs from those of previous studies, we used the lovastatin-treated Arabidosis as a model for examining the effects of MVA pathway deficiency and the Arabidopsis mutant *cla1-1*, a null mutant of the gene for DX synthase as a model of MEP pathway deficiency.

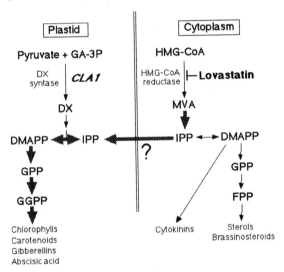

Figure 1. Pathways of isoprenoid biosynthesis in higher plants

DMAPP, Dimethylallyl diphosphate; DX, 1-deoxy-D-xylulose-5-phosphate; FPP, farnesyl diphosphate; GA-3P, glyceraldehyde 3-phosphate; GPP, geranyl diphosphate; GGPP, geranylgeranyl diphosphate; HMG-CoA, 3-hydroxy-3-methylglutaryl-coenzyme A; IPP, isopentenyl diphosphate; MVA, mevalonic acid

3. Results and Discussion

3.1. *MVA restores plastid development in cla1 mutant*

We applied 1 μM lovastatin to Arabidopsis seeds cultured in liquid MS medium and cultured under light (16-h light/8-h dark photoperiod) or continuous dark. For both the light- and the dark-grown plants, the lovastatin caused wild-type cultures to develop a dwarf phenotype exhibiting short hypocotyls and roots and contracted cotyledons. In contrast, the *cla1-1* plants exhibited a severe albino phenotype in addition to growth inhibition and contracted cotyledons.

Upon MVA treatment, the white cotyledons of both the light- and the dark-grown *cla1-1* mutant plants became pale-yellow (Fig. 2). Furthermore, the dwarf phenotype and contracted cotyledons of light-grown *cla1-1* mutants were partially reversed. Fluorescence microscopy revealed that application of MVA to *cla1-1* plants in the light induced significant recovery of chlorophyll autofluorescence. Electron microscopy showed MVA to *cla1-1* plants partially recovered formation of grana by stacking of thylakoid membranes in chloroplasts in the light and formation of crystalline prolamellar bodies and plastoglobuli in etioplasts in the dark. The most likely candidates for the MVA-derived compounds involved in these plastid development processes are chlorophyll in the chloroplasts and carotenoids in the etioplasts. Thus, the flow of MVA-derived compounds to the MEP pathway, and the subsequent formation of functional products from these compounds, occurs in both light- and dark-grown plants.

Furthermore, the application of lovastatin to *cla1-1* plants induced a severe dwarf phenotype, caused depigmentation of cotyledons and completely blocked chlorophyll synthesis. Initially, *cla1-1* plants seemed to have a limited ability to synthesize small amounts of chlorophyll. An explanation for these observations is that MVA-derived products constantly flow from the MVA pathway to the MEP pathway in limited amounts and this compensation mechanism may allow limited growth of lovastatin-treated wild-type plants (MVA pathway deficient) and *cla1-1* plants (MEP pathway deficient).

3.2. *BL and GA restore inhibition of cell elongation or endoreduplication of lovastatin-treated Arabidopsis*

Lovastatin-treatment induces inhibition of cell elongation. The application of BL or GA restored cell size of hypocotyls of lovastatin-treated plants in the light, while ZA restored none of all. Although also dark-grown lovastatin-treated plants exhibit short hypocotyl phenotype, application of BL, GA, ZA or combination of these hormones did not restore hypocotyl length. We found that lovastatin-treatment induces inhibition of endoreduplication of hypocotyls of dark-grown Arabidopsis. Application of BL or GA canceled the inhibition of endoreduplication. In contrast, ZA-treatment was not effective to endoreduplication level. In these contexts, it may be a possible assumption that MVA flows to GA in addition to BL, and do not flows to ZA.

248

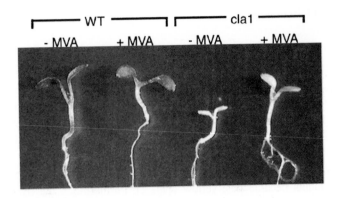

Figure 2. Morphological recovery of *cla1-1* plants by MVA-treatment

4. Aknowledgements

We would like to thank Professor Patricia León (Universidad Nacional Autónoma de México, México) for providing seeds of *cla1-1* mutant. This work was supported by the Ministry of Education, Culture, Sports, Science and Technology of Japan (Grant-in-Aid no. 13740475).

5. References

Eisenreich, W., Schwarz, M., Cartayrade, A., Arigoni, D., Zenk, M.H. and Bacher, A. (1998) The deoxyxylulose phosphate pathway of terpenoid biosynthesis in plants and microorganisms. Chem. Biol. 5, R221-R233.

Estévez, J.M., Cantero, A., Reindl, A., Reichler, S., León, P. (2001) 1-Deoxy-D-xylulose-5-phosphate synthase, a limiting enzyme for plastidic isoprenoid biosynthesis in plants. J. Biol. Chem. 276, 22901-22909.

Lichtenthaler, H.K. (1999) The 1-deoxy-D-xylulose-5-phosphate pathway of isoprenoid biosynthesis in plants. Ann. Rev. Plant Mol. Biol. 50, 47-65.

Rohdich, F., Hecht, S., Gärtner, K., Adam, P., Krieger, C., Amslinger, S., Arigoni, D., Bacher, A., Eisenreich, W. (2002) Studies on the nonmevalonate terpene biosynthetic pathway: metabolic role of IspH (LtB) protein. Proc. Natl. Acad. Sci. USA 99, 1158-1163.

THE STUDY OF THE MECHANISM OF CELL ELONGATION USING AN INHIBITOR OF 3-HYDROXY-3-METHYLGLUTARYL-COA REDUCTASE

M. SUZUKI, N. NAGATA, Y. KAMIDE, S. YOSHIDA AND T. MURANAKA
Laboratory for Biochemical Resources, Plant Science Center, RIKEN
2-1, Hirosawa, Wako, Saitama, 351-0198, Japan

1.   Abstract

We are investigating the role of 3-hydroxy-3-methylglutaryl-CoA reductase on the development of plant body by using the HMGR-inhibitor, lovastatin. Then, it was shown that lovastatin inhibits cell elongation of *Arabidopsis thaliana*. DNA microarray to examine the alteration of gene expression with or without lovastatin comprehensively was performed in order to investigate what kinds of genes cause the inhibition of cell elongation. As a result, it was found that 38 and 20 genes were down-regulated by lovastatin under light and dark condition respectively, while few genes were up-regulated. However, the alteration of the expression of genes involved in the biosynthesis of brassinosteroids or gibbellins is not detected by DNA microarray. Then, RT-PCR to investigate the time course expression of these genes after lovastatin treatment was performed. As a result, it was found that the expression level of the *GA2* gene increased at 4h under light condition and of the *GA5* gene decreased at 4h and 12h under light condition. The expression level of these genes did not alter under dark condition. These data suggests that lovastatin treatment cause some influence on the gene expression in MEP pathway.

2.   Introduction

Plant isoprenoids are synthesized through mevalonate (MVA) pathway in cytoplasm or through 2-*C*-methyl-D-erythritol-4-phosphate (MEP) pathway in plastid. Many steroid compounds including brassinosteroids (BR) are synthesized through the MVA pathway via isopentenyl diphosphate (IPP). 3-hydroxy-3-methylglutaryl-CoA reductase (HMGR) is a key enzyme in the MVA pathway. On the other hand, IPP is also synthesized through MEP pathway in plastid. Chlorophylls, carotenoids and some phytohormones (gibberellins (GA) and abscisic acids) are

N. Murata et al. (eds.), Advanced Research on Plant Lipids, 249–252.
© 2003 *Kluwer Academic Publishers. Printed in the Netherlands.*

synthesized in this pathway.

It had been shown that the inhibition of HMGR affects cell division of tobacco BY-2 cells and reduce contents of cytokinins in BY-2 cells (Crowell et al. 1992, Laureys et al. 1998). Therefore, it was understood that the inhibition of HMGR lead to the inhibition of biosynthesis of cytokinins. However the influence of HMGR-inhibitors on the growth of whole plant has not been well understood. Here, we examined how the treatment of HMGR-inhibitor, lovastatin, affects the growth of *Arabidopsis thaliana* plants.

3. Results and Discussion

Arabidopsis seedling grown with lovastatin for 6 days showed dwarf phenotype. Under light condition, lovastatin inhibits the elongation of hypocotyl and roots, enlargement of cotyledons, and development of the inflorescence. Under dark condition, lovastatin inhibits the elongation of hypocotyl and roots. This means that the inhibition of HMGR cause the inhibition of cell elongation.

Then, to investigate the mechanism of the inhibition of cell elongation by lovastatin, the alteration of gene expression with or without 1μM lovastatin was analyzed by DNA microarray (GeneChip). Total RNA was extracted from 6 days-old seedling after 12h incubation with or without lovastatin. Up-regulated genes are shown in the Table. 1. Seven and 15 genes were up-regulated under light and dark condition respectively. No genes were commonly up-regulated between light and dark.

TABLE 1. A list of lovastatin up-regulated genes.

Light	dark
Phytohormones	**Phytohormones**
ethylene responsive element binding factor	GAST1
ACC synthase	IAA5
Cell wall	auxin-induced protein like protein
putative endchinase	transport inhibitor response 1
peroxidase	**Cell wall**
putative peroxidase	putative extensin
peroxidase ATP23a	**Transcription**
Unknown/unclassified	DNA-binding protein
glutathione transferase	AtMYB8
hypothetical protein	**Stress response/disease**
	RD29B
	putative disease resistance protein
	Unknown/unclassified
	glutamine-dependent asparagine synthase
	VBVXE09 3
	similar to HSR201
	tapetum-specific A9
	putative non-LTR retroelement Rtase
	hypothetical protein

Down-regulated genes are shown in the Table. 2. Thirtyeight and 20 genes were down-regulated under light and dark condition respectively. Some genes (shown by *italic*) were commonly down-regulated between light and dark. Many of these genes encode cell wall protein and peroxidase. Therefore, these genes may regulate cell elongation directly.

TABLE 2. A list of lovastatin down-regulated genes.

Light	dark
Phytohormones	
jasmonate inducible protein isolog	**Cell wall**
Cell wall	*arabinogalactan-protein*
arabinogalactan-protein	*xyloglucan endtransglycosylase*
xyloglucan endtransglycosylase	*extensin-like protein*
extensin-like protein	*extensin-like protein*
extensin-like protein	**Transcription**
β-expansin	AP2 domain containing protein
putative pectinesterase protein	putative transcription factor
putative pectine methyesterase	putative transcription factor
putative extensin with leusine-rich repeat	**Peroxidase**
Transcription	*peroxidase ATP11a*
CALIFLOWER	*putative peroxidase*
Peroxidase	*putative peroxidase*
peroxidase ATP11a	*prxr7*
putative peroxidase	*prxr11*
putative peroxidase	**Signal transduction**
prxr7	similar to LIM17
prxr11	**Metabolism**
prxr4	putative glycerol-3-phosphate permease
peroxidase ATP21a	similar to mandelonitril lyase
P450	**Unknown/unclassified**
putative cytochrome P450 protein	*extA*
putative cytochrome P450 protein	similar to actin binding protein
Stress response/diseasse	CAF protein
Drought-inducible cystein proteinase RD21a	*unknown*
putative disease resistance response protein	
putative disease resistance response protein	
nodulin-26-like protein	
Signal transduction	
member of blue copper protein family	
receptor protein kinase like	
similar to copper-binding protein	
Unknown/unclassified	
extA	
pEARLI 1-like protein	
NWMU4-2s albumine 4 precursor	
gemine-like protein	
unknown	
unknown	
unknown	
hypothetical protein	
hypothetical protein	
putative protein	

Unexpectedly, no genes involved in the biosynthesis of BR or GA exist in this list. Therefore, the time course expression of genes in MVA and MEP pathways were analyzed by RT-PCR. Only the *GA2* (Yamaguchi et al. 1998) and *GA5* (Xu et al. 1995) genes expression was affected by lovastatin under light condition (data not shown). The *GA2* gene expression was activated after 4h from lovastatin treatment. The *GA5* gene expression was repressed after 4h and 12h from lovastatin treatment. These genes expression were not affected by lovastatin under dark condition. This means that lovastatin treatment cause some influence on the gene expression in MEP pathway.

4. Conclusion and perspective

The inhibition of HMGR by lovastatin cause the inhibition of cell elongation under light and dark condition. Then, lovastatin regulated genes were identified by DNA microarray. No genes involved in the biosynthesis of BR and GA exist in the list. However, it was shown that lovastatin affects some gene expression in MEP pathway under only light condition. These results suggest that signal transduction may exist from MVA pathway to MEP pathway and different mechanisms to elongate cells exist between under light and dark condition and.

Two genes encoding HMGR (*HMG1* and *HMG2* (Learned et al. 1989, Caelles et al. 1989)) exist in the Arabidopsis genome. To elucidate the regulatory mechanism of MVA pathway, the knowledge of the function of each gene should be well understood. Therefore, we are screening the KO mutant of the *HMG1* and *HMG2* gene. Now, we obtain three lines of *hmg1* mutant and one line of *hmg2* mutant. Near future, the role of HMGR on the regulation of MVA pathway and the relationship between MVA and MEP pathway will be elucidated.

5. References

Caelles, C., Ferrer, A., Balcells, L., Hegardt, F.G. and Boronat, A. (1989) Isolation and characterization of a cDNA encoding *Arabidopsis thaliana* 3-hydroxy-3-methylglutaryl coenzyme A reductase. Plant Mol. Biol. 13, 627-638.

Crowells, N.D. and Salaz, S.M. (1992) Inhibition of growth of cultured tobacco cells at low concentrations of lovastatin is reversed by cytokinin. Plant Physiol. 100, 2090-2095.

Laureys, F., Dewitte, W., Witters, E., Montagu, V.M., Inze, D. and Onckelen, V.H. (1998) Zeatin is indispensable for G$_2$-M transition in tobacco BY-2 cells. FEBS L. 426, 29-32.

Learned, R.M. and Fink, G.R. (1989) 3-Hydroxy-3-methylglutaryl-coenzyme A reductase from *Arabidopsis thaliana* is structurally distinct from the yeast and animal enzymes. Proc. Natl. Acad. Sci. USA 86, 2779-2783.

Xu, Y.L., Li, L., Wu, K., Peeters, A.J., Gage, D.A. and Zeevaart, J.A. (1995) The GA5 locus of *Arabidopsis thaliana* encodes a multifunctional gibberellin 20-oxidase: molecular cloning and functional expression. Proc. Natl. Acad. Sci. USA 92, 6640-6644.

Yamaguchi, S., Sun, T., Kawaide, H. and Kamiya, Y. (1998) The GA2 locus of *Arabidopsis thaliana* encodes *ent*-kaurene synthase of gibberellin biosynthesis. Plant Physiol. 116, 1271-1278.

CHARACTERISATION OF ENZYMES INVOLVED IN TOCOPHEROL BIOSYNTHESIS

R. SADRE, H. PAUS, M. FRENTZEN and D. WEIER
RWTH Aachen, Institute for Biology I, Worringerweg 1, 52056 Aachen, Germany

1. Introduction

Tocopherols are lipid-soluble antioxidants collectively known as vitamin E. By scavenging singlet oxygen and free radicals they have an important role in stabilising unsaturated fatty acids and in serving as nutraceuticals for human nutrition.

Tocopherols are synthesised only in photosynthetic organisms. The pathway requires the flux of intermediates from the isoprenoid as well as the aromatic compound synthesis. The committed step in this pathway is catalysed by a membrane bound prenyltransferase transferring a prenyl group from phytyl pyrophosphate (phytyl-PP) to homogentisate, so that 2-methyl-6-phytylplastoquinol is formed. The chromanol headgroup is subsequently formed by the activity of a tocopherol cyclase.

2. Experimental

Synechocystis sp. PCC6803 cultures were maintained according to Castenholz (1988). The open reading frames (ORF) *slr1736* and *slr1737* were amplified from *Synechocystis* genomic DNA and used to create *Synechocystis* null mutants by homologous recombination of the chloramphenicol cassette-disrupted genes into the *slr1736* and *slr1737* wild-type locus, respectively, as described by Williams et al. (1988).

Synechocystis lipids were extracted and utilised for HPLC analysis on a normal phase column with 98,5% hexane/1,5% isopropanol as mobile phase. Tocopherols were detected by fluorescence using 290 nm excitation and 325 nm emission and identified by cochromatography with standards.

N. Murata et al. (eds.), Advanced Research on Plant Lipids, 253–256.
© 2003 *Kluwer Academic Publishers. Printed in the Netherlands.*

For expression studies the ORFs *slr1736* and *slr1737* were cloned into an expression vector and transformed into *E. coli*. After 2 h induction with isopropyl-β-D-thiogalactopyranoside, *E. coli* cells were harvested, resuspended in buffer (25 mM Bis-Tris Propane-HCl, pH 7.0, 5 mM DTT, 200 μM Pefabloc) and disrupted by sonification. The 5000×g supernatant was then used for separating membranes and soluble proteins by ultracentrifugation at 150,000×g for 1h.

Standard prenyltransferase assays contained in a total volume of 100 μl 25 mM Bis-Tris Propane-HCl buffer, pH 7.0, 0.5 μM [³H]-homogentisate, 0.4 μM phytyl-PP, 1 mM $MgCl_2$ and *E. coli* membrane fractions (up to 4 μg protein). After 20 min at 25°C lipophilic products were extracted, separated by TLC in petrolether/diethylether (10:1, v/v), visualized with a Bioimager FLA 3000, identified by cochromatography with authentic lipids and the labelling was quantified by scintillation counting.

3. Results and Discussion

3.1. Identification of genes involved in tocopherol biosynthesis

Database searches were carried out to identify genes encoding enzymatic activities involved in tocopherol biosynthesis from *Synechocystis* sp. PCC6803. The ORFs *slr1736* and *slr1737* representing a putative polyprenyltransferase and a putative tocopherol cyclase were cloned. To verify that the identified ORFs are required for tocopherol biosynthesis, null mutants were generated. The tocopherol levels of the wild-type and the mutants were then analysed by HPLC.

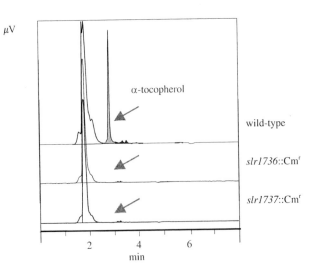

Figure 1. HPLC chromatographic analysis of lipid extracts from wild-type *Synechocystis* and the two mutants *slr1736*::Cm^r and *slr1737*::Cm^r.

Synechocystis wild-type contains predominantly α-tocopherol while β- and γ-tocopherol were detectable in trace amounts only (about 1% of total). In contrast to the wild-type, the *slr1736*::Cm[r] and *slr1737*::Cm[r] mutants were found to be devoid of tocopherol (Fig.1) consistent with the assumption that the respective ORFs encode enzymes involved in tocopherol synthesis.

Since *slr1736* and *slr1737* are directly contiguous ORFs in the *Synechocystis* genome, prenyltransferase activities were determined in *Synechocystis* wild-type and the respective mutants by measuring the incorporation rates of [³H]-homogentisate into chloroform soluble products using phytyl-PP as prenyl donor.

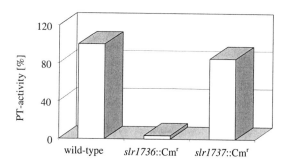

Figure 2. Prenyltransferase (PT) activity of wild-type *Synechocystis* and the two mutants *slr1736*::Cm[r] and *slr1737*::Cm[r]. Enzyme assays were carried out with membrane fractions isolated from *Synechocystis* cells.

The enzymic assays showed that the tocopherol deficient phenotype of the *slr1736* null mutant is due to the inactivation of the prenyltransferase. These data coincide with recent publications (Collakova and DellaPenna, 2001; Schledz et al., 2001, Savidge et al., 2002). On the other hand, the *slr1737* null mutant displayed prenyltransferase activity similar to the wild-type (Fig. 2). Hence the phenotype of the *slr1737* null mutant depends on another enzymic activity, namely a tocopherol cyclase. This is consistent with recent data of Subramaniam et al. (2001) demonstrating that the intermediate 2,3-dimethyl-5-phytylplastoquinol accumulated in a *slr1737* null mutant.

3.2. Expression of slr1736 *and* slr1737 *in E. coli*

The *Synechocystis* ORFs *slr1736* and *slr1737* from *Synechocystis* were cloned and expressed in *E. coli*. Our analyses showed that *slr1736* encodes a prenyltransferase located in the bacterial membranes. In contrast to *slr1736*, *slr1737* was expressed in *E. coli* primarily as a soluble protein. These subcellular localisations correlate with the hydropathy profiles of the two proteins. We are currently purifying the cyclase as recombinant His-tagged protein for further protein and enzyme studies.

The prenyltransferase of *Synechocystis* expressed in *E. coli* showed highest activities at pH 7, required divalent cations for activity and was more active with Mg^{2+} than with Mn^{2+}. The prenylation rates in dependence on the substrate concentrations gave highest enzymatic activity at about 0.5 μM homogentisate (Figure 3A) and 0.4

256

μM phytyl-PP (Figure 3B). The apparent K_M value for phytyl-PP was determined at about 0.1 μM.

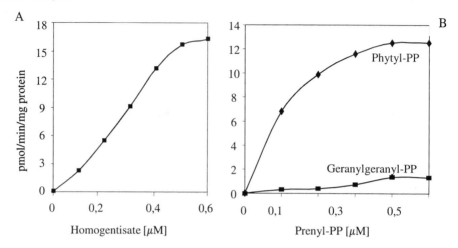

Figure 3. Prenylation rates as a function of the concentrations of homogentisate using 0.5 μM phytyl-PP as prenyl donor (A). Incorporation rates of [^{3}H]-homogentisate (0.5μM) as a function of the concentrations of phytyl-PP and geranylgeranyl-PP (B).

As depicted in Figure 3B, the prenyltransferase shows a pronounced specificity for phytyl-PP in comparison to geranylgeranyl-PP. This prenyl specificity is consistent with the vitamin E composition of *Synechocystis* lacking tocotrienols (Fig.1) and the recent data of Collakova and DellaPenna (2001). Hence, the *Synechocystis* prenyltransferase *slr1736* represents the homogentisate phytyltransferase involved in tocopherol biosynthesis.

Acknowledgements

This project is supported by the Bundesministerium für Bildung, Wissenschaft, Forschung und Technologie (Förderkennzeichen 0312252G/6)

4. References

Castenholz, R. W. (1988) Methods in Enzymology, 167, 68-92

Collakova, E. and DellaPenna, D. (2001) Isolation and Functional Analysis of Homogentisate Phytyltransferase from *Synechocystis* sp. PCC 6803 and Arabidopsis, Plant Physiology, 127, 1113-1124

Savidge, B., Weiss, J. D., Wong, Y.-H. H., Lassner, M. W., Mitsky, T. A., Shewmaker, C. K., Post-Beitenmiller, D. and Valentin, H. E. (2002) Isolation and Characterization of Homogentisate Phytyltransferase Genes from *Synechocystis* sp. PCC 6803 and Arabidopsis, Plant Physiology, 129, 1-12

Schledz, M., Seidler, A., Beyer, P. and Neuhaus G. (2001) A novel phytyltransferase from *Synechocystis* sp. PCC 6803 involved in tocopherol biosynthesis., FEBS Letters, 499, 15-20

Subramaniam et al. (2001) WO 01/79472 patent application.

Williams J. G. K. (1988) Methods in Enzymology, 167, 766-778

Chapter 7:

Lipase and Lipid Catabolism

TWO NOVEL TYPES OF ARABIDOPSIS PHOSPHOLIPASE D
Oleate-stimulated PLDδ and Ca²⁺-independent PLDζ1

CHUNBO QIN,[1] WEIQI LI,[1] YUEYUN HONG,[1] WENHUA ZHANG,[1]
TARA WOOD,[1] MAOYIN LI,[2] RUTH WELTI[2], XUEMIN WANG,[1]
[1]*Department of Biochemistry,* [2]*Division of Biology, Kansas State University*
Manhattan, KS 66506, USA Fax: (785) 532-7278; E-mail: wangs@ksu.edu

1. Introduction

Phospholipase D (PLD) is often the most prevalent lipolytic enzyme in plant tissues. PLD-mediated hydrolysis may lead to production of signaling messengers, alteration of membrane lipid composition, or membrane degradation. This enzyme is subjected to complex control in the cell to achieve diverse cellular functions. In Arabidopsis, 12 *PLD* genes are identified; they are grouped tentatively into five classes, *PLDα(1,2,3,4)*, *β(1,2)*, *γ(1,2,3)*, *δ*, and *ζ(1,2)* based on gene architecture, sequence similarity (Fig. 1A), domain structure (Fig.1B), biochemical properties, and the order of cDNA cloning (Qin and Wang, 2002). In addition, two *PLDδ* cDNA variants, δa and δb, have been cloned, which may result from alternative splicing (Gardiner et al, 2001; Katagiri et al, 2001; Wang and Wang, 2001). Thus, the total number of PLD isoforms in Arabidopsis is greater than 12.

Analysis of the PLD sequences yields two evolutionarily divergent groups (Qin and Wang, 2002). The major group that includes PLDαs, βs, γs, and δ is characterized by having a C2 domain, a Ca²⁺-dependent phospholipid binding structural fold. PLDζs constitute the other group of PLDs that have Phox homology (PX) and pleckstrin homology (PH), but not C2, domains. The PX domain was originally identified in the protein p47[phox], a component of phagocyte NADPH oxidase. This protein module contains a conserved proline-rich motif and can bind both phosphoinositides and proteins with the SH3 domain. PH domains are composed of an approximately 120 amino acid sequence found in more than 100 proteins. PH domains bind to phosphoinositides, but the specificity may vary.

Three types of PLD, α, β, and γ, have been characterized previously in Arabidopsis. PLDα is the most common plant PLD, and its activity is independent of phosphatidylinositol 4,5-bisphosphate (PIP₂) when assayed at millimolar concentrations of Ca²⁺. PLDβ and PLDγ are PIP₂-dependent and most active at micromolar levels of Ca²⁺ (Wang, 2000). Two new types of PLDs, PLDδ and ζ1, were identified in the past year. The oleate-stimulated PLDδ is associated with plasma membrane (Wang and Wang, 2001) and binds to microtubules (Gardiner et al, 2001). The PX/PH containing PLDζ1 does not require divalent cations for activity (Qin and Wang, 2002). Some of the major biological and catalytic properties of the new PLDs will be briefly described in this paper.

N. Murata et al. (eds.), Advanced Research on Plant Lipids, 259–262.

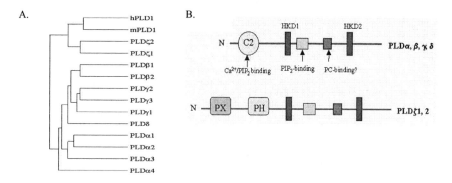

Figure 1. Sequence similarities and domain structures of Arabidopsis PLDs. A. Grouping of PLD using a dendrogram of the catalytic HKD sequences generated by the PILEUP program in GCG. hPLD, human PLD; mPLD, mouse PLD. B. Major domains of the C2 PLDα, β, γ, and δ and the PX/PH PLDζs. All PLDs have conserved the catalytic, duplicated HKD motifs (HxKxxxxD) and a putative PC-binding, "IYIENQFF" motif.

2. Oleate-stimulated PLDδ

2.1. *Oleate-stimulated activity is tightly associated with the plasma membrane*

Oleate-stimulated PLD activity was detected in the microsomal fraction, but not in the soluble fraction of Arabidopsis leaves (Wang and Wang 2001). This PLD activity did not result from any of the previously characterized isoforms, PLDα, β or γ (Fig. 2A). A distinct Arabidopsis PLD cDNA, designated as PLDδ was identified and active protein was expressed in *E. coli*. The purified PLDδ showed no activity toward phosphatidylcholine (PC) in the absence of oleic acid, but became highly active in the presence of oleic acid (Fig. 2A). The oleate activation of PLDδ was dose-dependent, with the maximal stimulation at 0.5 mM. Other unsaturated fatty acids, such as linoleic and linolenic acids, increased the PLD activity to a lesser extent (50%) than oleic acid. In contrast, the saturated fatty acids, palmitic and stearic acids, had no stimulatory effect (Wang and Wang, 2001). These results indicated that the oleate stimulation was not caused by oleate acting as a nonspecific surfactant in the assay. Ca^{2+} was required for oleate-activated PLD activity, and the optimal Ca^{2+} concentration was approximately 100 µM. Immunoblotting with PLDδ-specific antibody indicated PLDδ was located only in the plasma membrane fractions of Arabidopsis, which is in agreement with the distribution of the oleate-activated PLD activity.

2.2. *Oleate promotes PC binding*

To gain insight into the mechanism of activation of PLDδ by oleate, lipid binding assays and site-directed mutagenesis were carried out (Wang and Wang, 2001). Without oleic acid, no PC binding to PLDδ was detected, but inclusion of oleate greatly stimulated PC binding. In contrast, substitution of oleate with stearate produced no stimulation of PC binding. In addition, a mutation at R399 decreased oleate binding by PLDδ and resulted in an inactive PLDδ. The oleate-promoted binding of PC by PLDδ may be relevant to the association of this enzyme with cellular membranes and its substrates.

2.3. *PLDδ is expressed highly in senescent tissues*

PLDδ protein was detectable in all tissues examined by immunoblotting, and its amounts relative to total proteins in roots, flowers, and stems were higher than those in leaves and siliques. The amount of PLDδ was greater in senescent than young leaves. Oleate-activated PLD activity was distributed in the same pattern as the PLDδ protein. RNA blotting with a *PLDδ*-gene specific probe showed that *PLDδ* mRNA levels were much higher in old leaves, stems, flowers, and roots than in young leaves and siliques (Wang and Wang, 2001). Because free fatty acid increases are characteristic of leaf senescence, the high oleate-activated PLDδ protein and mRNA levels in old leaves raise interesting questions about the role of PLDδ in leaf development and senescence.

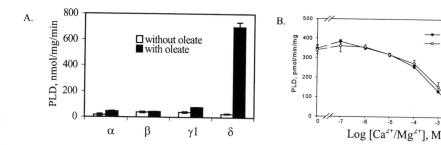

Figure 2. A. Oleate-activated PLD activities of PLDα, β, γ, and δ. PLDs were expressed in *E. coli* from the cDNAs of Arabidopsis PLDβ, γ1, and δ and castor bean PLDα. B. Divalent cation-independent activity of PLDζ1. 2 mM EGTA and 2 mM EDTA were added to the reactions with no Ca^{2+} or Mg^{2+}.

3. Ca^{2+}-independent PLDζ1

3.1. *PLDζs contain PX/PH, but not C2 domains*

A novel group of Arabidopsis PLDs, PLDζ1 and ζ2, has recently been identified. The two PLDζs both contain a PX domain and a PH domain near the N-terminus, but not a C2 domain (Fig.1B), which is present in all the previously cloned plant PLDs, but not in animal or fungal PLDs. The overall domain structures and sequences of the PLDζs are more similar to mammalian PLDs than to the other Arabidopsis PLDs (Fig.1A), as PX/PH domains are also present in the mammalian ones (Qin and Wang, 2002).

3.2. *No divalent cation is required for PLDζ1 activity*

A full-length cDNA for Arabidopsis PLDζ1 has been identified and used to express catalytically active PLD in *E. coli* (Qin and Wang, 2002). In contrast to all the C2-containing PLDs that require either micromolar or millimolar levels of Ca^{2+}, the PX/PH-containing PLDζ1 does not require Ca^{2+} or any divalent cation for activity. Its highest activity occurred in the zero to nanomolar concentrations of Ca^{2+}/Mg^{2+}. The activity decreased gradually in the micromolar range and dropped rapidly when cation concentrations approached millimolar levels (Fig. 2B).

3.3. *PLDζ1 selectively hydrolyzes PC*

Previously characterized Arabidopsis PLDs can cleave several common membrane lipids (Wang, 2000). For instance, PLDα hydrolyzes PC, phosphatidylethanolamine (PE) and phosphatidylglycerol (PG) equally well, and PLDβ and γ can hydrolyze PC, PE and phosphatidylserine (PS). On the other hand, of the several common substrates tested so far, PLDζ1 hydrolyzes only PC well, but has negligible activity toward PE, PS, and PG (Qin and Wang, 2002). This PC-selective activity is distinctively different from other cloned Arabidopsis PLDs, but similar to the mammalian ones. In addition, PLDζ1 requires PIP_2 for its PC-hydrolyzing activity; surfactants, such as Triton X-100, cannot substitute for PIP_2. However, unlike the PIP_2-dependent PLDβ and γ, PLDζ1 doesn't require PE in the substrate vesicles.

4. Closing remarks

PLD was first discovered and cloned in plants. Arabidopsis has 12 PLD genes, whereas so far only two PLD genes are identified in mammals and one PLD gene in yeast. The biochemical properties, domain structures, and genomic organization of plant PLDs are much more diverse than those of other organisms (Qin and Wang, 2002). This is in contrast to the other phospholipase families; more diversity in structures and regulatory properties has been observed for PI-PLC and PLA_2 in animals than in plants (Wang, 2001). The current knowledge of the biochemical and molecular properties of the new PLDs indicates further that PLDs in plants are regulated and activated differently. The revelation of the diverse PLDs opens new doors to studying their cellular functions.

Acknowledgements

This work was supported by grants from National Science Foundation and U.S. Department of Agriculture. This is contribution 02-430-B from the Kansas Agricultural Experiment Station.

References

1. Gardiner, J.C., Harper, J.D.I., Weerakoon, N.D., Collings, D.A., Ritchie, S., Gilroy, S., Cyr, R.J. and Marc, J. (2001) A 90-kD phospholipase D from tobacco binds to microtubules and the plasma membrane. Plant Cell. 13, 2143-2158.
2. Katagiri, T., Takahashi. S. and Shinozaki, K. (2001) Involvement of a novel Arabidopsis phospholipase D, AtPLDdelta, in dehydration-inducible accumulation of phosphatidic acid in stress signalling. Plant J. 26, 595-605.
3. Qin, C. and Wang, X. (2002) The Arabidopsis phospholipase D family: characterization of a Ca^{2+}-independent and phosphatidylcholine-selective PLDζ1 with distinct regulatory domains. Plant Physiol. 128, 1057-1068.
4. Wang, C. and Wang, X (2001) A novel phospholipase D of Arabidopsis that is activated by oleic acid and associated with the plasma membrane. Plant Physiol. 127, 1102-1112.
5. Wang, X (2001) Plant phospholipases. Annu. Rev. Plant Physiol. Plant Mol. Biol. 52, 211-231.

FUNCTIONAL ANALYSIS OF *ARABIDOPSIS* LIPASE GENES

A.K. PADHAM, M.T. KAUP, C.A. TAYLOR, M.K.Y. LO,
J.E. THOMPSON
Department of Biology, University of Waterloo
200 University Ave. West
Waterloo, Ontario, Canada
N2L 3G1

1. Abstract

Three lipase genes containing the ten amino acid lipase consensus sequence, [LIV]-X-[LIVFY]-[LIAMVST]-G-[HYWV]-S-X-G-[GSTAC], have been isolated from *Arabidopsis thaliana*. One of these genes, designated LIP-1, appears to encode a senescence- induced galactolipase. LIP-1 is up-regulated in parallel with diacylglycerol acyltransferase (DGAT1) in senescing leaves suggesting that their cognate proteins work in concert. A second lipase gene, designated LIP-2, appears to encode a plastid lipase that is also up-regulated in senescing leaves. LIP-2 protein is localized in the stroma of chloroplasts and is likely a triacylglycerol (TAG) lipase. The third lipase gene, designated LIP-3, is a cytosolic lipase expressed in developing rosette leaves that is strongly induced by UV irradiation, a treatment known to simulate the effects of pathogen ingression on gene expression.

2. Senescence-induced Galactolipase (LIP-1)

Senescence is the terminal phase of development of a plant or plant organ. In *Arabidopsis,* senescence of rosette leaves is initiated coincident with the onset of bolting between weeks 4 and 5 after planting. The chloroplast is the first organelle to exhibit functional deterioration in senescing leaves. Symptoms of this deterioration include loss of chlorophyll, depletion of photosynthetic proteins and the formation of stromal plastoglobuli enriched in catabolites arising from the dismantling of thylakoid membranes. Galactolipids, specifically monogalactosyldiacylglycerol (MGDG) and digalactosyldiacylglycerol (DGDG), are the dominant lipids of thylakoids accounting for more than 60% of total polar lipids in photosynthesizing tissues. De-esterification of galactolipids is a pronounced feature of natural leaf senescence and is also induced in the event of environmental stress (Sahsah et al., 1998). Proteins with galactolipid-hydrolyzing activity have been purified from leaves of bean (Anderson et al., 1974) and wheat (O'Sullivan et al., 1987). These enzymes proved capable of de-esterifying fatty acids from both *sn* positions of MGDG and DGDG and were also able to utilize phospholipids, but not triacylglycerol, as substrate. Accordingly, they have been classified as galactosyl diglyceride acyl hydrolases or galactolipases. Recently, a drought-induced patatin-like gene was isolated from cowpea and its cognate protein shown to

N. Murata et al. (eds.), Advanced Research on Plant Lipids, 263–266.

have galactolipid-hydrolyzing activity (Matos et al., 2000). However, this galactolipid-hydrolyzing protein is not predicted to have a chloroplast targeting sequence, and its subcellular localization has not been established.

The LIP-1 gene isolated from *Arabidopsis* appears to encode a senescence-induced galactolipase. Expression of the gene is strongly up-regulated in senescing rosette leaves and is also induced when leaves are treated with the stress hormones, ABA and ethylene. LIP-1 protein is present in all chlorophyll-containing tissues, but not in non-photosynthesizing tissues such as roots and seeds, and is localized in the stroma and on thylakoid membranes of chloroplasts. Recombinant LIP-1 protein also proved capable of de-esterifying fatty acids from galactolipids, although this was not always seen because not all preparations of recombinant LIP-1 protein are enzymatically active. In addition, the rosette leaves of transgenic plants expressing antisense LIP-1 under the regulation of a constitutive promoter exhibit delayed senescence. These observations are consistent with the contention that LIP-1 contributes to the dismantling of thylakoid membranes in senescing leaves by cleaving fatty acids from galactolipids (Fig. 1).

2.1 *LIP-1 and Diacylglycerol Acyltransferase (DGAT1) are up-regulated in parallel in Senescing Leaves*

Free fatty acids derived from the membranes of senescing leaves are converted to phloem-mobile sucrose by β-oxidation, the glyoxylate cycle and gluconeogenesis. This enables the carbon equivalents of membrane fatty acids to be translocated out of the senescing leaf to other parts of the plant, in particular developing seeds. Of particular interest is the finding that diacylglycerol acyltransferase (DGAT1) (EC 2.3.1.20), the enzyme that catalyzes the terminal acylation step in the biosynthesis of TAG, is up-regulated in parallel with LIP-1 in senescing rosette leaves of *Arabidopsis*. DGAT1 is present in most plant organs including leaves, petals, fruits, anthers and developing seeds (Hobbs et al., 1999). In seeds, TAG is thought to be synthesized within the membranes of the endoplasmic reticulum and subsequently released into the cytosol in the form of oil bodies which bleb from the cytoplasmic surface of the endoplasmic reticulum (Huang, 1992).

DGAT1 has been quite extensively studied in *Arabidopsis thaliana*. The gene is found on chromosome II, approximately 17.5 ± 3cM from the *sti* locus and 8 ± 2cM from the *cp2* locus. It has been established that the *Arabidopsis* EST clone, E6B2T7, corresponds to the DGAT1 gene, and the full-length cDNA for DGAT1 (~2.0kb) has been sequenced (Hobbs et al., 1999). Though TAG formation in seeds is believed to occur in the ER, there have been several reports indicating that purified chloroplast envelope membranes from leaves are also capable of synthesizing this storage lipid (Siebertz et al., 1979; Martin and Wilson, 1983; Martin and Wilson, 1984). Moreover, TAG is known to be present in plastoglobuli, which are lipid bodies localized in the stroma of chloroplasts (Martin and Wilson, 1984).

When gel blots of RNA isolated from rosette leaves at various stages of development were probed with the *Arabidopsis* EST clone, E6B2T7, which has been annotated as DGAT1, a steep increase in DGAT1 transcript levels was evident in the senescing leaves coincident with the accumulation of TAG. The increase in DGAT1 transcript correlated temporally with enhanced levels of DGAT1 protein detected immunologically. There was also a large increase in both the number and size of plastoglobuli in senescing chloroplasts, which are known to contain TAG. Two lines of evidence indicated that the TAG of senescing leaves is synthesized in chloroplasts and sequesters fatty acids released from the catabolism of thylakoid galactolipids. First, TAG isolated from senescing leaves proved to be enriched in hexadecatrienoic acid (16:3) and linolenic acid (18:3), which are normally present in thylakoid galactolipids. Second, DGAT1 protein in senescing leaves was found to be associated with chloroplast membranes.

These findings collectively indicate that diacylglycerol acyltransferase plays a role in senescence by sequestering fatty acids de-esterified from galactolipids into TAG, which is then temporarily stored in plastoglobuli. Indeed, this would appear to be an intermediate step in the conversion of thylakoid fatty acids to phloem-mobile sucrose during leaf senescence and could account for the large increase in both size and abundance of plastoglobuli in senescing chloroplasts. Lipase-mediated de-esterification of fatty acids from the polar lipids of membrane bilayers is an autocatalytic process inasmuch as released fatty acids perturb the structure of bilayers, rendering the complex lipids therein more accessible to lipase action. Moreover, free fatty acids act as detergents and can be quite toxic at high concentrations. Thus the formation of TAG following de-esterification of galactolipids in senescing chloroplasts may be a means of storing free fatty acids in a metabolically inert form until they can be further catabolized under controlled conditions.

3. Senescence-induced Plastid Lipase (LIP-2)

A second lipase gene isolated from *Arabidopsis* (LIP-2) is also up-regulated in senescing rosette leaves. LIP-2 encodes a protein that has a chloroplast targeting sequence and is localized in the chloroplast stroma. Moreover, it is present in both green and non-green tissues of *Arabidopsis,* viz., leaves, stems, siliques and roots, indicating that it may be a general plastid lipase. Antisense suppression of LIP-2 expression using a constitutive promoter yields transgenic plants that are severely stunted. This phenotype is consistent with the contention that LIP-2 protein mobilizes stored lipid, presumably TAG, into free fatty acids that are necessary for growth and development. Its role in senescing chloroplasts is likely to be that of hydrolyzing TAG within plastoglobuli. The ensuing free fatty acids would then be converted to phloem-mobile sucrose and translocated out of the senescing leaf tissue (Fig. 1).

4. Cytosolic Lipase (LIP-3)

A third lipase gene isolated from *Arabidopsis* (LIP-3) is expressed throughout growth and development of rosette leaves and is localized in the cytosol. This gene is strongly up-regulated when plants are exposed to UV-B irradiation, a treatment that simulates pathogen ingression in terms of its effects on gene expression. Recombinant LIP-3 protein de-esterifies phospholipids and only minimally hydrolyzes galactolipids and TAG, indicating that it is a phospholipase. The expression of LIP-3 is down-regulated with the onset of senescence. Accordingly, LIP-3 is clearly distinguishable from a senescence-induced cytosolic lipase that has been isolated from *Arabidopsis* and carnation petals (Hong et al., 2000; He and Gan, 2002) .

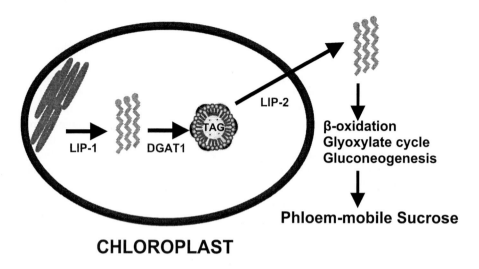

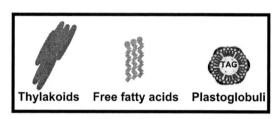

Figure 1. Model illustrating the proposed roles of LIP-1 (galactolipase), DGAT1 and LIP-2 (plastid lipase) in senescing chloroplasts. LIP-1 is a putative galactolipase that de-esterifies fatty acids from galactolipds. These fatty acids are converted to TAG by DGAT1 and temporarily stored in plastoglobuli. TAG fatty acids are de-esterified by LIP-2 and metabolized to phloem-mobile sucrose.

4. References

Anderson, M.M., McCarty, R.E. and Zimmer, E.A. (1974) The role of galactolipids in spinach chloroplast lamellar membranes. I Partial purification of a bean leaf galactolipase and its action on subchloroplast particles. Plant Physiol. 53, 699-704.

He, Y. and Gan, S. (2002) A gene encoding an acyl hydrolase is involved in leaf senescence in *Arabidopsis*. Plant Cell 14, 805-815.

Hobbs, D.H. and Hills M.J. (2000) Expression and characterization of diacylglycerol acyltransferase from *Arabidopsis thaliana* in insect cell cultures. Biochem. Soc. Trans. 28, 687-689.

Hong, Y., Wang, T.W., Hudak, K.A., Schade, F., Froese, C.D. and Thompson, J.E. (2000) An ethylene-induced cDNA encoding a lipase expressed at the onset of senescence. Proc. Natl. Acad. Sci. USA 97, 8717-8722.

Huang, A.H.C. (1992) Oil bodies and oleosins in seeds. Annu. Rev. Plant Physiol. Plant Mol. Biol. 43, 177-200.

Martin, B.A. and Wilson, R.F. (1983) Properties of diacylglycerol acyltransferase from spinach leaves. Lipids 18, 1-6.

Martin, B.A. and Wilson, R.F. (1984) Subcellular localization of TAG synthesis in spinach leaves. Lipids 19, 117-121.

Matos, A.R. d'Arcy-Lamet, A., Franca, M., Zuily-Fodil, Y. and Pham-Thi, A.T. (2000) A patatin-like protein with galactolipase activity is induced by drought stress in *Vigna unguiculata* leaves. Biochem. Soc. Trans. 28, 779-781.

O'Sullivan, J.N., Warwick, N.W.M. and Dalling, M.J. (1987) J. Plant Physiol. 131, 393-404.

Sahsah, Y., Campos, P., Gareil, M., Zuily-Fodil, Y. and Pham-Thi, A.T. (1998) Enzymatic degradation of polar lipids in *Vigna unguiculata* leaves and influence of drought stress. Physiol. Plant. 104, 577-586.

Sakaki, T., Kondo, N. and Yamada, M. (1990) Pathway for the synthesis of triacylglycerols from monogalactosyldiacylglycerols in ozone-fumigated spinach leaves. Plant Physiol. 94, 773-780.

Siebertz, H.P., Heinz, E., Linscheid, M., Joyard, J. and Douce, R. (1979) Characterization of lipids from chloroplast envelopes. Eur. J. Biochem. 101, 429-438.

TRANSCRIPTIONAL ACTIVATION OF JASMONATE BIOSYNTHESIS ENZYMES IS NOT REFLECTED AT PROTEIN LEVEL

I. STENZEL[1], B. HAUSE[1], I. FEUSSNER[2] AND C. WASTERNACK[1]
[1]*Institute of Plant Biochemistry, Dept. of Natural Products*
Weinberg 3, D-06120 Halle, Germany
[2]*Institute of Plant Genetics and Crop Plant Research (IPK)*
Corrensstrasse 3, D-06466 Gatersleben, Germany

1. Introduction

Jasmonic acid (JA) and its precursor 12-oxo phytodienoic acid (OPDA) are lipid-derived signals in plant stress responses and development (Wasternack and Hause, 2002). Within the wound-response pathway of tomato, a local response of expression of defense genes such as the proteinase inhibitor 2 gene (*PIN2*) is preceded by a rise in JA (Herde et al., 1996; Howe et al., 1996) and ethylene (O'Donnell et al., 1996). Mutants affected in JA biosynthesis such as *def1* (Howe et al., 1996) or *spr-2* (Li et al., 2002) clearly indicated that JA biosynthesis is an ultimate part of wound signaling. It is less understood, however, how the rise in JA is regulated.

Except for the β-oxidation steps, all enzymes of JA biosynthesis have been cloned in tomato. Four cDNAs coding for different lipoxygenase (LOX) forms (Heitz et al., 1997), two allene oxide synthases (AOS; Howe et al., 2000; Sivasankar et al., 2000), one allene oxide cyclase (AOC; Ziegler et al., 2000) and a homologue of the *Arabidopsis* OPDA reductase 3 (OPR3; Acc.-No. AJ278332) were found. Most of them were shown to be transcriptionally up-regulated following wounding, and a feed-forward regulation of JA biosynthesis was suggested (Laudert and Weiler 1998; Sivasankar et al., 2000). However, treatment of tomato leaves with OPDA or JA leading to transcriptional up-regulation of LOX or AOS was not accompanied with endogenous formation of JA within 24 h as analyzed by isotopic dilution analysis (Miersch and Wasternack, 2000). Interestingly, in an unwounded tomato leaf the AOC protein, essential for the establishment of the correct stereochemistry within the JA structure, is confined to parenchymatic cells around the vascular bundles (Hause et al., 2000), where the generation of the wound signal systemin is located (Jacinto et al., 1997). Such a tissue-specific occurrence of AOC suggests that a wound-induced rise of JA may occur spatially and temporally regulated.

N. Murata et al. (eds.), Advanced Research on Plant Lipids, 267–270.

2. Materials and methods

Lycopersicon esculentum cv Moneymaker grown for 6 weeks was used in all experiments. Growth, treatments, northern, immunoblot and immunohisto-chemical analyses were performed as described (Hause et al., 2000).

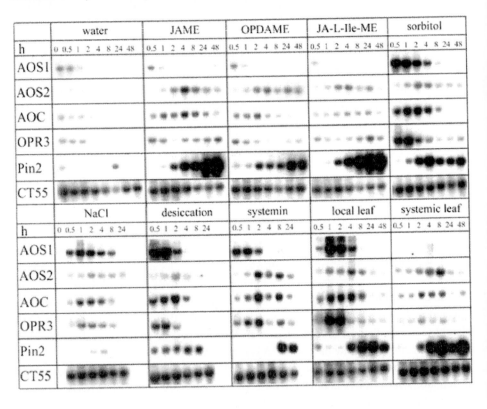

Figure 1. mRNA accumulation in leaves of 6 week-old tomato plants following different treatments. Total RNA (20 µg/lane) was loaded. Loading control was performed with a cDNA of the ATP/ADP carrier (CT55).

3. Results and discussion

As a type of comparative analysis, we analyzed kinetics of AOC mRNA accumulations and compared them with that of mRNAs coding for AOS1, AOS2, and OPR3 as well as for the wound response marker *PIN2*. Various treatments including stresses known to lead to endogenous rise of JA were used (Fig. 1). These data were compared to the kinetic of AOC protein accumulation (Fig. 2A) as well as with its tissue-specific occurrence (Fig. 2B).

Compounds such as the methyl esters of JA (JAME), OPDA (OPDAME) and the amino acid conjugate JA-L-Ile-ME exhibit transient mRNA accumulation of AOS1, AOS2, AOC and OPR3, which precedes *PIN2* mRNA accumulation. A distinct time course differing slightly among the JA biosynthesis enzymes (e.g. AOS1 before AOS2) was also found upon stress-treatment such as floating of leaflets on sorbitol, NaCl, or desiccation. The strong local wound response was accompanied only from a weak but significant systemic response, contrasting from the strong systemic *PIN2* expression. The kinetics of mRNA accumulation of JA biosynthesis enzymes clearly precedes that of the JA response marker *PIN2* suggesting a causal link. However, the transient rise of JA following wounding or other stresses applied to tomato leaves is already detectable between 20 and 40 min. Immunoblot analysis revealed constitutive occurrence of AOC protein in water floated or JAME-, systemin- or sorbitol-treated leaves (Fig. 2A). Moreover, in the immunocytochemical inspection, the AOC protein was clearly confined only to vascular bundles (Fig. 2B, left), even under conditions of transient AOC mRNA accumulation (Fig. 2B, right).

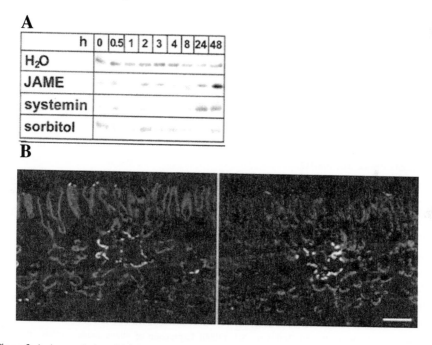

Figure 2. A: Accumulation of AOC protein in excised leaves of 6 weeks old tomato plants cv. Moneymaker following different treatments. B: Immunocytochemical analysis of AOC protein in cross-sections of excised leaves of 6 weeks old tomato plants cv. Moneymaker treated with water (left) or 100 nM systemin (right) *via* the petiole for 3 h.

These data suggest that there is no causal link between transient mRNA accumulation and the protein level. Recently, we could show that transgenic tomato plants overexpressing AOC constitutively were able to form JA only upon wounding (Stenzel et al., submitted). These and further data argue for regulation of JA biosynthesis in tomato by substrate availability and preferential generation of JA in main veins of a leaf. It will be interesting to analyze the reason for the additionally occurring transient mRNA accumulation described here.

4. References

Hause, B., Stenzel, I., Miersch, O., Maucher, H., Kramell, R., Ziegler, J. and Wasternack, C. (2000) Tissue-specific oxylipin signature of tomato flowers: allene oxide cyclase is highly expressed in distinct flower organs and vascular bundles. Plant J., 24, 113-126.

Heitz, T., Bergey, D.R. and Ryan, C.A. (1997) A gene encoding a chloroplast-targeted lipoxygenase in tomato leaves is transiently induced by wounding, systemin, and methyl jasmonate. Plant Physiol., 114, 1085-1093.

Herde, O., Atzorn, R., Fisahn, J., Wasternack, C., Willmitzer, L. and Pena-Cortes, H. (1996) Localized wounding by heat initiates the accumulation of proteinase inhibitor II in abscisic acid-deficient plants by triggering jasmonic acid biosynthesis. Plant Physiol., 112, 853-860.

Howe, G.A., Lee, G.I., Itoh, A., Li, L. and DeRocher, A.E. (2000) Cytochrome P450-dependent metabolism of oxylipins in tomato. Cloning and expression of allene oxide synthase and fatty acid hydroperoxide lyase. Plant Physiol., 123, 711-724.

Howe, G.A., Lightner, J., Browse, J. and Ryan, C.A. (1996) An octadecanoid pathway mutant (JL5) of tomato is compromised in signaling for defense against insect attack. Plant Cell, 8, 2067-2077.

Jacinto, T., McGurl, B., Franceschi, V., Delano-Freier, J. and Ryan, C.A. (1997) Tomato prosystemin promoter confers wound-inducible, vascular bundle-specific expression of the β-glucuronidase gene in transgenic tomato plants. Planta, 203, 406-412.

Laudert, D. and Weiler, E.W. (1998) Allene oxide synthase: a major control point in *Arabidopsis thaliana* octadecanoid signalling. Plant J., 15, 675-684.

Li, L., Li, C., Lee, G.I. and Howe, G.A. (2002) Distinct roles for jasmonate synthesis and action in the systemic wound response of tomato. Proc. Natl. Acad .Sci. USA, 99, 6416-6421.

Miersch, O. and Wasternack, C. (2000) Octadecanoid and jasmonate signaling in tomato (*Lycopersicon esculentum* Mill.) leaves: Endogenous jasmonates do not induce jasmonate biosynthesis. Biol. Chem., 381, 715-722.

O'Donnell, P.J., Calvert, C., Atzorn, R., Wasternack, C., Leyser, H.M.O. and Bowles, D.J. (1996) Ethylene as a signal mediating the wound response of tomato plants. Science, 274, 1914-1917.

Sivasankar, S., Sheldrick, B. and Rothstein, S.J. (2000) Expression of allene oxide synthase determines defense gene activation in tomato. Plant Physiol., 122, 1335-1342.

Wasternack, C. and Hause, B. (2002) Jasmonates and octadecanoids - signals in plant stress responses and development. Prog. Nuc. Acid Res., 72, 165-220.

Ziegler, J., Stenzel, I., Hause, B., Maucher, H., Hamberg, M., Grimm, R., Ganal, M. and Wasternack, C. (2000) Molecular cloning of allene oxide cyclase. J. Biol. Chem., 275, 19132-19138.

MEMBRANE LOCALIZATION OF *ARABIDOPSIS* ACYL-CoA BINDING PROTEIN ACBP2

H-Y. LI AND M-L. CHYE
Department of Botany, The University of Hong Kong.
Pokfulam Road, Hong Kong, China.

Abstract

Cytosolic acyl-CoA binding proteins bind long-chain acyl-CoAs and act as intracellular acyl-CoA transporters and pool formers. Recently, we have characterized *Arabidopsis thaliana* cDNAs encoding novel forms of ACBP, designated ACBP1 and ACBP2, that contain a hydrophobic domain at the N-terminus and show conservation at the acyl-CoA binding domain to cytosolic ACBPs. We have previously demonstrated that ACBP1 is membrane-associated in *Arabidopsis*. Here, Western blot analysis using anti-ACBP2 antibodies on *A. thaliana* protein show that ACBP2 is located in the microsome-containing membrane fraction and in the subcellular fraction containing large particles (mitochondria, chloroplasts and peroxisomes), resembling the subcellular localization of ACBP1. To further investigate the subcellular localization of ACBP2, we fused ACBP2 translationally in-frame to GFP. Using particle gene bombardment, ACBP2-GFP and ACBP1-GFP fusion proteins were observed transiently expressed at the endoplasmic reticulum and at the plasma membrane in onion epidermal cells. GFP fusions with deletion derivatives of ACBP1 or ACBP2 lacking the transmembrane domain were impaired in membrane targeting. Our investigations also showed that when the transmembrane domain of ACBP1 or that of ACBP2 was fused with GFP, the fusion protein was targeted to the plasma membrane, thereby establishing their role in membrane targeting. We conclude that ACBP2, like ACBP1, is also a membrane protein that likely functions in membrane-associated acyl-CoA transfer / metabolism.

Introduction

We have isolated and characterized cDNAs and their corresponding genes encoding ACBPs from *A. thaliana*, designated ACBP1 (Chye, 1998; Chye *et al.*, 1999) and ACBP2 (Chye *et al.*, 2000) that are distinct from cytosolic ACBPs (Færgeman *et al.*, 1997) by the presence of an N-terminal hydrophobic domain. We have shown that ACBP1 accumulates in developing seeds (Chye, 1998). Our results from immunoelectron microscopy suggest that ACBP1 likely participates in intermembrane lipid transport from the endoplasmic reticulum (ER) *via* vesicles to the plasma

N. Murata et al. (eds.), Advanced Research on Plant Lipids, 271–273.

272

membrane where it could maintain an acyl-CoA pool (Chye *et al.*, 1999). We have further shown that *Escherichia coli* expressed recombinant ACBP1 and ACBP2 bind acyl-CoA and mutations in the acyl-CoA binding domain of ACBP2 disrupt palmitoyl-CoA binding function (Chye, 1998; Chye *et al.*, 2000). Here we show that ACBP2 is a membrane-associated protein using (a) Western blot analysis with anti-ACBP2 antibodies on subcellular fractions of *Arabidopsis* protein and (b) GFP-tagged *in vivo* tests of transient expression in onion epidermal cells. We further verify that the N-terminal transmembrane domain mediates membrane targeting in ACBP2 and ACBP1.

Results

Localization of ACBP2 in the membrane-associated subcellular fraction
Western blot analysis of *A. thaliana* subcellular fractions using anti-ACBP2 antibodies showed a cross-reacting band of apparent molecular mass 38 kDa in total protein (Figure 1a, lane 1), in the microsome-containing membrane fraction (lane 2) and in the fraction containing large particles (lane 4). The cross-reacting band was not seen in the cytosol fraction (lane 3) and the nuclear fraction (lane 5).

Co-localization of ACBP2-GFP with the plasma membrane
We used the GFP vector, pGFP-2, in which GFP is expressed from the CaMV 35S promoter (Kost *et al*, 1998). The ACBP2-GFP fusion consists of amino acids 1 to 346 of ACBP2 fused translationally in-frame to GFP (Figure 2a), while its deletion derivative lacking the putative transmembrane, designated ACBP2ΔM-GFP, consists of amino acids 34 to 346 of ACBP2 fused to GFP (Figure 2b). As a positive control, we constructed an ACBP1-GFP fusion protein consisting of amino acids 1 to 197 of ACBP1 fused translationally in-frame to GFP (Figure 2d) as we have previously shown that ACBP1 is membrane-associated by immunoelectron microscopy (Chye *et al.*, 1999). We also included a deletion derivative of ACBP1 amino acids 41 to 197 consisting of amino acids 41 to 197, that lacks the transmembrane domain, designated ACBP1ΔM-GFP (Figure 2e).

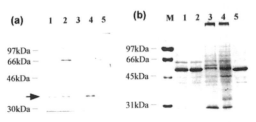

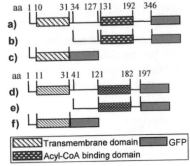

Figure 1. (a) Western blot analysis on subcellular fractions of silique-bearing *Arabidopsis* harvested 7 days after flowering using antipeptide antibodies against ACBP2. Total whole plant protein (lane 1), membrane (lane 2), cytosol (lane 3), large particles including mitochondria, chloroplasts and peroxisomes (lane 4), nuclei (lane 5) subcellular fractions. (b) A gel identically loaded as (a) stained with Coomassie blue. M, molecular mass markers.

Figure 2. Construction of ACBP2-GFP and ACBP1-GFP fusions. (a) pACBP2-GFP (b) pACBP2ΔM-GFP (c) pACBP2M-GFP (d) pACBP1-GFP (e) pACBP1ΔM-GFP and (f) pACBP1M-GFP.

Following particle bombardment of onion epidermal cells, GFP-images of single optical sections were obtained using confocal laser scanning microscopy (Figure 3). Using ACBP2-GFP, fluorescence was observed predominantly at the plasma membrane and to a lesser extent, at the ER (Figure 3a). ACBP1-GFP was also expressed predominantly at the plasma membrane and to a lesser extent, at the ER (Figure 3d). With ACBP2ΔM-GFP (Figure 3b) and ACBP1ΔM-GFP (Figure 3e), fluorescence was seen in the nucleus and the cytoplasm, like control plasmid pGFP-2 (Figure 3g). The signal detected in the nucleus is possibly due to diffusion of GFP through nuclear pore complexes (von Arnim et al., 1998). Figure 4a shows that the fluorescence observed due to transient expression of the ACBP2-GFP is associated with the retracted plasma membrane following plasmolysis. A transmitted light image of the same plasmolysed cell (Figure 4b) confirms that the ACBP2-GFP did not co-localize with the cell wall. Upon plasmolysis using pACBP1-GFP, we again observed co-localization of green fluorescence with the retracted plasma membrane (Figure 4c) as confirmed by a transmitted light image of the same cell (Figure 4d).

The transmembrane domain of ACBP2 or ACBP1 functions in membrane targeting
With ACBP2M-GFP in which the transmembrane domain of ACBP2 (amino acids 1 to 31) was fused to GFP (Figure 2c), green fluorescence was observed predominantly at the plasma membrane (Figure 3c). Similar fluorescence at the plasma membrane (Figure 3f) were obtained using the ACBP1M-GFP (Figure 2f). We concluded that the transmembrane domain in ACBP2 and that in ACBP1 primarily targets the protein to the plasma membrane.

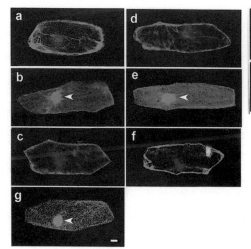

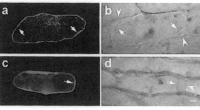

Figure 4. Transient expression of ACBP2-GFP and ACBP1-GFP fusion proteins in onion epidermal cells after plasmolysis.
(a) Epifluorescence of ACBP2-GFP (b) Transmitted light image of cell shown in (a). (c) Epifluorescence of ACBP1-GFP (d) Transmitted light image of cell shown in (c). Arrows indicate plasma membrane and arrowheads indicate cell wall. Dark spots in (b) and (d) are due to biolistic particles on the cell surface. Bar, 20 μm

Figure 3. Confocal images of ACBP2-GFP and ACBP1-GFP transiently expressed in onion epidermal cells. (a) ACBP2-GFP (b) ACBP2ΔM-GFP (c) ACBP2M-GFP (d) ACBP1-GFP (e) ACBP1ΔM-GFP (f) ACBP1M-GFP (g) GFP. Bar, 20 μm Arrowheads point nuclei.

274

Discussion

Our results of the membrane localization of ACBP2 by Western blot analysis and by using GFP fusions are consistent. Using GFP fusions, we showed that the transmembrane domain of ACBP1 and that of ACBP2 act as strong membrane targeting signals. The similar targeting pattern seen in both ACBP1M-GFP and ACBP2M-GFP may be reflected by the high conservation of amino acids (71% identity) within their transmembrane domains. The subcellular location of a protein plays a critical role in its function. That an N-terminal transmembrane domain is present in each of ACBP1 and ACBP2 makes them unique from cytosolic ACBPs. As yet, the biological functions of these ACBPs are not entirely clear. Like ACBP1, transport of ACBP2 to the plasma membrane could be mediated by ER-derived vesicles that are present in the subcellular fraction of large particles shown by Western blot analysis to be enriched in both ACBP1 (Chye, 1998) and ACBP2 (this study).

Acknowledgements

We thank Prof. N.-H. Chua for the gift of plasmid pGFP-2. This work is partially supported by the Research Grants Council of the Hong Kong Special Administrative Region, China (Project HKU7087/99M). H.-Y. Li was supported by a studentship from the University of Hong Kong.

References

Chye, M.-L. (1998) *Arabidopsis* cDNA encoding a membrane-associated protein with an acyl-CoA binding domain. Plant Mol. Biol. 38, 827–838.

Chye, M.-L., Huang, B.-Q. and Zee, S.-Y. (1999) Isolation of a gene encoding *Arabidopsis* membrane-associated acyl-CoA binding protein and immunolocalization of its gene product. Plant J. 18, 205-214.

Chye, M.-L., Li, H.-Y. and Yung, M.-H. (2000) Single amino acid substitutions at the acyl-CoA-binding domain interrupt ^{14}[C]palmitoyl-CoA binding of ACBP2, an *Arabidopsis* acyl-CoA-binding protein with ankyrin repeats. Plant Mol. Biol. 44, 711-721.

Færgeman, N.J. and Knudsen, J. (1997) Role of long-chain fatty acyl-CoAs esters in the regulation of metabolism and cell signaling. Biochem. J. 323, 1-12.

Kost, B., Spielhofer, P. and Chua, N.-H. (1998) A GFP-mouse talin fusion protein labels plant actin filaments *in vivo* and visualizes the actin cytoskeleton in growing pollen tubes. Plant J. 16, 393–401.

Pfanner, N., Orci, L., Glick, B.S., Amherdt, M., Arden, S.R., Maihotra, V. and Rothman, J.E. (1989) Fatty acyl-Coenzyme A is required for budding of transport vesicles from Golgi cisternae. Cell 59, 95-102.

Smith, J.A., Krauss, M.R., Borkird, C. and Sung, Z.R. (1988) A nuclear protein associated with cell divisions in plants. Planta 174, 462-472.

von Arnim, A.G., Deng, X.W. and Stacey, M.G. (1998) Cloning vectors for the expression of green fluorescent proteins in transgenic plants. Gene 221, 35-43

SHIFT IN FATTY ACID AND OXYLIPIN PATTERN OF TOMATO LEAVES FOLLOWING OVEREXPRESSION OF THE ALLENE OXIDE CYCLASE

H. WEICHERT[1], H. MAUCHER[2], E. HORNUNG[1], C. WASTERNACK[2] AND I. FEUSSNER[1]

[1]Institute of Plant Genetics and Crop Plant Research (IPK)
Corrensstrasse 3, D-06466 Gatersleben, Germany
[2]Institute of Plant Biochemistry, Dept. of Natural Products
Weinberg 3, D-06120 Halle, Germany

1. Introduction

Polyunsaturated fatty acids (PUFAs) are a source of numerous oxidation products, the oxylipins. In leaves, α-linolenic acid (α-LeA) is the preferential substrate for lipid peroxidation reactions. This reaction may be catalyzed either by a 9-lipoxygenase (9-LOX) or by a 13-LOX and oxygen is inserted regioselectively as well as stereospecifically leading to formation of **13S**- or **9S**-hydroperoxy octadecatrienoic acid (13-/9-HPOT; Brash, 1999). At least, seven different enzyme families or reaction branches within the LOX pathway can use these HPOTs or other hydroperoxy PUFAs leading to (i) keto-PUFAs (LOX); (ii) epoxy hydroxy-PUFAs (epoxy alcohol synthase, EAS); (iii) octa-decanoids and jasmonates (allene oxide synthase, AOS); (iv) leaf aldehydes and leaf alcohols (hydroperoxide lyase, HPL); (v) hydroxy PUFAs (reductase); (vi) divinyl ether PUFAs (divinyl ether synthase, DES); and (vii) epoxy- or dihydrodiol-PUFAs (peroxy-genase, POX; Fig. 1; Feussner and Wasternack, 2002).

Such a multiple use of free HPOTs suggests a regulatory role of its level and degree of consumption in the different branches, which can be recorded by targeted metabolic profiling of oxylipins. Indeed, initial work showed a preferential shift of oxy-lipins into the reductase branch following salicylate treatment of barley leaves (Weichert et al., 1999). In germinating cucumber cotyledons mainly 13-LOX products were detected and were found to be directed preferentially into the reductase as well as the HPL branch (Weichert et al., 2002), whereas in *Arabidopsis thaliana* 12-oxo phyto-dienoic acid (OPDA) and jasmonic acid (JA), both formed by the AOS branch, accumulated preferentially (Stenzel et al., submitted). In potato suspension cultures, the reductase, the EAS as well as the DES branch are specifically stimulated upon elicitor

N. Murata et al. (eds.), Advanced Research on Plant Lipids, 275–278.

treatment (Göbel et al., 2001). Moreover, data are accumulating that the total amount of free oxylipins is at least in some plants much lower than that of oxylipins within the fraction of esterified fatty acids. Based on the initial observation that a specific LOX of lipid bodies in cucumber uses esterified PUFAs as a substrate, thereby initiating storage lipid mobilization (Feussner et al., 2001) and upon improved analysis, the predominant occurrence of oxylipins in their esterified form became obvious. In case of *A. thaliana*, esterified 13-H(P)OT exceeded free 13-H(P)OT by four orders of magnitude (Stenzel et al., submitted), whereas cucumber seedlings contain about 100-fold more esterified **13S** –**h**ydro**p**eroxy **o**ctadeca**d**ienoic acid (13-HPOD) than free 13-HPOD (Weichert et al., 2002). Also OPDA can occur predominantly in its esterified form (Stelmach et al., 2001). Thus, metabolic profiling of oxylipin pattern needs comparison of profiles of both, free and esterified oxylipins.

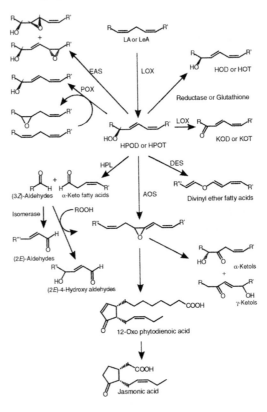

Figure 1. The LOX pathway.

2. Materials and methods

Lycopersicon esculentum cv Lukullus and homocygous lines carrying the full-length cDNA of AOC (Ziegler et al.,2000) in sense orientation under the control of the 35S promoter were used (Stenzel et al. submitted). Oxylipin profiling of non-esterified fatty acid derivatives was performed as described (Weichert et al., 1999). For analysis of esterified oxylipins, lipid extracts were transmethylated with sodium methoxide and oxylipins were analyzed as methyl esters.

3. Results and discussion

To inspect whether transgenetically altered branches of the LOX pathway may cause a shift of compounds between the branches and/or between the free and esterified amount of oxylipins, we have chosen the AOS branch. Here, the <u>a</u>llene <u>o</u>xide <u>c</u>yclase (AOC)-catalyzed step is of special importance due to its establishment of the correct stereo-chemistry as well as its regulatory role (Stenzel et al., submitted). AOC was cloned from tomato (Ziegler et al., 2000) and found to be a single copy gene which is specifically expressed in all vascular bundles (Hause et al., 2000). Here, we analyzed metabolic profiles of free and esterified oxylipins in leaves of wild type and transgenic tomato plants overexpressing the AOC under the control of the 35S promoter (OE-AOC plants). In freshly harvested leaves of 6 week-old plants fatty acids as well as hydro-peroxy, hydroxy and keto PUFA derivatives, all of them in the free and esterified form, were analyzed. Data exhibiting a shift between wild-type and OE-AOC plants were selected and are shown in Fig. 2: Among the esterified fatty acids only the α-LeA (18:3) exhibited a weak rise in OE-AOC plants (Fig. 2A), whereas these plants showed less free fatty acids (Fig. 2C). However, the most dramatic changes appeared in case of esterified 13-HOT and esterified 13-HOD (Fig. 2B). Whereas both compounds were undetectable in their esterified form in wild-type leaves, these oxylipins accumulated to remarkable levels in the leaves of OE-AOC plants. In addition, for their free forms a significant decrease was found (Fig. 2D). These data may show for the first time that a block in the AOS branch of the LOX pathway lead to an accumulation of metabolites of the reductase pathway in the membrane fraction and not in the fraction of free fatty acid derivatives. Further analysis will be needed to see whether the accumulation of oxylipins within membranes serves as a temporal storage form or has other physiological functions.

4. References

Brash, A. R. (1999) Lipoxygenases: Occurrence, functions, catalysis, and acquisition of substrate. J. Biol. Chem., 274, 23679-23682.

Feussner, I., Kühn, H. and Wasternack, C. (2001) The lipoxygenase dependent degradation of storage lipids. Trends Plant Sci., 6, 268-273.

Feussner, I. and Wasternack, C. (2002) The lipoxygenase pathway. Annu. Rev. Plant Physiol. Plant Mol. Biol., 53, 275-297.

278

Göbel, C., Feussner, I., Schmidt, A., Scheel, D., Sanchez-Serrano, J., Hamberg, M. and Rosahl, S. (2001) Oxylipin profiling reveals the preferential stimulation of the 9-lipoxygenase pathway in elicitor-treated potato cells. J. Biol. Chem., 276, 6267-6273.

Hause, B., Stenzel, I., Miersch, O., Maucher, H., Kramell, R., Ziegler, J. and Wasternack, C. (2000) Tissue-specific oxylipin signature of tomato flowers: allene oxide cyclase is highly expressed in distinct flower organs and vascular bundles. Plant J., 24, 113-126.

Stelmach, B.A., Müller, A., Hennig, P., Gebhardt, S., Schubert-Zsilavecz, M. and Weiler, E.W. (2001) A novel class of oxylipins, sn1-O-(12-Oxophytodienoyl)-sn2-O-(hexadecatrienoyl)-mono-galactosyl diglyceride, from *Arabidopsis thaliana*. J. Biol. Chem., 276, 12832-12838.

Weichert, H., Kolbe, A., Kraus, A., Wasternack, C. and Feussner, I. (2002) Metabolic profiling of oxylipins in germinating cucumber seedlings - lipoxygenase-dependent degradation of triacyl-glycerols and biosynthesis of volatile aldehydes. Planta, in press.

Weichert, H., Stenzel, I., Berndt, E., Wasternack, C. and Feussner, I. (1999) Metabolic profiling of oxylipins upon salicylate treatment in barley leaves - preferential induction of the reductase pathway by salicylate. FEBS Lett., 464, 133-137.

Ziegler, J., Stenzel, I., Hause, B., Maucher, H., Hamberg, M., Grimm, R., Ganal, M. and Waster-nack, C. (2000) Molecular cloning of allene oxide cyclase. J. Biol. Chem., 275, 19132-19138.

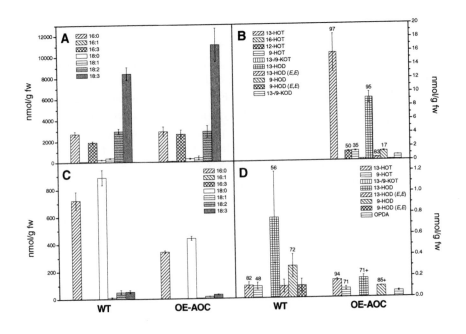

Figure 2. Oxylipin profiling of tomato leaves. (A) Esterified fatty acids, (B) esterified oxylipins, (C) free fatty acids, and (D) free oxylipins. Amount of *S* enatiomers is given above the column. The data show the mean of 2 independent experiments.

REACTION CHARACTERIZATION OF PHOSPHOLIPASEA₂ ON HYDROLYSIS OF PHOSPHOLIPID IN W/O MICROEMULSION

K. YAMAZAKI, M. IMAI and I. SUZUKI

1866 Kameino, Fujisawa, Kanagawa-pref. 252-8510   JAPAN

Laboratory of Food Chemical Engineering

Department of Food Science and Technology

Graduate School of Bioresource Sciences

Nihon University

ABSTRACT

Phosphatidylcholine (PC) from soybean was hydrolyzed using W/O microemulsions. In this system, PC played the role of the substrate and amphiphile. Isooctane (2-2-4 trimethylpentane: C_8*) was used as the major solvent, and several alcohols, i.e., 1-butanol (C_4), 1-pentanol (C_5), 1-hexanol (C_6) and 1-octanol (C_8) were employed as co-solvents to create suitable hydrophobic conditions. The hydrophobicity of the mixed solvent system was determined by the logP. A lower Michaelis-Menten constant, almost the same level as ca. 5 mM, was obtained in systems in which alcohols with an even number of carbons were employed as a co-solvent. On the other hand, the C_5 and the mixed C_4 and C_6 alcohols systems showed larger Michaelis constants. The maximum reaction rate (V_{max}) was ca. 0.03[(mM-fatty acid)(mg-PLA$_2$)$^{-1}$(s)$^{-1}$] in the whole range of solvent compositions.

1. INTRODUCTION

Phospholipids are widely distributed in nature as a main ingredient of biomembranes. They can be found in brain, nerves, internal organs, blood, chicken eggs and seeds. In the body,

N. Murata et al. (eds.), Advanced Research on Plant Lipids, 279–282.
© 2003 *Kluwer Academic Publishers. Printed in the Netherlands.*

phospholipids maintain functions, such as metabolism, neural transmission and immunity. At present, phospholipids are utilized in the food and cosmetics industries as biocompatible amphiphiles of natural origin[3]. The purpose of this investigation is to characterize the reactions of phospholipaseA$_2$ using water immiscible substrates, and to propose a more suitable solvent design through kinetic parameters determined by hydrolysis of PC.

2. MATERIALS and METHODS

In this investigation, PC (soybean lecithin [Epikuron 200, Lucas Meyer (Hamburg)], (phosphatidylcholine: 95%)) was used as the substrate and amphiphile. The enzyme used was phospholipaseA$_2$ (PLA$_2$) from hog pancreas (568 U/mg, 14 kDa, Fluka (Buchs)). Isooctane:C$_8$ was used as the major solvent, and 1-butanol:C$_4$, 1-pentanol:C$_5$, 1-hexanol:C$_6$, 1-octanol:C$_8$ were used as co-solvents[1]. They were analytical grade and purchased from Wako Pure Chemical Ltd (Osaka). All test solvents were prepared at a molar ratio of main solvent to co-solvent of 11:1. Hydrophobicities of mixed solvent systems were elucidated by using the logP[2]. The solvent mixtures prepared in this study have eight different hydrophobicities with a range of logP from 3.816 to 3.991. Figure 1 shows the order of hydrophobicities of each test solvent. As the aqueous phase, 0.1 M Tris-HCl buffer (pH=8.0 at 313 K) was used. The enzymatic reaction was performed at 313 K[3].

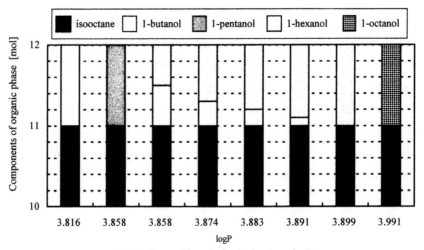

Fig. 1 Composition of mixed solvents vs. logP

3. RESULTS and DISCUSSION

Hydrolysis production (FA: fatty acid) was obtained in all experimental systems. Figure 2 shows a typical time course of hydrolysis of PC (30 mM) in the mixed solvent of C_8^*:C_6 of 11:1. This hydrolysis has a slight delay at the very beginning of the reaction. The length of the lag phase decreased, changed with components of the organic solvent. As hydrophobicity became lower, the length of the lag phase became shorter (1 min-5 min). The initial hydrolysis rate determination was performed over 0-10 min periods during which the rates remained linear. The yield of FA was elucidated when the hydrolysis production profile leveled off. Regardless of the components of the organic solvents, the yields were almost constantly over 80 %.

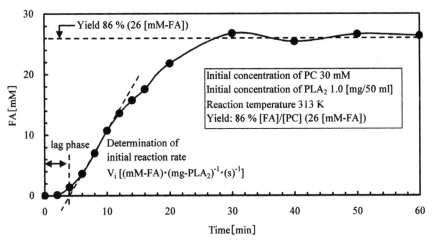

Fig. 2 Time course of hydrolysis of PC=30mM by PLA$_2$
in molar ratio mixture solvent of isooctane:1-hexanol=11:1

Figure 3 shows the kinetic analysis as $[S_0] \cdot V_i^{-1}$ vs. $[S_0]$. The intercept on abscissa means a negative Michaelis constant, and the slope of the correlation line represents a reciprocal of the maximum reaction rate. The K_m changed remarkably in a limited range of logP ($3.82 \leq logP \leq 3.90$). In this range, the systems that used C_5 and mixed C_4 and C_6 alcohols were included. In particular, the maximum Michaelis-Menten constant (K_m=42mM) was measured in the isooctane-1-pentanol (C_8^*:C_5=11:1, logP=3.858) system. The minimum Michaelis-Menten constant (K_m=5 mM) was obtained in the systems employing alcohols with an even number of carbons as co-solvents. The maximum reaction rate (V_{max}=0.03 [(mM-FA)(mg-PLA$_2$)$^{-1}$(s)$^{-1}$]) was almost constant within the experimental range of logP.

282

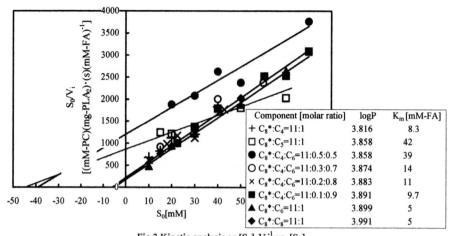

Fig.3 Kinetic analysis as $[S_0] \cdot V_i^{-1}$ vs. $[S_0]$
LogPs were evaluated according to ref 2). K_m values presented in Fig. 3 were obtained.

4. CONCLUSION

The hydrolysis of phospholipid by phospholipaseA$_2$ in W/O microemulsions was characterized in various organic solvents. Suitable hydrophobic conditions were discussed in relation to the reaction parameters and hydrophobicity index logP. The systems, that used alcohols with even numbers of carbon atoms, 1-butanol (C_4), 1-hexanol (C_6) and 1-octanol (C_8), as co-solvents were favorable to hydrolysis and the logPs were 3.816, 3.899 and 3.991, respectively. On the other hand, K_m values obtained when C_5 and the mixed C_4 and C_6 alcohols were used as co-solvents, whose logPs were in very limited range, seemed to be inappropriate for hydrolysis. The maximum reaction rate (V_{max}=0.03 [(mM-FA)(mg-PLA$_2$)$^{-1}$(s)$^{-1}$]) was almost constant throughout the experimental range of logP.

REFFERENCES

1) Garti, N., D. Lichtenberg and T. Silberstein, (1997) The hydrolysis of phosphatidylcholine by phospholipaseA$_2$ in microemulsion as microreactor. Colloids and Surfaces, A, 128, 17-25.

2) Lanne, C., S. Boeren, K. Vos and C. Veeger, (2001) Rules for Optimization of Biocatalysis in Organic Solvents. Biotech. and Bioeng. 30, 81-87.

3) Morgado, M.A.P., J.M.S. Cabral and D.M.F. Prazeres, (1996) PhospholipaseA$_2$-catalyzed hydrolysis of lecithin in a continuous reversed-micellar membrane bioreactor. J. Am. Oil. Chem. Soc. 73, 337-346.

MOLECULAR CHARACTERIZATION OF RICE α-OXYGENASE
A comparison with mammalian prostaglandin H synthases

T. KOEDUKA, K. MATSUI, Y. AKAKABE, and T. KAJIWARA
Department of Biological Chemistry, Faculty of Agriculture,
Yamaguchi University,
Yoshida 1677-1, Yamaguchi city, Yamaguchi 753-8515, Japan

Abstract

Long-chain fatty acids can be metabolized to C_{n-1} aldehydes by α-oxidation in plants. The reaction mechanism of the enzyme has not been elucidated. In this study, the gene encoding rice α-oxygenase was expressed in *E. coli* and purified to apparently homogenous state. Site-directed mutagenesis revealed that His-158, Tyr-380, and Ser-558 were essential for the α-oxygenase activity. These residues are conserved in prostaglandin H synthases (PGHS) and known as a heme-ligand, a source of a radical species to initiate oxygenation reaction, and a residue involved in substrate binding, respectively. The rice α-oxygenase activity was inhibited by imidazole but hardly inhibited by nonsteroidal anti-inflammatory drugs, which are known as typical PGHS inhibitions. In addition, peroxidase activity could not be detected with α-oxygenase when palmitic acid 2-hydroperoxide was used as a substrate. From these findings catalytic resemblance between α-oxygenase and PGHS seems to be evident although there still are differences in their substrate recognitions and peroxidation activities.

Product analysis and substrate specificity

When the purified recombinant enzyme was reacted with linoleic acid at 25°C, formation of (8Z,11Z)-heptadecadienal, which is a fatty aldehyde with one methylene length shorter than the substrate, could be detected. 2-Hydroperoxy fatty acids are chemically unstable, and have a half-life time of about 30 min in an aqueous buffer at 23°C (Hamberg et al., 1999). Thus, in order to detect the unstable fatty acid 2-hydroperoxide, fatty acid as a substrate was incubated with the purified enzyme on ice, and the product was immediately esterified with 9-anthryldiazomethane (ADAM). The resulting ADAM esters were analyzed by HPLC. When palmitic acid was incubated with the recombinant purified enzyme on ice, 2-hydroperoxide of palmitic acid was detected as a major product. When palmitic acid 2-hydroperoxide was incubated with the recombinant enzyme under the presence of guaiacol, which is widely used as a

N. Murata et al. (eds.), Advanced Research on Plant Lipids, 283–286.
© 2003 *Kluwer Academic Publishers. Printed in the Netherlands.*

cosubstrate for peroxidase activity of PGHSs, little increase in the amount of 2-hydroxide was detected. In the case of PGHSs, the turnover rates of oxygenase pathway and peroxidase pathway are almost stoichiometrically same. In contrast, in the case of rice α-oxygenase, fatty acid 2-hydroperoxide was the major product at least under the reaction condition employed here, therefore, the rice enzyme has little need to provoke peroxidase pathway.

With linoleic acid as a substrate there is no change in the absorbance between 210 to 350 nm including absorption at 234 nm corresponding to the formation of a conjugated diene hydroperoxy moiety during the reaction. This indicated that rice α-oxygenase could not abstract the bisallylic proton to form conjugated diene hydroperoxide but exclusively abstracted α-proton. When various chain-length of fatty acids, from C10 to 20, were used as the substrates, the highest activity could be detected with the polyunsaturated substrates having double-bonds such as linoleic acid, α-linolenic acid, and γ-linolenic acid (Fig. 1).

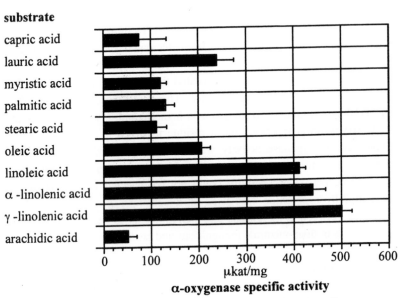

Figure 1. **Effect of the chain-length on substrate for α-oxygenase activity.** α-Oxygenase activity was determined at 25°C as the rate of oxygen consumption, using a Clark type oxygen electrode (Yellow Springs instruments). The reaction was initialized by adding an appropriate amount (50 μg) of the purified enzyme to 0.1 M Hepes buffer (pH 7.2) including 500 nmol of the substrate with gum arabic. The error bars represent the mean ± S.D. ($n=3$).

It seemed that the activities depend on the solubility of substrates. However, it must be noticed that all of them were available as the substrates for α-oxygenase.

Inhibition of rice α-oxygenase

As noted with plant α-oxidation system (Galliard et al., 1976), imidazole was a potent reversible inhibitor for the rice α-oxygenase and the *Ki* value was determined to be

0.046 mM (Table 1). It is well known that imidazole can bind to Fe in most heme proteins (Psylinakis et al., 2001), which results in the inactivation of the function of the heme protein. On the contrary, no inhibition of ram seminal PGHS has been reported with even at high imidazole levels as much as more than 500 mM (Tsai et al., 1993 although addition of imidazole changed the absorption spectrum of PGHS, which indicated that imidazole could bind to the PGHS. This was again a big difference in the properties of rice α-oxygenase and mammalian PGHSs. However, they were not potent inhibitors for rice α-oxygenase activity. It may suggest that the conformation of substrate binding site of rice α-oxygenase is somewhat different from that of mammalian PGHSs.

Site-directed mutagenesis

Through comparison of the rice α-oxygenase sequence with those of mammalian PGHSs, we found that most residues reported to be essential to the catalytic activity of PGHSs are conserved in the rice sequence. In order to determine whether they also function as essential residues to support the α-oxidation activity, site-directed mutageneses were carried out. The results are summarized in Fig. 2.

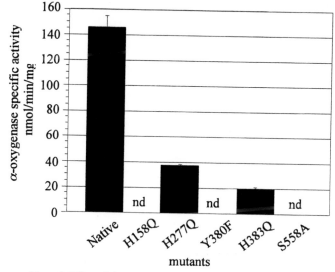

Figure 2. **Effect of site-directed mutagenesis on α-oxygenase activity.**
The activity was determined by HPLC. The error bars represent the
mean ± S.D. (*n*=3). nd, not detected.

Within the five mutated proteins, H158Q, Y380F, and S558A, showed no detectable activity with either the assay following oxygen consumption or formation of fatty aldehydes. His-158 could be the heme ligand of α-oxygenase, and the inhibition of α-oxygenase activity by imidazole indicates that the residue takes a role as one of the heme ligands. Tyr-380 and Ser-558 were also important residues for the activity. The sequences around the tyrosine residues in PGHS and α-oxygenase are highly conserved.

The fact that Y380F showed no activity indicates that the tyrosine residue is essential to the oxygenase probably through formation of tyrosyl radical. It was interesting to know that Ser-558 is essential for the activity of rice α-oxygenase.

References

Hamberg, M., Sanz, A., Castresana, C. (1999) Alpha-oxidation of fatty acids in higher plants. Identification of a pathogen inducible oxygenase (piox) as an alpha-dioxygenase and biosynthesis of 2-hydroperoxylinolenic acid . *J. Biol. Chem.* **274**, 24503-13.

Galliard, T., Matthew, J. A. (1976) The enzymic formation of long chain aldehydes and alcohols by alpha-oxidation of fatty acids in extracts of cucumber fruit (Cucumis sativus). *Biochim. Biophys. Acta* **424**, 26-35.

Psylinakis, E., Davoras, E. M., Ioannidis, N., Trikeriotis, M., Petrouleas, V., Ghanotakis, D. F. (2001) Isolation and spectroscopic characterization of a recombinant bell pepper hydroperoxide lyase. *Biochim. Biophys. Acta* **1533**, 119-27.

Tsai, A. L., Kulmacz, R. J., Wang, J. S., Wang, Y., Van Wart, H. E., Palmer, G. (1993) Heme coordination of prostaglandin H synthase. *J. Biol. Chem.* **268**, 8554-8563.

THE LIPOXYGENASE PATHWAY IN MYCORRHIZAL ROOTS OF *MEDICAGO TRUNCATULA*

M. STUMPE[1], I. STENZEL[2], H. WEICHERT[1], B. HAUSE[2] AND I. FEUSSNER[1]

[1]*Institute of Plant Genetics and Crop Plant Research (IPK) Corrensstrasse 3, D-06466 Gatersleben, Germany*
[2]*Institute of Plant Biochemistry, Weinberg 3, D-06120 Halle/Saale, Germany*

1. Introduction

Mycorrhizas are by far the most frequent occurring beneficial symbiotic interactions between plants and fungi. Species in >80 % of extant plant families are capable of establishing an arbuscular mycorrhiza (AM). In relation to the development of the symbiosis the first molecular modifications are those associated with plant defense responses, which seem to be locally suppressed to levels compatible with symbiotic interaction (Gianinazzi-Pearson, 1996). AM symbiosis can, however, reduce root disease caused by several soil-borne pathogens. The mechanisms underlying this protective effect are still not well understood. In plants, products of the enzyme lipoxygenase (LOX) and the corresponding downstream enzymes, collectively named LOX pathway (Fig. 1B), are involved in wound healing, pest resistance, and signaling, or they have antimicrobial and antifungal activity (Feussner and Wasternack, 2002). The central reaction in this pathway is catalyzed by LOXs leading to formation of either 9- or 13-**h**ydro**p**eroxy **o**cta-deca(**di**/**t**rien)oic acids (9/13-HPO(D/T); Brash, 1999). Thus LOXs may be divided into 9- and 13-LOXs (Fig. 1A). Seven different reaction branches within this pathway can use these hydroperoxy polyenoic fatty acids (PUFAs) leading to (i) keto PUFAs by a LOX; (ii) epoxy hydroxy-fatty acids by an **e**poxy **a**lcohol **s**ynthase (EAS); (iii) octadecanoids and jasmonates via **a**llene **o**xide **s**ynthase (AOS); (iv) leaf aldehydes and leaf alcohols via fatty acid **h**ydro**p**eroxide **l**yase (HPL); (v) hydroxy PUFAs (reductase); (vi) divinyl ether PUFAs via **d**ivinyl **e**ther **s**ynthase (DES); and (vii) epoxy- or dihydrodiol-PUFAs via **p**er**ox**ygenase (POX; Feussner and Wasternack, 2002). AOS, HPL and DES belong to one subfamily of P450-containing enzymes, the *CYP74* family (Feussner and Wasternack, 2002). Here, the involvement of this *CYP74* enzyme family in mycorrhizal roots of *M. truncatula* during early stages of AM symbiosis formation was analyzed.

287

N. Murata et al. (eds.), Advanced Research on Plant Lipids, 287–290.
© 2003 *Kluwer Academic Publishers. Printed in the Netherlands.*

288

2. Materials and methods

Cultivation and inoculation of *M. truncatula*. cv. Jemalong aas well as the propagation of the AM fungus *Glomus intraradices* Schenck & Smith was performed as described (Maier et al., 1995). Expression and analysis of recombinant *CYP74* enzymes was carried out as before (Matsui et al., 2000). Oxylipin profiling followed a protocol described by Weichert et al. (1999; 2002).

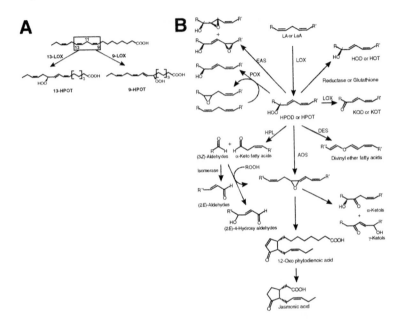

Figure 1. A: The positional specificity of LOXs. B: The LOX pathway.

3. Results and discussion

Three cDNAs coding for enzymes of the *CYP74* family were isolated from a cDNA library prepared from mycorrhizal roots of *M. truncatula* cv. Jemalong A17 (provided by Dr. M. Harrison, Ardmore, USA) by hybridization with PCR products obtained from *M. truncatula* leaf mRNA by RT-PCR using primers from EST sequence data. Due to homologies to other *CYP74*s the sequences were named MtAOS and MtHPL1,2, respectively, and were deposited into GenBank (Acc.-No.: CAB54847-9). For biochemical identification cDNAs were expressed in *E. coli* and the recombinant proteins were analyzed for products formed by incubation with radioactive 13-HPOT (Fig. 2A, B). As shown in Fig. 2A MtAOS could be identified as an AOS by formation of α-ketol and it was specific for 13-LOX-derived products (not shown). MtHPL1,2, however, formed ω-oxo fatty acids (Fig. 2B). Both enzymes showed no preference against a positional isomer as shown by aldehyde formation after incubation of MtHPL1,2 with a mixture of 9-HPOD and 13-HPOT (Fig. 2C). Thus MtAOS could be identified as a 13-AOS (*CYP74A*), and MtHPL1,2 belong to the family of 9/13-HPLs (*CYP74C*).

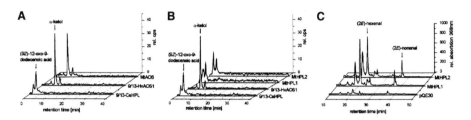

Figure 2. Product analysis of MtAOS (A) and MtHPL1,2 (B, C). A, B: Recombinant enzymes were incubated with 13-[$^{14}C_1$]-HPOT and products were analyzed as described by radio-HPLC (Matsui et al., 2000). C: Aldehyde formation was detected by HPLC via the 2.4-dinitrophenyl hydrazones after incubation with a mixture of 9-HPOD and 13-HPOT, respectively (Weichert et al., 2002).

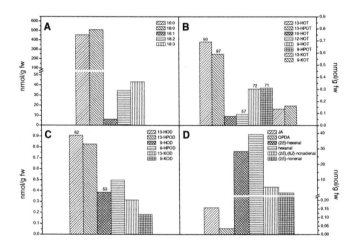

Figure 3. Oxylipin signature of mycorrhizal roots of *M. truncatula*. A: Free fatty acids, B: Hydro(pero)xides of 18:3, C: Hydro(pero)xides of 18:2, D: Oxylipins of the AOS pathway (JA: jasmonic acid, OPDA: 12-oxo phytodienoic acid) and of HPL pathway. The amount of *S* enatiomers is given above the column.

Since an enzyme representing the AOS branch and two enzymes representing the HPL branch of the LOX pathway had been isolated, the oxylipin signature of mycorrhizal roots was analyzed in order to find out whether both pathways may be active in this tissue and if one of these pathways may be the preferred one. The substrates of the LOX reaction, free 18:2 and 18:3, respectively, were found to accumulate to higher amounts than all oxylipins detected, indicating that the initial substrate for the LOX reaction may not be limiting in this tissue (Fig. 3A). At the level of the next reaction step, the LOX reaction, a strong preponderance of 13-H(P)O(D/T)s could be detected showing a specific excess of 13-LOX-derived products over 9-LOX-derived hydro(pero)xy PUFAs (Fig. 3B, C). They may serve as substrates for the following reactions, i.e. catalyzed by *CYP74*-enzymes. This was substantiated by the presence of JA and OPDA, respectively, which are specific metabolites of the 13-AOS pathway. However, the HPL pathway may be the dominant reaction branch of the LOX pathway in this tissue, because HPL-derived aldehydes accumulated in a more than 200-fold excess over that of the octadecanoids (Fig. 3D), as well as in a 50-fold excess over the products of the reductase reaction (HO(D/T)s; Fig. 3B, C). In addition, within the group of aldehydes the 13-HPL-derived metabolites hexanal and (2*E*)-hexenal were found preferentially (Fig. 3D), indicating that the specific formation of 13-LOX-derived substrates determines the specificity of this reaction although the isolated HPLs from *M. truncatula* seem to have no substrate preference. Future analysis will show whether the HPL pathway is specifically induced during early stages of AM symbiosis formation in mycorrhizal roots, i.e. by jasmonates, and how this fungus is protected against these toxic aldehydes, while the growth of other fungi seem to be inhibited by these substances (Croft et al., 1993).

4. References

Brash, A.R. (1999) Lipoxygenases: Occurrence, functions, catalysis, and acquisition of substrate. J. Biol. Chem., 274, 23679-23682.

Croft, K.P.C., Juttner, F. and Slusarenko, A.J. (1993) Volatile products of the lipoxygenase pathway evolved from *Phaseolus vulgaris* (L.) leaves inoculated with *Pseudomonas syringae* pv *phaseolicola*. Plant Physiol., 101, 13-24.

Feussner, I. and Wasternack, C. (2002) The lipoxygenase pathway. Annu. Rev. Plant Physiol. Plant Mol. Biol., 53, 275-297.

Gianinazzi-Pearson, V. (1996) Plant cell responses to arbuscular mycorrhizal fungi: Getting to the roots of the symbiosis. Plant Cell, 8, 1871-1883.

Maier, W., Peipp, H., Schmidt, J., Wray, V. and Strack, D. (1995) Levels of a terpenoid glycoside (blumenin) and cell wall-bound phenolics in some cereal mycorrhizas. Plant Physiol., 109, 465-470.

Matsui, K., Ujita, C., Fujimoto, S., Wilkinson, J., Hiatt, B., Knauf, V., Kajiwara, T. and Feussner, I. (2000) Fatty acid 9- and 13-hydroperoxide lyases from cucumber. FEBS Lett., 481, 183-188.

Weichert, H., Stenzel, I., Berndt, E., Wasternack, C. and Feussner, I. (1999) Metabolic profiling of oxylipins upon salicylate treatment in barley leaves - preferential induction of the reductase pathway by salicylate. FEBS Lett., 464, 133-137.

Weichert, H., Kolbe, A., Kraus, A., Wasternack, C. and Feussner, I. (2002) Metabolic profiling of oxylipins in germinating cucumber seedlings - lipoxygenase-dependent degradation of triacylglycerols and biosynthesis of volatile aldehydes. Planta, in press.

PURIFICATION AND IMMUNOLOGICAL PROPERTIES OF GALACTOLIPASE FROM *PHASEOLUS VULGARIS* LEAVES

T. SAKAKI[1], N. NAKAJIMA[2] and Y. HAMAYA[1]

[1]*Hokkaido Tokai University, Sapporo, 005-8601, Japan*
[2]*National Institute for Environmental Studies, Tsukuba, 305-0053, Japan*

1. Introduction

Galactolipid-hydrolyzing enzymes (galactolipases) have been reported to occur in leaves, seeds, and tubers of some plant species (Huang, 1987).   In leaf cells, at least one of these enzymes is associated with chloroplast membranes where the substrate for the enzyme is abundant (Anderson et al., 1974; O'Sullivan et al., 1987).   Because galactolipase is found to be active in the isolated chloroplasts, it is assumed that the enzyme should exist *in vivo* in an inactive form.   We showed that galactolipids such as mono- and digalactosyldiacylglycerol (MGDG and DGDG) were hydrolyzed in leaf cells when plants were subjected to stress conditions like ozone exposure (Sakaki et al., 1990).   The hydrolysis of galactolipids also increased in leaves in response to wounding (Conconi et al., 1996).   Thus it is possible that galactolipase activity is enhanced in leaf cells under such stress conditions.   To know the regulation of galactolipid hydrolysis in leaf cells, we purified galactolipase from kidney bean leaves and determined its enzymic and immunological properties.

2. Materials and Methods

Various plants were grown in a glasshouse or in the farm of Hokkaido Tokai University. Galactolipase activity was measured according to the method of Anderson et al. (1974) with modifications of the reaction mixture (1ml), which contained 100mM citrate-NaOH buffer (pH 6.0), 0.4% (w/v) Triton X-100, 1mM spinach MGDG as the substrate and enzyme.   One unit of enzyme was defined as described in Matsuda et al. (1979). Several kinds of column chromatography used for the purification of galactolipase were performed with the elution solvents as follows: Sephadex G-150 and HiPrep Sephacryl S-200, 50mM K-Pi buffer (pH 7) and 0.15M NaCl; DEAE-Toyopearl, 50mM K-Pi buffer (pH 7) with linear gradient of NaCl (0-0.3M or 0-0.11M); phenyl-Sepharose, 20mM K-Pi buffer (pH 7) with linear gradient of NaCl (2.0-0M) and ethylene glycol (0-

291

N. Murata et al. (eds.), Advanced Research on Plant Lipids, 291–294.
© 2003 *Kluwer Academic Publishers. Printed in the Netherlands.*

70%); butyl-Toyopearl, 50mM K-Pi buffer (pH 7) and 1mM EDTA with linear gradient of sodium cholate (0-30mM); Con-A Sepharose, 50mM K-Pi buffer (pH 7) and 0.25M NaCl with linear gradient of methyl glucoside (0-50mM). Native- and SDS-PAGE were carried out with separation gels of 7.5% and 10% polyacrylamide, respectively. After SDS-PAGE, proteins were transferred to a PVDF membrane and stained with rabbit anti-galactolipase and alkaline phosphatase-conjugated goat anti-rabbit IgG.

TABLE 1. Galactolipase activities in leaves of various cultivars of *Phaseolus vulgaris* L. and other plant species

Plant species (Common name)	Cultivar	Specific activity[1] (mU/mg protein)	
		Particulate	Soluble
Phaseolus vulgaris L. (Kidney bean)	Koshu-Shakugosun	343.5	96.6
	Kentucky Wonder	225.9	53.3
	Hon-Kintoki	102.9	9.6
	Shiro-Kinugasa	55.5	10.0
	Kurodane-Kinugasa	36.3	7.2
	Shin-Edogawa	0.94	0.31
	Edogawa	0.78	0.38
	Masterpiece	0.58	0.18
	Naga-Uzura	0.34	0.06
	Satsuki-Midori (No. 2)	0.16	0.05
	Yamashiro-Kurosando	0.12	0.07
Vigna angularis (Willd.) Ohwi & Ohashi (Azuki bean)	Doyo	1747	239
	Wase	1380	169
Glycine max Merr. (Soybean)	Vilocci	5.00	25.7
Vicia faba L. (Broad bean)	Otafuku	0.19	1.08
Helianthus annuus L. (Sunflower)	Kawa (No. 55)	0.04	0.12
Lactuca sativa L. (Lettuce)	White Pallace	0.08	0.09
Zea mays L. (Maize)	Yellow Dentcorn	0.20	ND
Lycopersicon esculentum Mill. (Tomato)	Fukuju (No.2)	ND	0.01
Spinacia oleracea L. (Spinach)	New Asia	ND	0.01
Raphanus sativus L. (Radish)	Minowase	ND	ND
Nicotiana tabacum L. (Tobacco)	Xanthi NC	ND	ND
Arachis hypogaea L. (Peanut)	Chiba-handachi	ND	ND

The youngest leaves fully developed were homogenized in 20 mM Tris/HCl buffer (pH 7.0) containing 0.4 M sucrose and fractionated by centrifugation at 20,000xg for 15 min. The resultant precipitate (particulate) and supernatant (soluble) fractions were assayed for the enzyme activity and the protein content. ND, not detected.

[1]Average of two separate samples.

3. Results and Discussion

To survey the occurrence of galactolipases in plants, the enzyme activity was assayed in the soluble and particulate fractions of leaf homogenates (TABLE 1).   We found that the activities in both fractions varied not only in plant species but also in cultivars of the same plant species.   Although azuki bean had the highest activity in both fractions, kidney bean plants were particularly interesting since the activity largely varied in eleven cultivars of this plant species (TABLE 1).   We selected the kidney bean cultivar, Koshu-Shakugosun, for further investigation of galactolipase and purified the enzyme from the particulate fraction, rich in chloroplast membranes, of the leaf homogenates.   Partial purification of leaf galactolipase from other kidney bean cultivar has already been reported by Anderson et al. (1974), followed by further purification to a single band on native-PAGE (Matsuda et al., 1979).

We purified the enzyme with two different procedures (TABLE 2).   First we tried to apply the affinity chromatography on palmitoylated gauze (Matsuda et al., 1979) after the step of heat treatment (TABLE 2), but the method did not work in our hands.   Thus we used common chromatographic methods, gel-filtration, anion-exchange and hydrophobic chromatography, and native-PAGE for further purification (Expt. 1 in TABLE 2).   The purified enzyme showed a single band on native-PAGE. The enzyme became unstable after the first ion-exchange chromatography as reported

TABLE 2.   Purification of galactolipase from particulate fractions of *Phaseolus vulgaris* (cv. Koshu-Shakugosun) leaves

Expt.	Steps	Total activity (Unit)	Total protein (mg)	Specific activity (Unit/mg protein)	Purification (-fold)
1	Particulate fraction	3986	15770	0.25	1.0
	Extract from acetone ppt[1]	2381	2937	0.81	3.2
	Heat extract[2]	2087	2274	0.92	3.7
	Sephadex G-150	2158	459	4.71	18.8
	DEAE-Toyopearl (1st)	1338	57.3	23.4	93.6
	Phenyl-Sepharose	780	19.3	40.4	162
	DEAE-Toyopearl (2nd)	670	3.8	176	704
	Native-PAGE	325	1.5	217	868
2	Heat extract[2]	881	313	2.81	3.7
	Butyl-Toyopearl	526	4.52	116	152
	DEAE-Toyopearl	261	0.755	346	453
	Con-A Sepharose	135	0.306	441	578
	HiPrep Sephacryl S-200	105	0.137	766	1005

[1] Acetone precipitates prepared from the particulate fraction were extracted with 50 mM K-Pi buffer (pH 7.0) containing 1 mM EDTA.

[2] Extracts from the acetone precipitates were heated at 65°C for 2 min.

by Anderson et al. (1974). In agreement with Matsuda et al. (1979), the purified enzyme hydrolyzed not only MGDG but also DGDG, SQDG, PC, PI, and PG, whereas it was inactive to TAG, indicating that the enzyme was classified into non-specific lipid acyl hydrolase (Huang, 1987). We prepared polyclonal antibodies against the purified enzyme with rabbit, which were shown to completely inhibit the crude and purified enzyme.

To obtain more stable galactolipase preparations, we tried to re-purify the enzyme with another procedure (Expt. 2 in TABLE 2). Hydrophobic chromatography on a butyl-Toyopearl column was found to be very effective for the purification and stability of the enzyme (TABLE 2). The purified enzyme showed a molecular mass of about 78 kDa by gel-filtration on HiPrep Sephacryl S-200, whereas it showed a predominant protein band of 38 kDa with a faint 53 kDa band by SDS-PAGE. The specific acitvity of the purified enzyme (766 U/mg) was 220- and 160-fold higher than those of the enzymes previously reported (Anderson et al., 1974; Matsuda et al., 1979). The N-terminal amino acid sequence of both polypeptides was determined.

Since the antibodies obtained with the first procedure reacted with many protein bands including 38 and 53 kDa polypeptides in the crude protein extracts from acetone precipitates, we purified the antibodies with a lectin-agarose column. The antibodies unabsorbed to the column reacted with only 38 kDa and 53 kDa polypeptides, suggesting that the original antibodies reacted not only with these polypeptides but also with polysaccharide chains bound to the enzyme. We further purified the antibodies with a PVDF membrane blotted with the 38 kDa band. The purified antibody inhibited galactolipase activity, indicating that at least the 38 kDa polypeptide was true galactolipase.

4. References

Anderson, M.M., McCarty, R.E. and Zimmer, E.A. (1974) The role of galactolipids in spinach chloroplast lamellar membranes. I. Partial purification of a bean leaf galactolipid lipase and its action on subchloroplast particles. Plant Physiol. 53, 699-704.

Conconi, A., Miquel, M., Browse, J.A. and Ryan, C.A. (1996) Intracellular levels of free linolenic and linoleic acids increase in tomato leaves in response to wounding. Plant Physiol, 111. 797-803.

Huang, A.H.C. (1987) Lipases. in P.K. Stumpf (ed.), The Biochemistry of Plants, Vol 4, Lipids: Structure and Function. Academic Press, Orlando, pp. 1-55.

Matsuda, H., Tanaka, G., Morita, K. and Hirayama, O. (1979) Purification of a lipolytic acyl-hydrolase from *Phaseolus vulgaris* leaves by affinity chromatography on palmitoylated gauze and its properties. Agric. Biol. Chem. 43, 563-570.

O'Sullivan, J.N., Warwick, N.W.M. and Dalling, M.J. (1987) A galactolipase activity associated with the thylakoids of wheat leaves (*Triticum aestivum* L.). J. Plant Physiol. 131, 393-404.

Sakaki, T., Kondo, N, and Yamada, M. (1990) Pathway for the synthesis of triacylglycerols from monogalactosyldiacylglycerols in ozone-fumigated spinach leaves. Plant Physiol. 94, 773-780.

HYDROLYSIS BEHAVIOR OF TRIGLYCERIDES BY *Rhizopus delemar* LIPASE IN REVERSE MICELLAR ORGANIC SOLVENTS

K.NAOE, Y. YAMADA, S. AWATSU, M. KAWAGOE
Department of Chemical Engineering, Nara National College of Technology
22 Yata, Yamato-Koriyama, Nara 639-1080, Japan
K. NAGAYAMA
Department of Material Science and Engineering, Kochi National College of Technology
200-1 Monobe, Nankoku, Kochi 783-8508, Japan
M. IMAI
Department of Food Science and Technology, College of Bioresource Science, Nihon University
1866 Kameino, Fujisawa, Kanagawa 252-8501, Japan

1. Introduction

The structure of a reverse micelle consists of an aqueous micro-domain facing the polar heads of the amphiphilic molecule that surrounds this core interacting with the bulk organic solvent, through the hydrophobic chains. The polar cores of the micelles have the ability to solubilize a significant amount of water. When the enzymes are microencapsulated, they are located in the interior aqueous phase of reverse micelles, and interacting with the micellar interface depending on the enzyme species. Enzymes have been traditionally used in aqueous media, but reverse micelles became attractive, especially, when substrates and/or products are hydrophobic and a low water content is desired. This applies specially to lipase reactions since the solubility of triglycerides is largely improved in organic solvents (Han and Rhee, 1986; Nagayama et al., 1996). Furthermore, since the minute size of water pools in reverse micelles on a nanometer scale, reverse micellar organic solvent offers enormous W/O interface area in an organic solvent. Owing to these features, the reverse micellar organic solvent is anticipated as reaction medium for industrial processes of lipase-catalyzed reaction using a hydrophobic substrate (Lee et al., 1998; Naoe et al., 2001).

In this study, the effect of system parameters (solvent type, water content, type of plant triglyceride) on hydrolysis activity of lipase in reverse micellar organic system was investigated.

N. Murata et al. (eds.), Advanced Research on Plant Lipids, 295–298.
© 2003 *Kluwer Academic Publishers. Printed in the Netherlands.*

2. Materials and Methods

Sodium bis(2-ethylhexyl) sulfosuccinate (Aerosol-OT, abbreviated as AOT), used as an amphiphilic molecule, was obtained from Nacalai Tesque, Inc. (Kyoto, Japan). The organic solvent isooctane (2,2,4-trimethylpentane) was purchased from Wako Pure Chemical Industries Ltd. (Osaka, Japan). Other solvents (n-heptane, n-octane, n-decane, and cyclohexane) were obtained from Kishida Chemical Co., Ltd. (Osaka, Japan). The triglycerides triolein and trilinolein, used as substrates, were purchased from Wako Pure Chemical Industries and Tokyo Kasei Ltd. (Tokyo, Japan), respectively. Lipase (*Rhizopus delemar*) was purchased from Seikagaku Kogyo (Tokyo, Japan). They were used without further purification.

Enzymatic activity of lipase (*Rhizopus delemar*) in this system was examined by its hydrolysis of triglyceride at 303 K. The enzymatic reaction was started by injecting lipase-containing buffer solution (6 mg-lipase/0.05 mL, 0.1 M acetate buffer, pH 5.6) into the pre-incubated AOT/solvent solution (39 mL) containing substrates and buffer solution at desired concentrations. The enzymatic activity was determined by measuring the increasing profile of free fatty acid produced according to the method previously described (Lowry and Tinsley, 1976). The water content of the reverse micellar organic phase was measured by Karl-Fisher titration.

3. Results and Discussion

3.1. *Effect of solvent on initial reaction rate of triolein hydrolysis*

Figure 1 shows the initial reaction rate of triolein hydrolysis in reverse micellar organic systems composed of different non-polar organic solvent at constant water content of the organic phase. The initial reaction rate depended on the organic solvent preparing the micellar phase. Highest initial reaction rate was obtained in the system prepared of isooctane, rather than those of straight chain alkanes. The cyclohexane system exhibited a lowest initial reaction rate. In the cases of straight chain alkane system, longer chain solvent showed higher initial reaction rate, indicating that the initial reaction rate was correlated with the hydrophobicity of solvent. The deviation of results in the isooctane and cyclohexane systems from those of the straight chain alkane systems implies that the structure of solvent may be a significant factor in the hydrolysis reaction by lipase in reverse micellar system, rather than the hydrophobicity of solvent.

3.2. *Effect of water content in micellar organic phase on initial reaction rate*

Figure 2 shows the effect of water content in reverse micellar organic phase on initial reaction rate of triolein hydrolysis by lipase in AOT/isooctane, n-heptane reverse

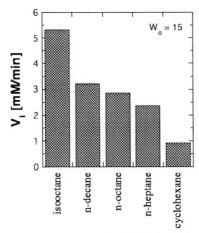

Figure 1. The initial reaction rate of hydrolysis of triolein by lipase in AOT reverse micellar systems prepared of different solvent.

micellar systems. Water content in micellar organic phase is expressed as the molar ratio of water to AOT in the organic phase, W_0. The initial reaction rate in the "bulky" isooctane system was strongly dependent on the water content W_0 and took a maximum at ca. 15 of W_0, while the initial reaction rate of the heptane system showed almost no change with the W_0 value. As the size of reverse micelle is correlated linearly with the W_0 value (Pileni et al., 1988), the W_0 change gives the change in the curvature of reverse micellar interface. The penetration of solvent molecule into the micellar interface may influence the reaction.

3.3. Effect of triglyceride structure on initial reaction rate of hydrolysis

Table 1. shows the initial reaction rate of hydrolysises of triolein and trilinolein by

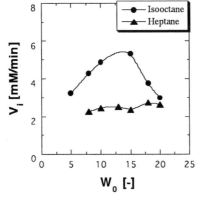

Figure 2. Effect of water content in micellar organic phase on initial reaction rate of triolein hydrolysis by lipase in AOT reverse micellar systems. Water content W_0 is molar ratio of water to AOT in organic phase. AOT conc. = 50 mM.

TABLE 1. Initial reaction rate of hydrolysis of triglycerides in
AOT/isooctane reverse micellar organic solution.

Triglyceride	V_i (mM/min)
Triolein	5.32
Trilinolein	3.99

$W_0 = 15$, AOT conc. $= 50$ mM, Triglyceride conc. $= 72$ mM

lipase in AOT/isooctane reverse micellar system. The initial reaction rate of triolein hydrolysis was higher than that of trilinolein. Trilinolein has more "bulky" structure with two double bonds than triolein with single one. This result implies that the penetration of triglyceride molecule into the micellar interface also influence the reaction in this system.

4. Conclusion

The effect of system parameters on hydrolysis activity of lipase in reverse micellar organic system was investigated. Among the solvents preparing the system, isooctane exhibited a highest initial reaction rate of triolein hydrolysis. In the isooctane system, the initial reaction rate depended on the water content. The initial reaction rate of triolein hydrolysis was higher than that of trilinolein. These results imply that the penetration of solvent and/or substrate into the micellar interface influence the reaction.

5. Nomenclatures

V_i: initial reaction rate of hydrolysis [mM/min]
W_0:   molar ratio of water to AOT in organic phase [mol-H_2O/mol-AOT]

6. References

Han, D. and Rhee, J. S., (1986) Characteristics of lipase-catalyzed hydrolysis of olive oil in AOT-isooctane reversed micelles: effect of water on equilibrium, Biotechnol. Bioeng., 28, 1250-1255.

Lee, K. K. B., Poppenborg, L. H., and Stuckey, D. C., (1998) Terpene ester production in a solvent phase using a reverse micelle-encapsulated lipase, Enzyme Microb. Technol., 23, 253-260.

Lowry, R. R. and Tinsley, I. J., (1976) Rapid colorimetric determination of free fatty acids, J. Am. Oil Chem. Soc., 53, 470-472.

Nagayama, K., Imai, M., Shimizu, M., and Doi, T., (1996) Kinetic study of lipase-catalyzed hydrolysis in AOT and lecithin microemulsion systems, Seibutsu Kogaku, 74, 255-260.

Naoe, K., Ohsa, T., Kawagoe, M., and Imai, M., (2001) Esterification by *Rhizopus delemar* lipase in organic solvent using sugar ester reverse micelles, Biochem. Eng. J., 9, 67-72.

Pileni, M. P., Zemb, T., and Petit, C., (1985) Solubilization by reverse micelles: solute localization and structure perturbation, Chem. Phys. Lett., 118, 414-420.

HYDROPEROXIDE LYASE AND LEAF ALDEHYDE FORMATION CAN BE GREATLY INCREASED IN LEAVES

WIPAWAN SIANGDUNG[1], HIROTADA FUKUSHIGE and DAVID HILDEBRAND
Department of Agronomy,
University of Kentucky, Lexington, KY 40546, U.S.A.

Introduction

2(*E*)-hexenal, leaf aldehyde, is used as flavor to impart a green character and freshness in foods and beverages. Leaf aldehyde is formed by the hydroperoxide lyase (HL) branch pathway of oxylipin pathway (Zimmerman and Coudron, 1979; Hatanaka et al., 1987; Vick and Zimmerman, 1987b). Linolenic acid (18:3), a major fatty acid in leaf lipids, is first oxidized by lipoxygenase (LOX) to produce 13-hydroperoxy 18:3 (13-HPOT). 13-HPOT is then cleaved by HL to produce the C_6 aldehyde, 3(*Z*)-hexenal and a 12-carbon oxo acid. 3(*Z*)-hexenal is then isomerized to leaf aldehyde or reduced to 3(*Z*)-hexenol, leaf alcohol (Fig. 1). Though this compound can been synthesized chemically, there are increasing demands for natural flavors as well as natural chemicals for fungicides and pesticides as more and more consumers are interested in

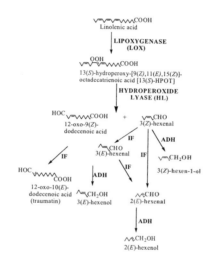

Figure 1. HL branch of the oxylipin pathway
IF = Isomerization Factor, ADH = Alcohol Dehydrogenase

natural products, organic production and 'green' industries (Whitehead et al., 1995). Also a more efficient plant-based production system will be more competitive with petroleum-based production making use from such a renewable source more economically attractive. Currently commercial production of natural leaf aldehyde is achieved from watermelon leaves because of the high yield of this compound from this source. This involves making large-scale homogenates of watermelon leaves with added linolenic acid or hydrolyzed linseed oil (Chou and Chin, 1994; Holtz et al. 2001). Hildebrand et al. (1990 and 1991) and Whitehead et al. (1995) proposed the usage of soybean LOX to produce 13-HPOT in combination with plant materials to generate C_6 aldehydes. Leaf aldehyde is currently valued as much as $100 per kg and the market for such compounds has increased > 70% in the past 10 years. This research reports the increased formation of leaf aldehyde from plant leaves by enhanced expression of the key enzyme involved in its formation and the usage of plant leaves as a factory for this compound.

N. Murata et al. (eds.), Advanced Research on Plant Lipids, 299–302.
© 2003 *Kluwer Academic Publishers. Printed in the Netherlands.*

Materials and Methods

The primers were designed based on the published sequence of *Arabidopsis thaliana HL* gene (Bate et al., 1998) with an added XhoI restriction site.
Forward primer: 5'ATAGCTCGAGATGTTGTTGAGAACGATG-3',
Reverse primer: 5'-GCGGCTCGAG TTA TTTAGCTTTAACAAC-3'.
The HL cDNA was amplified by Promega's Access RT-PCR System kit and cloned into pGEM-T Easy vector. HL coding region was introduced into the XhoI site of the expression vector pKYLX71:35S2 (Schardl et al., 1987). The pKYLX71:35S2-HL-KM construct was introduced into *Agrobacterium tumefaciens* AGV3850 by the freeze-and-thaw method (An et al., 1988). The desired transformants were confirmed by restriction enzyme analysis and electrophoresis before tobacco leaf infection. Leaves from *Nicotiana tabacum* tobacco seedlings (8 weeks old) were transformed with the leaf dip method.

About 0.2 g of leaves (transgenic, wild type lines of tobacco and watermelon) were frozen in liquid nitrogen, ground to a powder in a chilled motar, and pestle and then further homogenized in 0.5 mL of 50 mM potassium phosphate, pH 6.8, 3 mM EDTA, 3mM DTT, 0.5% Triton X-100, 1% Protease inhibitor cocktail (Sigma) and 5% polyvinylpolypyrolidone. After incubation for 30 min on ice, the homogenates were centrifuged at 13,000 rpm for 15 min at 4°C to remove cell debris. Then the supernatants were used in the enzyme assay step. The protein content was determined by a modified Lowry method. HL activities were assessed by the indirect assay described by Vick (1991). The reaction was started by adding 2 µL crude extract into 998 µL of a mixture of 0.2 mM 13-HPOT, 0.2mM NADH, 15 µg alcohol dehydrogenase (ADH). The decrease of NADH during the conversion of aldehydes to alcohols was followed at 340 nm as a measure of the activity of enzyme.

About 0.2 g of mature leaves were cut into small pieces, 50 µL 10% 18:3 (or 89.8 µmol/g F.W.) available from Sigma Co., 100 mg of soybean acetone powder and 2.0 mL deionized water. The mixture in the test tube was homogenized with a Tissuemizer for 1 min at room temperature. The homogenate 50 µL of 10 mM 2(*E*)-heptenal (in ethanol) then was added as an internal standard, 2 mL of saturated NaCl solution, and 2 mL of diethyl ether (EE) and vortexed. The tube was left for 1 min at room temperature before centrifuging. The centrifugation was conducted for a couple of minutes in SpeedVac for phase separation. The EE phase was collected into a new small test tube and dried over Na_2SO_4. The dried sample was transferred to a GC vial, and analyzed by GC within 24 h on a 0.2 mm x 14 m FFAP column. In this work, in addition to leaf aldehyde (2(*E*)-hexenal), 3(*Z*)-hexenal was found.

Results and Discussions

Constitutive expressing regenerated T_0 plants were selected on kanamycin.. Among 23 transgenic lines, two lines N8 and N24, had greatly increased HL activity comparable to wild type and controls. Particularly, line N24 had approximately 5-fold higher amounts or more HL activity than that of watermelon leaves (Fig. 2).

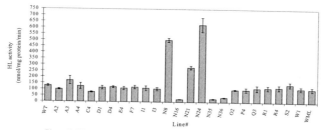

Figure 2. HL activity of transformed tobacco and watermelon (WML) leaves.
Note:WT= wild type, I1 and I3 = vector controls. Values represent means +/−SE of replicates (n=3-7)

Grinding green tissue with 18:3 (substrate of oxylipin pathway) increased C_6 aldehydes. When soybean acetone powder (LOX source) and green tissues were combined and ground with 18:3, the increase in C_6 aldehyde formation was much greater than using the green tissues and 18:3 alone (Fig. 3). N24 and watermelon leaves produced similar amounts of C_6 aldehydes after addition of 18:3.

N24 and N8 were found to consistently have higher leaf aldehyde levels than watermelon. Although the HL activity of N24 had 5 times greater than watermelon, the leaf aldehyde production was only 1.3 times when compare to watermelon suggesting LOX is a limiting factor for C_6 aldehyde and alcohol formation in green tissues.

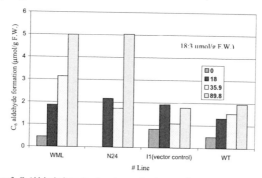

Figure 3. C_6 Aldehyde formation from transgenic tobacco and watermelon leaves homogenized with soybean meal and 18:3 in different amounts

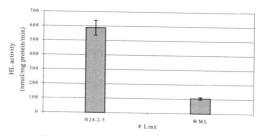

Figure 4. HL activity of T_1 transgenic tobacco and watermelon leaves

Figures 4 and 5 indicate that T_1 generation of N24 has the high HL activity corresponding to the amount of C6 aldehyde formation.

302

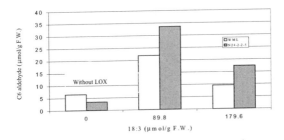

Figure 5. C_6 aldehyde formation from soybean acetone powder and 18:3 homogenized with watermelon and T_1 transgenic tobacco leaves compared with production without added LOX.

Conclusion

An *Arabidopsis* HL cDNA was over expressed in tobacco leaves using the strong constitutive promoter, CaMV 35S. This showed the high activity of HL in transformed tobacco leaves. Grinding fresh leaves with 18:3 and soybean meal resulted in the high production of aldehydes (3(Z)-hexenal and 2(E)-hexenal) Among transgenic lines analyzed one, the T_1 N24 plant, has approximately 5-6-fold higher HL activity than that of watermelon leaves. The aldehydes level of N24 (T_1) is about 34 µmol/g F.W whereas the level in watermelon is about 30 µmol/g F.W. This small difference in leaf aldehyde production between N24 and watermelon suggests a limiting level of LOX in these reactions.

References

An, G., P.R. Ebert, A. Mitra and S.B. Ha. 1988. Binary Vectors. In: "Plant Molecular Biology Manual", Gelvin, S.B., Schilperoort, R.A. and Verma, D.P.S. (Eds.), Kluwer Academic Publishers, Boston, pp. A3:1-19.

Bate, N.J., S. Sivasankar, C. Moxon, J.M.C. Riley, J.E. Thompson and S.J. Rothstein. 1998. Molecular characterization of an Arabidopsis gene encoding hydroperoxide lyase, a cytochrome P-450 that is wound inducible. Plant Physiol. 117: 1393-1400.

Chou, S.-R. and C.-K. Chin. 1994. Control of the production of *cis*-3-hexenal, lipid-derived flavor compound, by plant cell culture. In: "Lipids in Food Flavors.", Ho, C.-T. and Hartman, T.G. (Eds.), American Chemical Society, Washington, D.C., pp. 282-291.

Hatanaka, A., T. Kajiwara and J. Sekiya. 1987. Biosynthetic pathway for C6-aldehyde formation from linolenic acid in green leaves. Chem. Phys. Lipids 44: 341-361.

Hildebrand, D.F., T.R. Hamilton-Kemp, J.H. Loughrin, K. Ali and R.A. Andersen. 1990. Lipoxygenase 3 reduces hexanal production from soybean seed homogenates. J. Agric. Food Chem. 38: 1934-1936.

Hildebrand, D.F., T.R. Kemp and R.A. Andersen. 1991 "Method of reducing odor associated with hexanal production in plant products ". Patent issued May 21, 1991. - U.S. Patent No. 5,017,386.

Holtz, R.B., M.J. McCulloch, S.J. Garger, R.K. Teague and H.F. Phillips. 2001. Method for providing green note compounds. U.S. Patent No. 6,274,358.

Schardl, C.L., A.F. Byrd, G. Bension, M.A. Altschuler, D.F Hildebrand and A.G. Hunt. 1987. Design and construction of a versatile system for the expression of foreign genes in plants. Gene 6: 1-11.

Vick, B.A. and D.C. Zimmerman. 1987b. Pathways of fatty acid hydroperoxide metabolism in spinach leaf chloroplasts. Plant Physiol. 85: 1073-1078.

Vick, V.A. 1991. A spectrophotometric assay for hydroperoxide lyase. Lipids 26: 315-320.

Whitehead, I.M., B.L. Muller and C. Dean. 1995. Industrial use of soybean lipoxygenase for the production of natural green note flavor compounds. Cereal Foods World 40: 193-194, 196-197.

Zimmerman, D.C. and C.A. Coudron. 1979. Identification of traumatin, a wound hormone, as 12-oxo-*trans*-10-dodecenoic acid. Plant Physiol. 63: 536-541.

DAD1-LIKE LIPASE (DLL) GENES IN *Arabidopsis thaliana*
Their Expression Profiles and Physiological Significances

K. MATSUI, AND T. KAJIWARA

Department of Biological Chemistry, Faculty of Agriculture, Yamaguchi University, Yamaguchi 753-8515, JAPAN

1. Introduction

Octadecanoid, or oxylipin, is a name for compounds derived from fatty acids through at least one step of oxygenation reaction. Jasmonates, short-chain aldehydes, hydroxylated fatty acids, and so on, are included in this family of compounds, and they are thought to be involved in resistance responses of plants against various biotic and abiotic stresses (Blée, 1998). In general, it is believed that octadecanoids are formed from lipids through hydrolysis to form free fatty acids, oxygenation by lipoxygenase or α-dioxygenase to form fatty acid hydroperoxides, and subsequent reaction on the hydroperoxides by fatty acid hydroperoxide lyase, allene oxide synthase, and so on. In the octadecanoid pathway, little is known about the first, lipid-hydrolyzing step (Matsui et al., 1999). Recently, Ishiguro et al. (2001) identified a phospholipase gene (*DAD1*) that is involved in jasmonate synthesis in anthers of *Arabidopsis*. *DAD1* expresses specifically in anthers, but little transcripts can be found in vegetative tissues, such as leaves. *dad1* mutant still has the ability to form jasmonates in leaves, and the mutant plants can form the other octadecanoids, such as short chain aldehydes as well as their corresponding wild type plants do (Matsui et al., unpublished). Thus, it can be assumed that the other *DAD1-like lipases (DLLs)* are involved in the octadecanoid syntheses in vegetative tissues. By searching *Arabidopsis* genome, it was shown that there are at least 13 genes homologous to *DAD1*. In order to identify the lipase accountable for the respective branch of the octadecanoid pathway, in this study, we attempted to reveal expression profiles of them.

2. Materials and Methods

N. Murata et al. (eds.), Advanced Research on Plant Lipids, 303–306.

Arabidopsis thaliana (No-0) was grown with soil at 22 C under 14-h light/10 h dark photo regime. Wounding was done on rosette leaves of one-month old plants by using forceps. Cold treatment was carried out at 4 C under continuous light. For treatment with chemicals, rosette leaves were cut at the base of petioles, and put into small plastic tubes filled with respective test solution. Chemicals used are, ABA; abscisic acid (150 µM), ACC; 1-aminocyclopropane-1-carboxylic acid (1 mM), NO; sodium nitroprusside (1 mM), MV; methyl viologen (10 µM), RB; Rose Bengal (50 *m*M), SA; salicylic acid (2 mM). Total RNA was extracted by using TRIzol reagent (Invitrogen), and reverse transcribed with oligo dT primer by using RT-PCR kit (Invitrogen). PCR was performed with ExTaq polymerase (Takara). As a control, actin (*AAc1*) gene was also amplified. The amount of cDNA used for PCR was respectively determined so that almost the same degree of amplification could be observed with the actin gene.

3. Results and Discussion

3-1. *Structure of DAD1 like lipase (DLL)*
DAD1-like lipases (DLLs) have molecular masses of about 45-50 kDa and show similarity with eukaryotic lipases, such as *R. miehei* lipase. As a motif typical to DLLs, GHS(L/M)G are highly conserved within DLLs. The Ser residue in this motif seems to consist an active center, or catalytic triad, with His and Asp, which are also highly conserved in DLLs. Computer analyses with the ChloroP, TargetP, and PSORT programs indicated that they could be grouped into four subfamilies according to their possible subcellular localization, namely, plastidic (class I), cytoplasmic (class II), mitochondrial (class III), and extracellular (class IV) ones. A phylogenetic tree constructed with the thirteen DLLs plus DAD1 (Figure 1) showed that they could be further classified into subgroups. For example, DLL1 showed highest similarity with DAD1 (class I-a), while DLL2, 7 and 6 (class I-b) showed relatively high similarity each other.

3-2. *Expression of DLL*
In order to get insights on the physiological significances of *DLLs* in *Arabidopsis*, profiles of expressions of *DLL* (*DLL1, 2, 3, 4, 6,* and *7*) were investigated by RT-PCR. As controls, actin (*AAc1*), glutathione S-transferase (*GST1*), and patatin 1 gene (*Pat1*) were used. *GST1* is known to be induced by various biotic and abiotic stresses. Patatin is one of lipases, and it was reported that the corresponding gene in *Vigna* was induced under drought stress (Matos, et al., 2001). First, expression of them in various organs of mature *Arabidopsis* plants was analyzed. Almost all the genes investigated here were highly expressed in unopened buds, most prominently

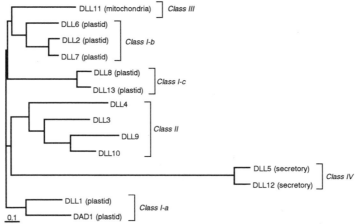

Figure 1. A phylogenetic tree of DLL (DAD1-like lipase) in *Arabidopsis*. According to the expected targeting site of each DLL, they are grouped into four classes, i.e., Class I for plastidic type, Class II for cytosolic type, Class III for mitochondrial type, and Class IV for secretory type.

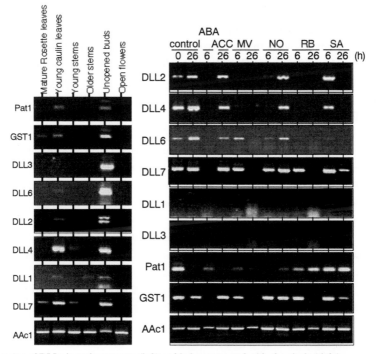

Figure 2. Expression of DLLs in various organs (left) and in leaves treated with chemicals (right).

ABA; abscisic acid (150 µM), ACC; 1-aminocyclopropane-1-carboxylic acid (1 mM), NO; sodium nitroprusside (1 mM), MV; methyl viologen (10 µM), RB; Rose Bengal (50 µM), SA; salicylic acid (2 mM).

with *DLL6* and *DLL3*, while their transcripts could be scarcely detected in the other vegetative organs. *DLL4* expresses highly in young leaves and almost the same profile could be observed with *Pat1*. *DLL7* expression was observed in all the organs examined. Accordingly, expression profiles of *DLL6* and *DLL3* were almost same, thus, they might be involved in the same physiological events, probably in reproductive ones, although their putative subcellular localizations are different each other. High expression of *DLL4* and *DLL2*, concomitant with *Pat1* could be observed with young organs, thus, they might take roles in early vegetative growth. In turn, rosette leaves were treated with ABA (drought-related phytohormone), ACC (ethylene precursor), MV (superoxide generator), NO (NO generator), RB (singlet oxygen generator), or SA (pathogenesis-related signal molecule). As shown in the right panel of Figure 2, expression of *DLL1* and *DLL3* could hardly detected in almost all the treatment. Interestingly, treatment with chemicals generate reactive oxygen species (i.e., MV and RB) suppressed expression of *DLL2* and *DLL4* while in the presence of the other type of reactive oxygen species, namely NO, their expression persisted. Among *DLLs*, *DLL7* and *DLL6* showed expression profile similar to *GST1*. Thus, they were thought to be involved in some sort of stress response. None of expression profile of *DLLs* showed similarity with *Pat1*.

This study showed that DLLs were differently regulated, which might suggest that they have their specific role in *Arabidopsis*. However, the information is still limited, and further study is now underway.

4. References

Blée, E. (1998) Phytooxylipins and plant defense reactions. Prog. Lipid Res. 37, 33-72.

Ishiguro, S., Kawai-Oda, A., Ueda, J., Nishida, I. and Okada, K. (2001) The *defective in anther dehiscence1* gene encodes a novel phospholipase A1 catalyzing the initial step of jasmonic acid biosynthesis, which synchronizes pollen maturation, anther dehiscence, and flower opening in Arabidopsis. Plant Cell, 13, 2191-2209.

Matos, A.R., d'Arcy-Lameta, A., França, M., Pétres, S., Edelman, L., Kader, J.-C., Zuily-Fodil, Y. and Pham-Thi, A. T. (2001) A novel patatin-like gene stimulated by drought stress encodes a galactolipid acyl hydrolase. FEBS Lett., 491, 188-192.

Matsui, K. Kurishita, S., Hisamitsu, A. and Kajiwara, T. (2000) A lipid-hydrolysing activity involved in hexenal formation. Biochem. Soc. Trans. 28, 857-860.

PEROXISOMAL ACYL-CoA THIOESTERASES FROM *Arabidopsis*

G. Tilton, J. Shockey, and J. Browse
Institute of Biological Chemistry
Washington State University
Pullman, WA 99164-6340
USA

Abstract

We have cloned two acyl-CoA thioesterases from the plant *Arabidopsis* that we have named *ACH1* and *ACH2*. Both genes code for proteins carrying peroxisomal targeting sequences (type 1). In an effort to characterize the activity and function of these proteins, we have used the following approaches:

- *In vitro* activity assays of recombinant 6His-tagged ACH2
- Quantitative RT-PCR
- Analysis of gene knock-out mutants

Results from these approaches showed that recombinant ACH2-6His is a long-chain thioesterase with preference for unsaturated fatty acyl-CoAs. *ACH1* and *ACH2* have similar expression profiles in various tissues of the plant. They are both induced during seedling germination when high levels of fatty acids are being oxidized, but also show substantial expression in flowers and leaves. T-DNA insertional mutants show no visible phenotype, germinate even in the absence of sucrose, and are fertile. These results indicate that ACH1 and ACH2 are not necessary for basic β-oxidation and are more likely involved in other specific reactions in the peroxisome.

Introduction

Acyl-CoA thioesterases catalyze the hydrolysis of acyl-CoAs, yielding free fatty acid and free CoA. These enzymes have been found in all classes of organisms, and in various subcellular locations. Peroxisomes often have multiple isoforms of acyl-CoA thioesterases, although their role in that

N. Murata et al. (eds.), Advanced Research on Plant Lipids, 307–312.
© 2003 *Kluwer Academic Publishers. Printed in the Netherlands.*

organelle has not been discovered. Peroxisomal acyl-CoA thioesterases are upregulated under conditions of high fatty acid oxidation such as fasting or exposure to hypolipidemic drugs in rats (Hunt et al. 1999 and 2002). Jones et al. (1999) discovered that a yeast acyl-CoA thioesterase is induced when grown in high-oleate media, and mutants grew much more slowly than wild-type. This indicates that the peroxisomal acyl-CoA thioesterases are somehow involved in metabolizing fatty acids, one of the principal metabolic processes that occurs in the peroxisome. In order to further study the role of peroxisomal acyl-CoA thioesterases we have cloned and characterized two such enzymes from *Arabidopsis*. This plant is useful for studying peroxisomal fatty acid metabolism because all of the β-oxidation reactions take place in the peroxisome. *Arabidopsis* relies heavily on seed oils for germination, during which time the fatty acids are broken down in the peroxisome and reassembled into sugars and other metabolites for vegetative growth. Our results indicated that the acyl-CoA thioesterases are indeed upregulated in tissues with high β-oxidation activity, but they are also induced in other tissues where β-oxidation is less active.

Experimental

Overexpression and Purification of ACH2-6His

The *ACH2* cDNA was cloned into pET-24d and introduced into BL-21 (DE3) *E. coli* cells. Cells were grown at 37°C in LB to an optical density of 0.5 - 0.6, then transferred to room temperature and induced with 800 μM IPTG. The cells were grown for 48 hours at room temperature before harvesting. Overexpressed ACH2-6His protein was purified using Ni^{2+}-affinity columns according to the Novagen His-Bind protocol. The purified protein was dialyzed against 50 mM KP_i, pH 6.0/20 mM NaCl and stored in 20% glycerol at -20°C.

In Vitro Activity Assays

ACH2-6His acyl-CoA thioesterase activity was measured at 412 nm in assay buffer (50 mM KP_i, pH 8.0, 200 μM DTNB) using 20 :M acyl-CoAs (Sigma) in a total volume of 750 μl. Optimal concentrations of BSA, determined earlier, were added to each reaction according to the chain length of the substrate. 2 μg of ACH2-6His was added to the reaction mixture to start each assay, and readings were taken every second for 1 minute. Specific activities were calculated using data points in the linear range the assay curve. The molar extinction coefficient of DTNB is $1.36 \times 10^4 \ M^{-1}cm^{-1}$.

Real Time RT-PCR

2-5 μg of DNAse-treated total RNA was used for reverse transcription using random hexamers and the Gibco Superscript RT II system. The RT reactions were RNAse-treated, then used as template for real-time PCR reactions with the Stratagene Brilliant

SYBR-Green QPCR Reaction Kit. Amplification was measured in real-time using the Stratagene Mx-4000 Multiplex quantitative PCR instrument. The amplification of *ACH1* and *ACH2* amplicons was normalized against the amplification of an *18S* control.

Analysis of T-DNA Insertional Mutants
ACH1 and *ACH2* T-DNA insertional mutants were identified by screening the University of Wisconsin T-DNA lines. Both mutants have inserts that prevent expression of a full-length transcript. Effects of the mutations on germination were analyzed by growing seeds in the dark on 1X MS salt plates/0.45 % phytagel with or without 1% sucrose. Wild-type and mutant growth was also compared by growing seeds in soil.

Results and Discussion

ACH2-6His Activity
ACH2-6His was active against a wide range of fatty acyl-CoA molecules (Fig. 1), and had the highest preference for long-chain unsaturated acyl-CoAs. This specificity is consistent with the types of fatty acyl-CoAs that would be expected to enter the peroxisome during seed germination since *Arabidopsis* seed oils contain mainly long-chain unsaturated fatty acids. This activity would allow ACH2 to be involved in a variety of processes such as transport of acyl-CoAs across the peroxisomal membrane, or in the first rounds of long-chain fatty acid β-oxidation. ACH2-6His does have activity against short-chain acyl-CoAs, although only at much higher concentrations (data not shown).

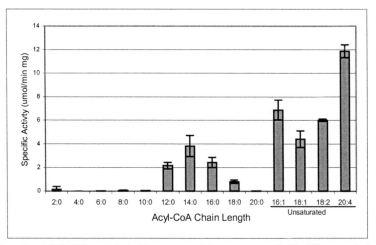

Figure 1   ACH2-6His acyl-CoA thioesterase activity was measured with different acyl-CoA chain lengths as in Experimental. Error bars represent standard deviation of three assays.

Real-Time RT-PCR

ACH1 and *ACH2* both have very similar expression patterns in all of the tissues we analyzed. Both genes are slightly upregulated during seed germination where the seed oils are rapidly being oxidized in the peroxisome (see Fig. 2). However, there is relatively more expression in flower and leaf tissue, indicating that the enzymes are involved in more than just basic fatty acid oxidation.

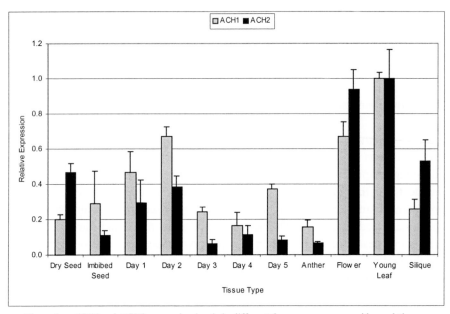

Figure 2. *ACH1* and *ACH2* expression levels in different tissues were measured by real-time quantitative PCR. The data is presented as percent expression of young leaf tissue, where the transcript level of both genes was highest. Error bars represent standard error of three assays.

T-DNA Mutant Analysis

In order to check whether ACH1 and ACH2 are important in the germination process, we germinated wild-type and mutant plants on media with or without sucrose. Seedlings with impaired fatty acid oxidation are often unable to grow properly without sucrose, presumably due to their inability to metabolize the seed oils for carbon (Hayashi et al. 1998, Eastmond et al. 2000). Accordingly, we germinated wild-type plants, single mutants (*ach1* or *ach2*), and a double mutant (*ach1ach2*) in the dark with our without 1% sucrose and compared their growth. We found that there was no difference in the germination and seedling establishment between the wild-type plant and the mutants, regardless of the presence or absence of sucrose (see Fig. 3)

1% Sucrose

No Sucrose

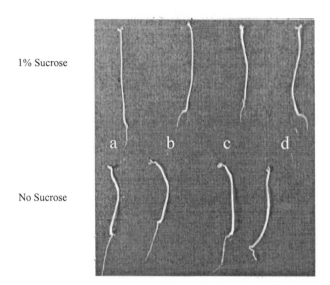

Figure 3. Etiolated seedlings were grown in 1% sucrose (top) or without sucrose (bottom). a) wild-type, b) *ach1* single mutant, c) *ach2* single mutant, d) *ach1ach2* double mutant.

All of these results would indicate that ACH1 and ACH2 do not play a crucial role in metabolizing the seed oils in germinating seedlings. However, we have not been able to determine whether the ACH1 and ACH2 represent all of the acyl-CoA thioesterase activity in the peroxisome. It is possible that other redundant enzymes exist that could replace the missing enzyme activity in the mutants. It is also possible that these enzymes have a very specific function that can only be detected under certain conditions such as environmental stress, defense response, or senescence. Hunt et al. propose that these enzymes may be important in maintaining proper Coenzyme A levels in the peroxisome, especially during times of high β-oxidation activity. If this is true in *Arabidopsis*, then it may be possible that the increased expression of these peroxisomal enzymes in young leaves and flowers is to divert CoA away from the peroxisome to other areas of the cell where it is needed, such as in lipid biosynthesis. Further study of the mutants will elucidate the conditions where acyl-CoA thioesterases are most important.

References

Hunt MC, Nousiainen SE, Huttunen MK, Orii KE, Svensson LT, Alexson SE (1999) Peroxisome proliferator-induced long chain acyl-CoA thioesterases comprise a highly conserved novel multi-gene family involved in lipid metabolism. J. Biol. Chem. 274(48):34317-26.

Hunt MC, Solaas K, Kase BF, Alexson SE (2002) Characterization of an acyl-coA thioesterase that functions as a major regulator of peroxisomal lipid metabolism. J. Biol. Chem. 277(2):1128-38.

Jones JM, Nau K, Geraghty MT, Erdmann R, Gould SJ. (1999) Identification of peroxisomal acyl-CoA thioesterases in yeast and humans. J. Biol. Chem. 274(14):9216-23.

Hayashi M, Toriyama K, Kondo M, Nishimura M. (1998) 2,4-Dichlorophenoxybutyric acid-resistant mutants of Arabidopsis have defects in glyoxysomal fatty acid beta-oxidation. Plant Cell 10(2):183-95

Eastmond PJ, Germain V, Lange PR, Bryce JH, Smith SM, Graham IA. (2000) Postgerminative growth and lipid catabolism in oilseeds lacking the glyoxylate cycle. Proc. Natl. Acad. Sci. U.S.A. 97(10):5669-74

Chapter 8:

Lipid Trafficking and Signaling

IDENTIFICATION AND CHARACTERIZATION OF A BINDING PROTEIN FOR *N*-ACYLETHANOLAMINES IN MEMBRANES OF TOBACCO CELLS

S. TRIPATHY[1], K. KLEPPINGER-SPARACE[1], R.A. DIXON[2], K.D. CHAPMAN[1]

[1]*Department of Biological Sciences, University of North Texas, Denton, TX 76203 USA*
[2]*The Samuel R. Noble Foundation, Plant Biology Division, Ardmore, OK 73402 USA*

1. Abstract

N-acylethanolamines (NAEs) are endogenous constituents of various plant tissues and are derived from the membrane phospholipid, *N*-acylphosphatidyl-ethanolamine (NAPE), by the action of phospholipase D. In earlier studies, *N*-myristoylethanolamine (NAE14:0) levels increased 10-50 fold in elicitor-treated tobacco leaves prompting us to speculate that plants can utilize an NAE signaling pathway following pathogen elicitor perception. Here we used [^{3}H]NAE14:0 in radioligand binding assays to identify a saturable, high affinity specific NAE-binding activity in suspension-cultured tobacco cells and in microsomes of tobacco leaves (apparent K_d of 75 nM and 35 nM, respectively). The mammalian cannabinoid (CB) receptor antagonist, AM281, reduced or eliminated the specific [^{3}H]NAE 14:0 binding. Maximal [^{3}H]NAE 14:0 binding activity was solubilized from leaf microsomes in 0.2 mM dodecylmaltoside (DDM) at a detergent-to-protein ratio of 0.4-to-1. Solubilization of functional NAE binding activity(ies) will facilitate the detailed biochemical characterization of what may be a plant counterpart of mammalian CB receptors.

2. Introduction

N-acylethanolamines (NAEs) are endogenous constituents of various plant and animal tissues (Chapman, 2000; Schmid et al., 1996). The formation and degradation of NAEs is subject to tight regulation in vertebrates and is termed the endocannabinoid signaling pathway. This pathway regulates a variety of physiological processes including neurotransmission, immune responses, embryo implantation and development, and cell proliferation (DiMarzo, 1998). The conservation of NAE occurrence and metabolism of NAEs in plant tissues (Chapman, 2000) suggests that important physiological processes in plants may also be regulated at the cellular level by NAEs.

NAE perception in animal systems is mediated by several types of membrane-bound cannabinoid (CB) receptors (Felder and Glass, 1998). Although still a matter of

N. Murata et al. (eds.), Advanced Research on Plant Lipids, 315–318.

intensive investigation, the predominant NAE that is considered to interact with mammalian CB receptors in vivo is *N*-arachidonylethanolamine (or anandamide). Manipulation of anadamide levels in vivo results in physiological and behavioral symptoms ascribed to classical cannabinoid drugs, and the pharmacological properties of anandamide binding to CB receptors in vitro are generally consistent with these in vivo functions (DiMarzo, 1998).

In previous studies with tobacco cell suspensions (Chapman et al, 1998), and later with leaves of intact plants (Tripathy et al., 1999), NAE14:0 was identified by GC-MS as an endogenous metabolite that was formed in response to fungal elicitor treatment. NAE14:0 levels increased 10-50 fold in tobacco leaves treated with xylanase or cryptogein protein elicitors. The levels of NAE14:0 measured in vivo after elicitor treatment were sufficient to activate *PAL2* expression in tobacco leaves, a known downstream response to these elicitors, and we hypothesized that an NAE signaling pathway, similar to the endocannabinoid signaling pathway of animal systems, participated in plant defense signaling. Here we identify saturable NAE14:0 binding proteins in plant membranes, and have solubilized functional NAE binding proteins from tobacco and *Arabidopsis thaliana* microsomal membranes.

3. Results

In tobacco cell suspensions, NAE14:0 accumulated in culture medium, not in cells (Tripathy et al, 1999), suggesting that NAEs were perceived at the cell surface. Specific [^{3}H]NAE 14:0 binding to tobacco cells (Fig 1, upper panel) supported this concept. Occupied NAE14:0 binding sites increased with increasing radioligand concentration, and this specific binding (total binding minus nonspecific binding) was reduced in the presence of the mammalian CB receptor antagonist, AM281 (Fig 1, upper panel). Similar binding results were obtained with microsomes isolated from tobacco leaves (Fig 1, lower panel), suggesting that the tobacco NAE14:0 binding protein(s) was (were) indeed membrane bound. Moreover, diminished NAE14:0 binding in the presence of AM281 in both preparations suggested some similarity of the plant NAE binding activity with CB receptors (NAE binding proteins) of animal tissues.

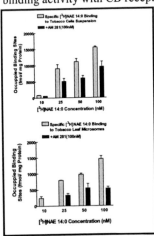

Figure 1. Specific binding of [^{3}H]NAE 14:0 to tobacco suspension cultured cells (derived from cv Xanthi) or to microsomes isolated from tobacco leaves (cv Xanthi). Specific binding was determined by subtracting nonspecific binding (assayed in the presence of 100-1000X excess of nonradioactive NAE 14:0) from total binding (no excess "cold" NAE14:0) at increasing concentrations of [^{3}H]NAE 14:0 in parallel assays (assayed in triplicate at each concentration). Typical saturation binding curves were obtained within the concentration range of 5 to 100 nM [^{3}H]NAE 14:0 for both cell suspensions (20 microgram protein per assay) and for leaf microsomes (50 microgram protein per assay). Occupied binding sites were estimated with Graphpad Prism 3.0 software. For binding of [^{3}H]NAE14:0 in the presence of AM281, 100nM AM281 (final concentration) was included in both total and nonspecific binding assays.

Conditions for the solubilization of functional [³H]NAE 14:0 binding activity from tobacco microsomal membranes were investigated. In preliminary experiments, a variety of detergents were tested in NAE14:0 binding assays, and only Triton X-100 and dodecylmaltoside (DDM) appeared to be compatible with binding assays at concentrations above their critical micelle concentration; i.e., other detergents eliminated or substantially inhibited binding activity compared with controls containing no detergent.

Ultracentrifugation of DDM-solubilized membrane proteins showed that considerable NAE 14:0 binding activity was indeed solubilized at concentrations near or above its CMC of 0.16 mM (Fig 2). Maximal activity was recovered in the soluble fraction after centrifugation in detergent treatments with 0.1, 0.2 and 0.4 mM DDM. Although NAE binding activity was observed in Triton X-100-treated membranes, little or no activity remained soluble after centrifugation (not shown).

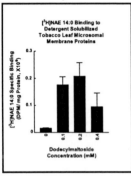

Figure 2. Detergent solubilization of active [³H]NAE-14:0 binding proteins from microsomes isolated from tobacco (*Nicotiana tabacum*, L. cv Xanthi) leaves. Microsomes (100 micrograms of protein) were treated with increasing concentrations of dodecylmaltoside (DDM) for 30 min on ice and centrifuged (150,000 xg) to separate proteins into detergent-solubilized fractions (supernatant) and insoluble fractions (pellet). Specific [³H]NAE 14:0 binding activity was determined in fractions and 0.2 mM DDM was reproducibly found to result in the highest solubilized NAE14:0 binding activity. Data presented are the means and standard deviation of 4 separate solubilization experiments.

Further experiments varying the detergent-to-protein ratio were conducted at 0.2 mM DDM, and a detergent-to-protein ratio of 0.4-to-1 was selected as the most effective for the reproducible isolation of detergent-solubilized NAE binding activity. These solubilization conditions also were effective for [³H]NAE14:0 binding activity in *Arabidopsis thaliana* microsomes, and a preliminary comparison of the biochemical properties for NAE binding activity from several plant sources is given in Table 1.

Table 1. Comparison of apparent NAE14:0 affinities (App Kd) and maximal binding (App Bmax) estimated from specific binding activities at increasing ligand concentration for various plant membrane sources (Nt, Nicotiana tabacum cells, microsomes and DDM-solubilized microsomal protein, and At, Arabidopsis thaliana leaf microsomes, and DDM solubilized microsomal proteins). Assays were conducted without (left two columns) or with (right two columns) 100 nM AM281. Parameters were estimated by fitting specific binding data to a single site binding equation (Graphpad Prism 3.0 software).

Protein Source	App Kd	App Bmax	App Kd With 100 nM antagonist AM281	App Bmax
	nM	dpm mg⁻¹ protein	nM	dpm mg⁻¹ protein
Nt Suspension cells	74	2991000	82	1888000
Nt leaf microsomes	35	196994	49	96554
Nt solubilized protein (leaf microsomes)	26	296376	ND	ND
At leaf microsomes	9	100302	ND	ND
At solubilized protein	13	600549	ND	ND

ND: Not determined

318

4. Conclusions

The evidence provided here supports the existence of a high affinity NAE14:0 binding protein(s) in membranes of higher plants. Moreover, conditions were identified to isolate detergent-solubilized functional protein preparations that will facilitate future biochemical approaches to purify and characterize this class of lipid binding proteins. Preliminary evidence indicates these NAE binding proteins may share functional similarities with NAE binding proteins in animal tissues which function as membrane receptors for endocannabinoid signal transduction (DiMarzo, 1998; Felder and Glass, 1998). These results continue to point to an evolutionarily-conserved, general NAE signaling pathway in plant and animal tissues.

5. Materials and Methods

[^{3}H]NAE14:0 was synthesized from uniformly labeled [^{3}H]myristic acid (49 Ci/mmol, NEN Life Sciences) following the method of Hillard and coworkers (1995) and purified by TLC. Binding assays were performed as developed previously for cannabinoid receptors (Hillard et al., 1995) using a multiscreen Whatman filtration system with BC Durapore 1.2 micron filters (Millipore) with a few modifications. Whole cell aliquots with 20-30 micrograms of protein were incubated for 30 min in 200 microL spent culture medium, pH 5.2-5.5, containing 1% w/v BSA and 5 mM DTT with shaking (140 rpm). Microsomes (150,000 xg, 60 min pellet of 10,000 xg, 30 min supt) were isolated as described previously (Chapman and Sriparameswaran, 1997) and resuspended in 1/10 of the original volume of homogenate. Microsomal proteins, 10-100 micrograms, were incubated as for cell suspensions except in 100 mM K-phosphate, pH 7.2, 10 mM KCl, 1 mM EDTA, 1 mM EGTA, 0.5 mM ascorbate, 1.5% w/v BSA, 5 mM DTT. Total binding was measured with undiluted radioligand, whereas nonspecific binding was measured with radioligand plus 100 to 1000X excess unlabeled NAE14:0. Binding assays were stopped by filtration and washed 3 times with equal volumes of ice cold incubation buffer to remove unbound ligand. Radioactivity remaining on filters was determined by liquid scintillation counting (Beckman model 3801), corrected for quenching and counting efficiency, and converted to DPM. Detergent solubilization of membrane proteins proceeded on ice for 30 min and was evaluated after centrifigation (150,000 xg, 60 min) by estimating binding activity and protein content in supernatant and pelleted fractions.

References

Chapman, K.D. (2000) Emerging physiological roles for *N*-acylphosphatidylethanolamine metabolism in plants. Chem. Phys. Lipids 108: 221-230.

Chapman, K.D., Sriparameswaran, A. (1997) Intracellular localization of *N*-acylphosphatidylethanolamine synthesis in cotyledons of cotton seedlings. Plant Cell Physiol. 38: 1359-1367.

Chapman, K.D., Tripathy, S., Venables, B., Desouza, A. (1998) *N*-Acylethanolamines: formation and molecular composition of a new class of plant lipids. Plant Physiol. 116: 1163-1168.

Chapman, K.D., Venables, B., Blair, R. Jr., Bettinger, C. (1999) *N*-Acylethanolamines in seeds: quantification of molecular species and their degradation upon imbibition. Plant Physiol. 120: 1157-1164.

DiMarzo, V. (1998) Endocannabinoids and other fatty acid derivatives with cannabimietic properties: biochemistry and possible physiological relevance. Biochim. Biophys. Acta 1392: 153-175.

Felder, C.C., Glass, M. (1998) Cannabinoid receptors and their endogenous agonists. Annu. Rev. Pharmacol. Toxicol. 38:179-200.

Hillard, C.J., Edgemond, W.S., Campbell, W.B. (1995) Characterization of ligand binding to the cannabinoid receptor of rat brain membranes using a novel method: application to anandamide. J. Neurochem. 64: 677-683.

Schmid, H.H.O., Schmid, P.C., Natarajan, V. (1996) The *N*-acylation-phosphodiesterase pathway and cell signaling. Chem. Phys. Lipids 80: 133-142.

Tripathy, S. Venables, B.J., Chapman, K.D. (1999) *N*-Acylethanolamines in signal transduction of elicitor perception: attenuation of alkalinization response and activation of defense gene expression. Plant Physiol. 121: 1299-1308.

LIPID TRANSFER PROTEINS :

A genomic approach to establish their suggested roles in lipid trafficking and signaling

J.C. KADER, F. GUERBETTE, C. VERGNOLLE A. ZACHOWSKI
*Laboratoire de Physiologie Cellulaire et Moléculaire, Unit 7632 CNRS /
Université Paris 6 , case 154, 4 place Jussieu, 75252 Paris cedex 05
(France)*

1. Introduction

Proteins capable of transferring lipids between membranes in vitro have been purified from a wide range of living organisms.The best known proteins of this category in higher plants are lipid transfer proteins (LTPs) which are able to transfer several different phospholipids between membranes and to bind fatty acids and acyl-CoA esters (Yamada, 1992 ; Kader, 1996, 1997). LTPs were initially supposed to participate to membrane biogenesis and to intracellulair lipid trafficking. However, no clear evidence of such roles have been demonstrated in vivo. Instead, LTPs are suggested to be involved in extracellular functions linked to defense reactions, in signaling mechanisms operating between pathogens and plants and in allergenic processes. In order to validate these hypotheses, it is of high interest to follow a genomic approach and to study the regulation of the expression of LTP genes by various abiotic and biotic stresses. These studies have been performed in Arabidopsis thaliana since its genome has been recently sequenced.

2. Structural properties of LTP

Plant LTPs form a very homogeneous class of proteins. They are small (below 10 kDa), abundant, stable and basic proteins Two families have been characterized , mainly in wheat, having molecular masses of 7 and 9 kDa. LTPs contain 8 strictly conserved cysteine residues and which form 4 intramolecular disulfide bonds established by the group of Yamada (1992). The 3D structure, based on X-ray diffraction and nuclear magnetic resonance, was determined both for LTPs isolated from various sources (wheat, maize, barley) (Charvolin et al, 1999 ; Douliez et al, 2001; Han et al , 2001). The structure shows a hydrophobic cavity delimited by 4 compacted helices. The volume of the hydrophobic cavity shows some variation depending on the size of the

N. Murata et al. (eds.), Advanced Research on Plant Lipids, 319–322.

bound ligands, thus explaining probably the non-specificity of LTP. This cavity accommodates various ligands from C10 to C18, but, as established by experiments with fluorescent lipids, with diverse specificity. The binding of acyl chains on LTP has been demonstrated by incubating maize protein using fluorescent analogs of fatty acids (anthroyl-oxy-stearic acid) (Guerbette et al, 1999). The increase in fluorescence indicated a binding of this fatty acid on maize LTP. It was found that various fatty acids or acyl-CoAs can compete with the fluorescent fatty acid, the most efficient being those having a chain length of 16 to 18 carbons.

3. Expression of LTP genes.

The expression pattern of LTP genes was found to be complex. It is characterized by a strong developmental, tissue and cell specificity and by an induction by environmental factors, such as extreme temperatures or drought stresses (Table 1). A strong specificity of expression in the epidermal cells has been observed in several plants : maize, carrot, Arabidopsis (Kader, 1996). High expression in tapetal cells has been observed in rapeseed (Kader, 1996) while fiber cells of cotton (Kader, 1996) and guard cells in Nicotiana (Smart et al, 2000) contain a highly expressed LTP gene. The induction of LTP genes by several abiotic and biotic stresses has been observed in several conditions : drought (Clark and Bohnert, 1999), low temperature (Dunn *et al*, 1998), abscisic acid (Kader, 1997) or cadmium (Hollenbach *et al*, 1997) induced the expression of LTP genes while the attack by fungal pathogens or virus led to the enhanced expression of LTP genes in barley and pepper, respectively (Garcia-Olmedo et al, 1998 ; Park et al, 2002). these observations are coherent with the postulated roles of LTPs in cutin biogenesis (Sterk *et al*, 1991). Another suggested role for LTP could be to facilitate pollen tube adhesion to the stylar matrix (Park et al, 2000).

Studies conducted on elicitins, small proteins secreted from various fungi and able to bind sterols, led to the conclusion that elicitins and LTP can compete for the same receptor located on plasma membrane from tobacco (Buhot et al, 2001). LTP can thus be considered as involved in cell to cell signalisation.

Surprisingly, LTPs have been identified as one of the major allergens in various fruits, seeds or pollen grains. Due to their extreme resistance to pepsin digestion, LTPs are potentially severe food allergens. Interestingly, some sequence similarities can be found between allergens and pathogenesis-related protein families among which LTP are classified (PR-14) (Hoffmann-Sommergruber, 2000).

4. Genomic studies on LTP

The LTP family is rather complex. We have identified in Arabidopsis thaliana among more than 100 ESTS encoding LTP-like proteins, 15 different genes. Analyzing the sequence of Arabidopsis, we have detected 40 sequences coding for LTP-like proteins, characterized by the cysteine motif (Vergnolle et al, 2000). The deduced molecular masses differ from 7 to 61 kDa. The expression of six of these genes have revealed that two of them are highly expressed in inflorescences while two others are induced by absicisic acid.

TABLE 1   Examples of pattern expression of lipid transfer protein (LTP) genes

Specificity or induction	Plant	Ref.
Epidermal cells	carrot	Sterk et al, 1991
	Arabidopsis	Ka der, 1996
	maize	
Tapetal cells	rape	Kader, 1996
Fiber cells	cotton	Kader, 1996
Style	Lily	Mollet et al, 2000
Salt	tomato	Kader, 1996
Low temperature	barley	Dunn et al, 1998
Drought	Arabidopsis	Clark and Bonhert, 1999
	Nicotiana	Smart et al, 2000
Abscisic acid	barley	Kader, 1997
cadmium	barley	Hollenbach et al, 1997
Fungal pathogens	barley	Garcia-Olmedo et al, 1998
Virus	pepper	Park et al, 2002

5. Perspectives for the future.

The complexity of the family of LTP suggests that they play probably several roles in plant cells. One unifying hypothesis is to propose that these proteins are involved in lipid signaling processes which are active in various conditions of abiotic and biotic stresses. Future efforts will concern the search for a function for these various proteins by studying the global expression of LTP genes in Arabidopsis with macro- or microarrays and by isolating mutants deficient in some of the LTP genes. This opens exciting perspectives for the elucidation of the role of LTP in intercellular lipid signaling linked to defense reactions.

6. References

Arondel, V., Vergnolle, C., Cantrel, C. and Kader J.C. (2000) Lipid transfer proteins are encoded by a small multigene family in Arabidopsis thaliana. Plant Sci.157,1-12.

Buhot, N., Douliez, J.P., Jacquemard, A., Marion, D., Tran, V., Maume, B.F., Milat, M.L., Ponchet, M., Mikes, V., Kader, J.C., and Blein, J.P.(2001) A lipid transfer protein binds to a receptor involved in the control of plant defence responses.FEBS Lett.509, 27-30.

Charvolin, D., Douliez, J.P., Marion, D., Cohen-Addad, C. and Pebay-Peyroula E. (1999) The crystal structure of a wheat nonspecific lipid transfer protein (ns-LTP1) complexed with two molecules of phospholipid at 2.1 A resolution. Eur.J.Biochem.264, 562-568.

Clark, A.M. and Bohnert, H.J. (1999) Cell-specific expression of genes of the lipid transfer protein family from Arabidopsis thaliana. Plant Cell Physiol. 40, 69-76.

Douliez, J.P., Jegou, S., Pato, C., Molle, D., Tran, V. and Marion, D. (2001) Binding of two mono-acylated lipid monomers by the barley lipid transfer protein, LTP1, as viewed by fluorescence, isothermal titration calorimetry and molecular modelling. Eur J Biochem. 268, 384-388.

Dunn, M.A., White, A.J., Vural, S. and Hughes, M.A. (1998)Identification of promoter elements in a low-temperature-responsive gene (blt4.9) from barley (Hordeum vulgare).,Plant Mol Biol. 38, 551-564.

Garcia-Olmedo, F., Molina, A., Alamillo, J.M. and Rodriguez-Palenzuela, P. (1998) Plant defense peptides. Biopolymers, 47, 479-91.

Guerbette, F, Grosbois, M., Jolliot-Croquin, A, Kader, J.C. and Zachowski, A. (1999) Comparison of lipid binding and transfer properties of two lipid transfer proteins from plants. Biochemistry 38, 14131-14137.

Han, G.W., Lee, J.Y., Song, H.K., Chang, C., Min, K., Moon, J., Shin, D.H., Kopka, M.L., Sawaya, M.R., Yuan, H.S., Kim, T.D., Choe, J., Lim, D., Moon, H.J. and Suh, S.W. (2001) Structural basis of non-specific lipid binding in maize lipid-transfer protein complexes revealed by high-resolution X-ray crystallography. J Mol Biol. 308, 263-278.

Hoffmann-Sommergruber, K. (2000) Plant allergens and pathogenesis-related proteins. What do they have in common? Int Arch Allergy Immunol.122, 155-166.

Hollenbach,,B.A., Schreiber, L., Hartung, W. and Dietz, K.J. (1997) Cadmium leads to stimulated expression of the lipid transfer protein genes in barley. Implications for the involvement of lipid transfer proteins in wax assembly. Planta, 203, 9-19.

Kader, J.C. (1996) Lipid transfer proteins in plants, Annu. Rev. Plant Physiol. Plant Mol. Biol. 47, 627-654

Kader, J.C. (1997) Lipid-transfer proteins : a puzzling family of plant proteins. Trends Plant Sci. 2, 66-70.

Park, C.J, Shin, R., Park, J.M., Lee, G.J., You, J.S. and Paek, K.H.(2002) Induction of pepper cDNA encoding a lipid transfer protein during the resistance response to tobacco mosaic virus.Plant Mol Biol. 48, 243-54

Park, S.Y., Jauh, G.Y., Mollet, J.C., Eckard, K.J., Nothnagel, E.A., Walling, L.L. and Lord, E.M.(2000) A lipid transfer-like protein is necessary for lily pollen tube adhesion to an in vitro stylar matrix.Plant Cell. 12, 151-64.

Smart, L.B., Cameron, K.D. and Bennett, A.B.(2000) Isolation of genes predominantly expressed in guard cells and epidermal cells of Nicotiana glauca. Plant Mol Biol. 42, 857-869.

Sterk, P., Booij, H., Scheleenkens, G.A., Van Kammen, A. and De Vries, S.C. (1991) Cell-specific expression of the carrot EP2 lipid transfer protein gene. Plant Cell 3 , 907-921

Yamada, M. (1992) Lipid transfer proteins in plants and microorganisms, Plant Cell Physiol. 33, 1-6

RAPID AND SLOW RESPONSE REACTIONS OF PLANTS ON EFFECT OF ANTIOXIDANT AMBIOL

N.V.BUDAGOVSKAYA[1] and V.I.GULIAEV[2]

[1]*Institute of Plant Physiology, Russian Academy of Sciences, Botanicheskaya 35, Moscow 127276, Russia;* [2]*All-Russian Institute of Genetics and Breeding of Fruit Plants, Russian Academy of Agricultural Sciences, Michurinsk, Russia*

Introduction

The use of a high sensitive method – laser interference auxanometry makes it possible to study the response reactions of the whole plant on effect of biologically active compounds in short time intervals (sec, min, hours). Ambiol, possessing antioxidant properties was used in the present work as a biologically active compound. Earlier (Budagovskaya 1997) we have demonstrated that growing of plants in prolonged (for many days) experiments in the presence of ambiol led to stimulation of growth of shoots and roots of the plant. This work presents the study of rapid (in first minutes after addition of ambiol to the root zone) and slow (few hours after addition of ambiol) response growth reactions of the shoots of plant on effect of the antioxidant.

Materials and methods

Plant growth rate was measured by laser interference auxanometer LINA-EM3D (Russia), equipped with a helium-neon laser (632.8 nm). Sensitivity of the auxanometer was 0.07 μm. Growth parameters, temperature of air and substrate at the root zone were registered every 10 sec. Plants of wheat cv. Yang mai (China), rice cv. Suxianggen 1 (China) and oat cv. Horizont (Russia) were used in experiments. Plants were grown in washed sand in Petri dishes, 9.5 cm in diameter. Three days before the start of experiment plants were placed by one in Petri dishes of 4 cm in diameter (wheat and rice plants) and 5 cm (oat plants, which had larger roots). In the course of growing plants were watered by tap water, the distilled water was used for control watering in experiments. Ambiol (2-methyl-4-dimethylaminomethyl-5-hydroxybenzimidazole) was synthesized at the Institute of Chemical Physics, Russia. Water solution of ambiol (10^{-5}-10^{-6} M) was introduced in 0.5 or 1.0 ml portions to the substrate at the root zone. Prior to addition of ambiol solution the 4 cm Petri dishes with sand contained 3 ml of water and 5 cm dishes – 5 ml. The measurement of the growth rate of the second leaf of plants was performed by auxanometer at 20-25°C and air humidity of 64%.

Results and discussion

A two phase response reaction of leaves was observed after addition of ambiol to the root zone of wheat, oat and rice plants. First phase – rapid change in their growth rate

N. Murata et al. (eds.), Advanced Research on Plant Lipids, 323–326.

(in first minutes after addition of ambiol), and second phase – slow, smooth change in the growth rate of leaves (less than 3 hours after addition of ambiol) (Fig. 1, 2, 3). The value and sign of the response reaction (acceleration or slowing down of the growth rate) depended on the initial growth rate of plants and amount of the ambiol added. At low initial growth rate of leaves (Fig. 1 and 2) a significant increase in their growth rate was observed after addition of ambiol. At high initial growth rate of leaves (Fig. 3, curve 1) stimulation of their growth by ambiol was expressed less both for rapid and slow response reactions. Repeated additions of ambiol to the experimental plant with initially higher growth rate (Fig. 3, curve 2; ambiol was added on the next day after its first addition) increased its concentration, what, apparently, promoted the change in the sign of the response reaction of the plant - a decrease in the growth rate of the leaf was observed. Fig. 3, curve 2 shows the dynamics of the change in growth rate of the leaf after two additions of ambiol on the second day of experiment. The next addition of ambiol (ten-fold concentrated solution than in previous additions) led to a still greater fall in the growth rate of the leaf (not shown on curve 2, Fig.3).

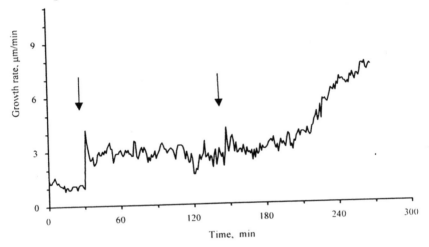

Fig. 1. Effect of ambiol (10^{-6} M, 0.5 ml) on the growth of the second leaf of 10-day-old wheat plant. ↓ - the moment of addition of ambiol.

Repeated addition of ambiol to the plant with low initial growth rate (Fig. 1 and 2) did not cause a decrease in its growth activity as for the plant with high growth rate (Fig. 3, curve 2), but on the contrary enhanced it. Repeated addition of ambiol (1.0 ml; 10^{-5} M) to oat plant on the 4-th day after initial addition increased the growth rate of second leaf from 10 μm/min to 17 μm/min (curve not shown).

Two phase response reaction of plants (existence of rapid and slow components) on addition of ambiol was observed for all experimental plants (Fig. 1-3). Apparently, the rapid response of the second leaf of plant on addition of ambiol to the root zone is possible as result of transfer of electric signal from roots to leaves; the slow response may be due to metabolic changes, connected with *de novo* biosynthesis. Possibility of a rapid informational connection between spatially separated plant organs involving excitation potentials was demonstrated in the work of Opritov (1970). The time interval between the rapid and slow response of plants (about 3 hours) is, apparently, needed for synthesis and accumulation of necessary substances in the proper amount for going to a new higher functional level. It results in increase of growth rate of plants.

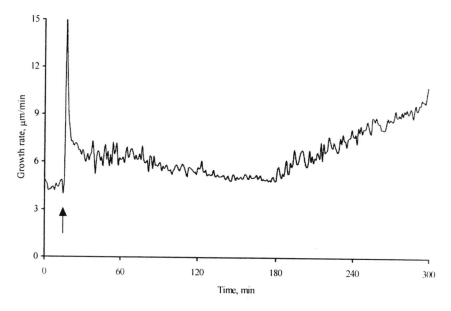

Fig. 2 . Effect of ambiol (10^{-5} M, 1.0 ml) on the growth of the second leaf of the 9-day-old oat plant. ↑ - the moment of addition of ambiol.

It is noteworthy that ambiol significantly increased the growth rate of plants, grown under deficiency of mineral nutrition. For plants, growing in washed sand, a decrease in growth rate, caused by deficiency of mineral compounds, was observed after exhaustion of nutritional substances of the seed. For wheat and oat plants the decrease in growth rate occurred earlier than for rice plants of the same age, differing by a smaller size. As demonstrated in a number of works under deficiency of mineral nutrition formation of active forms of oxygen was enhanced (Cakmak and Marschner 1988a), the membrane permeability increased (Cakmak and Marschner 1988b) and the ratio of saturated to non-saturated fatty acids was changed (Rivera and Penner 1978). That is why in our experiments addition of exogenous antioxidant ambiol to plants under conditions of deficiency of mineral nutrition may provoke improvement of the state of membranes and increase in functional activity of plants. Earlier we have demonstrated the change infunctional activity of isolated chloroplasts (photophosphorylation and light induced proton uptake) under effect of antioxidants (Budagovskaya 1994). Natural antioxidants cysteine and ascorbate enhanced the initially low activity of chloroplasts and decreased the high activity. A similar regularity in effect of antioxidant ambiol (including the effect of its concentration) on the integral process – plant growth is also seen among the results of the present work. All these data may point to a possible participation of antioxidants in regulation of functional activity of plants. Apparently, under normal conditions there is a certain balance of free radical oxidation (FRO) and antioxidation (AO) due to activity of antioxidizing enzyme systems and concentration of non-enzymatic antioxidants. Under unfavorable conditions

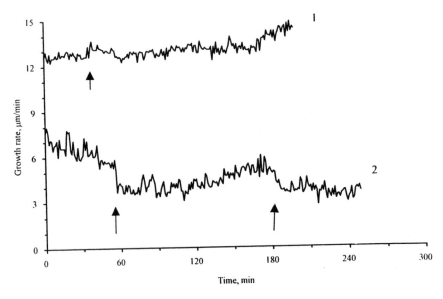

Fig. 3. Effect of ambiol (10^{-6} M, 0.5 ml) on the growth of the second leaf of 15-day-old rice plant. ↑ - the moment of addition of ambiol.

the equilibrium FRO/AO may be shifted to enhancement of free radical oxidation, what will lead to disturbances in functional activity of plants. In this case addition of exogenous antioxidants will promote a stimulating effect. At high functional activity of plants, when free radical oxidation and antioxidation are well balanced, addition of exogenous antioxidants will shift the equilibrium FRO/AO from the optimal level. This will cause a decrease in functional activity of plants.

References

Budagovskaya. N.V. 1994. Antioxidants and free radical oxidation as regulators of ATP biosynthesis in chloroplasts. Proceedings of the Royal Society of Edinburgh 102B, 237-241.

Budagovskaya, N.V. (1997) Effect of calcium channels blocker and growth regulators on the state of photoautotrophic and geterotrophic plant tissues. The FASEB Journal 11, 1435.

Cakmak, I. and Marschner, H. (1988a) Enhanced superoxide radical production in roots of zinc-deficient plants. J. Exp. Bot. 39, 1449-1460.

Cakmak, I. and Marschner, H. (1988b) Zinc-dependent changes in ESR signals, NADPH oxidase and plasma membrane permeability in cotton root s. Phys. Plant. 73, 182-186.

Opritov, V.A. and Kalinin, V.A. (1970) Modification of the energy state and functional activity of the conducting system of fodder beat during propagation of excitation wave. Fiziologia Rastenii 17, 769-775 [in Russian].

Rivera, C.M. and Penner, D. (1978) Effect of calcium and nitrogen on soybean (Glicine max) root fatty acid composition and uptake of linuron. Weed Sci. 26, 647-650.

LOCALIZATION OF POLLEN TUBE NpCDPK7 PROTEIN IN GENERATIVE CELL MEMBRANE MEDIATED BY AMINO TERMINAL MYRISTOYLATION.

[1]MASAAKI K WATAHIKI, RICHARD M PARTON
[3*]ANTHONY J TREWAVAS
The University of Edinburgh, Institute of Cell and Molecular Biology, King's Buildings, Edinburgh, Scotland, United Kingdom.
[1]m.watahiki@ed.ac.uk, [3]trewavas@ed.ac.uk, [*]Corresponding author

Abstract

Calcium dependent protein kinases (CDPKs) are a group of protein kinases whose primary function is the interpretation of Ca^{2+} signals. We have investigated the localization of a calcium dependent protein kinase (CDPK) in the germinating pollen grain and growing pollen tubes. Pollen specific CDPK of *Nicotiana plumbaginifolia*, NpCDPK7, has 52 N-terminal amino acid residues containing a putative myristoylation site and 34 amino acids in the C-terminal domain which includes a putative microbody targeting sequence. Green florescent protein (GFP) was introduced between the N-terminal domain and kinase domain of NpCDPK7 and pollen grains were transformed with this construct by biolistic gun. In pollen tubes, GFP florescence was observed on the surface of the generative cell (containing the two sperm cells), on the plasma membrane and in granules in the ungerminated pollen grain. Mutation or deletion analysis of the N-terminal domain supports the localization of NpCDPK7 protein as determined by the N-terminal region myristoylation. NpCDPK7 may be involved in localized calcium signaling, sensing calcium transients directly under the plasma membrane.

Introduction

Calcium is one of the essential minerals for pollen tube growth and an important signaling component. Recent imaging techniques have let us visualize the oscillatory process of tip growth and the tip high calcium gradient of growing pollen tubes revealing calcium concentration is well coordinated with pulsate growth and reorientation (Messerli and Robinson, 1997, Malho and Trewavas, 1996, Parton et al., 2001). Calcium dependent protein kinase (CDPK) is one of the protein kinases found in plants and protozoan and is a large gene family, in *Arabidopsis* 42 CDPK sequences are reported (Harmon et al., 2001). CDPKs are activated by increased calcium concentration and function as direct transducers of calcium signaling. To investigatetheir role in the calcium signaling in growing pollen tubes, CDPK isozymes

N. Murata et al. (eds.), Advanced Research on Plant Lipids, 327–330.

were cloned from *Nicotiana plumbaginifolia* and tissue specificity was studied as well as localization in the growing tube and the effects of over expression.

Materials and Methods

Nicotiana Plumbaginifolia and *Nicotiana tabacum* were grown under glass. Pollen was collected and used for experiments immediately or frozen with liquid nitrogen and stored at −80 °C. *N. Plumbaginifolia* pollen was cultured in germination medium (Read et al., 1993) for 1, 2, 4 or 8h and then combined for extraction of mRNA. A pollen specific expression vector was constructed by exchanging cauliflower mosaic virus 35S promoter of pART7 vector with the pollen specific promoter of LAT52 gene (Twell et al., 1991), named as pART8. 500 ng of DNA was co-precipitated with 0.6 mg of gold particles (Biorad, Uk) and transformation by biolistic gun was performed according to Kost (Kost et al., 1999). GFP florescence was observed by laser confocal scanning microscope (MRC600, Biorad) and wide field fluorescence microscope (Diaphoto, Nikon) using planapo X60 (N.A. 1.2) or planapo X40 (N.A. 0.95). Domain search analysis of NpCDPK7 was performed in PROSITE (www.expacy.ch/prosite/). Antibody was raised in rabbits and affinity purified.

Results

cDNA of NpCDPK7 consists of 2102bp and encodes 528 amino acids. NpCDPK7 sequence has the closest similarity (80% amino acid identity) to *Arabidopsis* CPK7 (Hrabak et al., 1996) and *Fragaria* CDPK isozyme, MAX17 (accession no. AF035944). Domain search of NpCDPK7 revealed two putative myristoylation signals in the first 52 amino acids and one microbody targeting sequence in the carboxyl terminus. The calmodulin-like domain has three EF-hand motifs as AtCPK7. Anti-NpCDPK7 antibody was raised against a peptide which commenced at the 2nd amino acid, glycin, to the 18th amino acid, lysine and gave high anti-NpCDPK7 specificity. The Predicted size of 59 KDa NpCDPK7 protein was detected in pollen (Fig. 1). The expression of NpCDPK7 is observed from late stage of pollen maturation in the anther and reached at maximum level in dehydrated pollen grain. The level of expression was not changed during pollen tube growth. Expression was not seen in leaf or petal. This indicates NpCDPK7 is expressed in pollen specifically.

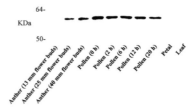

Figure 1. Expression of NpCDPK7 protein in anther, pollen, petal and leaf. Pollen was incubated in germination medium from 0h to 20h. Protein was extracted from the tissues and 10 μg of the total proteins are subjected to SDS-PAGE.

To investigate the localization of NpCDPK7, smGFP was introduced between the N-terminal domain and kinase domain of NpCDPK7 (NpCDPK7-GFP). The fusion gene was over expressed under the control of the LAT52 gene pollen specific promoter (Twell et al., 1991). Wild type pollen of *N. tabacum* germinates from 30 min to 90 min after imbibition. Over expression of NpCDPK7-GFP inhibits pollen germination (Figure 3) and GFP localization was seen in generative cell membrane, plasma membrane and some small, roughly spherical structures (Figure 2, top). Over expression of NpCDPK7 lacking GFP also inhibited germination (Figure 3). These results indicate that NpCDPK7-GFP protein may have similar activity to wild type NpCDPK7 protein and reveals the native localization of endogenous NpCDPK7. Knock-out of the kinase domain by mutation of the ATP binding site (NpCDPK7-K82M-GFP) allowed partial germination of pollen. NpCDPK7-K82M-GFP protein localized on the generative cell surface, and plasma membrane, but only weakly in the nucleus and cytosol (Figure 2, middle). Plasma membrane localization of NpCDPK7-GFP and NpCDPK7-K82M-GFP suggests that the amino-terminal myristoylation signal may contribute to membrane localization. Fusion protein of first 52 amino acids of NpCDPK7 and GFP were sufficient to identify the plasma membrane localization and generative cell localization (Figure 2, bottom). Mutational analysis of the myristoylation signal abolished membrane localization (data not shown).

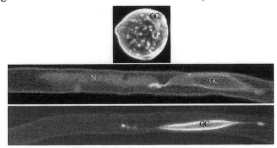

Figure 2. GFP fluorescence of pollen or growing pollen tube. Pollens were transformed with NpCDPK7-GFP (top), NpCDPK-K82M-GFP (middle) or NV2smGFP (bottom) and incubated for 6h in germination medium. GFP fluorescence from nucleus (N), generative cell (GC) and plasma membrane.

Pollen germination was inhibited by the over expression of NpCDPK7 or NpCDPK7-GFP fusion protein (Figure 3). There is a possibility that germination inhibition could be caused by over expression of active kinase and subsequent inappropriate phosphorylation. To check this possibility, NpCDPK7-ΔNV which lacked the first 52 amino acids was examined. GFP fusion of NpCDPK7-ΔNV localized in the cytoplasm (data not shown). NpCDPK7-ΔNV which had an active kinase domain but was not localized in the membrane did not inhibit germination of pollen although tube growth was aberrant after germination (Figure 3). This result indicates that only over expression of active kinase, which targets the native position of NpCDPK7 inhibits germination. K252a, kinase inhibitor (10 µM) also inhibited germination (Figure 3).

Discussion

NpCDPK7 expression was specific in the late maturation stage of pollen. Localization of NpCDPK7 in the plasma membrane implies that the calcium signal to which it

330

respond is localized to the plasma membrane with NpCDPK7 acting on locally available substrates. Over expression of NpCDPK7 inhibits germination while mis-localized active kinase does not. It is possible that its represents a dominant negative effect on plasma membrane. Inhibition of germination by K252a would support this. Overall these results indicate that NpCDPK7 may contribute to the germination process of pollen. The hydration process is one of the critical event for pollen germination. Aquaporin is an intrinsic membrane protein and transmits water molecules through lipid bilayers. One isozyme of CDPK, CDPKα activates aquaporin *in vivo* (Weaver et al., 1991). The target of NpCDPK7 could be aquaporin involved in the hydration process of dry pollen. NpCDPK7 also localizes in the generative cell membrane and subsequent sperm cell membrane (data not shown). During pollen development in *Nicotiana*, nucleus divides asymmetrically and one side of nucleus (generative cell nucleus) is surrounded by two layer of cell membrane on the plasma membrane. The origin of the outer layer of the generative cell is the plasma membrane. Localization of myristoylated NV2smGFP in the generative cell membrane and plasma membrane implicate a similar lipid composition between them. The possible role of NpCDPK7 in generative cell cytokinesis is currently under investigation.

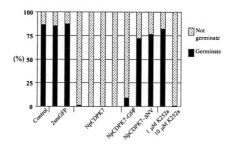

Figure 3. Germination rate of *N.tabacum* pollen. Pollens are transformed with cDNA of 2smGFP, NpCDPK7, NpCDPK7-GFP and NpCDPK7-ΔNV or incubated with K252a. Control is WT pollen. Germination rate was examined 6h after imbibition.

Harmon AC, Gribskov M, Gubrium E, Harper JF. (2001) The CDPK superfamily of protein kinases. New Phytol. 151, 175-183.
Hrabak,E.M., Dickmann,L.J., Satterlee,J.S., Sussman, M.R. (1996) Characterization of eight new members of the calmodulin-like domain protein kinase gene family from *Arabidopsis thaliana*. Plant Mol. Biol. 31, 405-412.
Kost B, Lemichez E, Spielhofer P, Hong Y, Tolias K, Carpenter C, Chua NH. (1999) Rac homologues and compartmentalized phosphatidylinositol 4,5-bisphosphate act in a common pathway to regulate polar pollen tube growth. J. Cell Biol. 145, 317-330.
Malho R., Trewavas AJ. (1996) Localized apical increases of cytosolic free calcium control pollen tube orientation. Plant Cell 8, 1935-1949.
Messerli M., Robinson KR. (1997) Tip localized Ca2+ pulses are coincident with peak pulsatile growth rates in pollen tubes of *Lilium longiflorum*. J. Cell Sci. 110, 1269-1278.
Parton RM., Fischer-Parton S., Watahiki MK., Trewavas AJ. (2001) Dynamics of the apical vesicle accumulation and the rate of growth are related in individual pollen tubes. J. Cell Sci. 114, 2685-2695.
Read SM., Clarke AE., Bacic A. (1993) Stimulation of growth of cultured Nicotiana tabacum W38 pollen tubes by poly (ethylene glycol) and Cu(II) salts. Protoplasma 177, 1-14.
Twell D, Yamaguchi J, Wing RA, Ushiba J, Mccormick S. (1991) Promoter analysis of genes that are coordinately expressed during pollen development reveals pollen-specific enhancer sequences and shared regulatory elements. Genes & Dev. 5, 496-507.
Weaver, C.D., Crombie, B., Stacey, G., and Roberts, D.M. (1991) Calcium-dependent phosphorylation of symbiosome membrane proteins from nitrogen-fixing soybean nodules. Plant Physiol. 95, 222-227.

GENE-ENGINEERED RIGIDIFICATION OF MEMBRANE LIPIDS ENHANCES THE COLD INDUCIBILITY OF GENE EXPRESSION IN *Synechocystis*

M. INABA[1,2], S. FRANCESCHELLI[1,3], I. SUZUKI[1,4], B. SZALONTAI[5], Y. KANESAKI[1], D. A. LOS[6], H. HAYASHI[2,7] and N. MURATA[1,4]

[1]*Department of Regulation Biology, National Institute for Basic Biology, Myodaiji, Okazaki 444-8585, Japan;* [2]*Satellite Venture Business Laboratory, Ehime University, Matsuyama 790-8577, Japan;* [3]*Department of Pharmaceutical Sciences, University of Salerno, Via Ponte Don Melillo, 84084 Fisciano, Italy;* [4]*Department of Molecular Biomechanics, School of Life Science, Graduate School for Advanced Studies, Myodaiji, Okazaki 444-8585, Japan;* [5]*Institute of Biophysics, Biological Research Center, Szeged, Hungary;* [6]*Institute of Plant Physiology, Russian Academy of Sciences, Botanicheskaya Street 35, 127276 Moscow, Russia;* [7]*Department of Chemistry, Ehime University, Matsuyama 790-8577, Japan.*

1. Introduction

The exposure of *Synechocystis* sp. PCC 6803 (hereafter *Synechocystis*) cells to low temperature induces the expression of a specific set of genes that includes the *desA*, *desD* and *desB* genes for acyl-lipid desaturases. Previously we produced a series of mutants of *Synechocystis* in which the extent of unsaturation of fatty acids is modified in a stepwise manner. One of these mutants, *desA⁻/desD⁻*, synthesizes only a saturated C16 fatty acid and a monounsaturated C18 fatty acid regardless of the growth temperature, while the wild type synthesizes di-unsaturated and tri-unsaturated C18 fatty acids in addition to the monounsaturated C18 fatty acid (Tasaka et al. 1996). This system is suitable to examine the contribution of membrane fluidity/rigidity to the cold-inducible gene expression.

2. Experimental procedures

Cells of wild-type and *desA⁻/desD⁻* strains of *Synechocystis* were grown photoautotrophically at 34°C under illumination from incandescent lamps at 70 μE m⁻² s⁻¹ in BG-11 medium with aeration by air that contained 1% CO_2 (Ono and Murata,

N. Murata et al. (eds.), Advanced Research on Plant Lipids, 331–334.

1981). At the early exponential phase of the growth (the optical density at 750 nm 0.3 to 0.4), 50-ml aliquots were withdrawn and incubated for 30 min under the same conditions except that the temperature was varied as indicated. Plasma and thylakoid membranes were isolated from *Synechocystis* cells that had been grown at 34°C as described (Murata and Omata 1988).

FTIR spectra were recorded at a spectral resolution of 2 cm^{-1} with an FTIR spectrometer (Szalontai et al. 2000). The DNA microarray analysis was performed as described previously (Suzuki et al. 2001; Kanesaki et al. 2002). The microarray covers 3079 genes (97% of the total genes) in *Synechocystis*.

3. Results and Discussion

3.1. FTIR analysis of the physical state of plasma membranes

FTIR spectra from plasma membranes from wild-type and *desA⁻/desD⁻* cells show the shift in frequency of the $v_{sym}CH_2$ band as a function of temperature. The concordant data were calculated from FTIR spectra that had been recorded as the temperature was increased in steps of 2-3°C. Thus, the curves represent the ordered state of the respective membranes. A lower frequency of $v_{sym}CH_2$ stretching in the 2,852 cm^{-1} region at a given temperature corresponds to a more ordered (rigidified) state. The results indicate that the plasma membranes from *desA⁻/desD⁻* cells were more rigid than those from wild-type cells, in particular at 22°C, the temperature at which we examined the expression of cold-regulated genes. Our FTIR analysis indicated that lipids in plasma membranes were rigidified with decreases in temperature and that rigidification was enhanced by the saturation of fatty acids upon inactivation of fatty acid desaturases.

3.2. DNA microarray analysis of effects of the membrane rigidification on the gene expression under isothermal conditions (34°C).

We examined effects of the membrane rigidification on the expression of individual genes when wild-type and *desA⁻/desD⁻* cells were grown at 34°C. Most data points appeared between two lines, which correspond to the ratios of the expression in *desA⁻/desD⁻* cells to that in wild-type cells being 0.5 and 2, respectively. This indicates that the expression levels of most genes under isothermal conditions were unaffected by the rigidification.

3.3. DNA microarray analysis of effects of the membrane rigidification on the cold-inducible gene expression

We examined effects of the incubation of cells at 22°C for 30 min, after growth at 34°C, on the expression of individual genes. In the case of both wild-type and *desA⁻/desD⁻* cells, many of the data points lie between two lines, which correspond to the ratios of the expression in cold shock-treated cells (22°C cells) to that in untreated cells (34°C cells) being 0.5 and 2, respectively. This indicate that, in both types of cell, expression of the majority of genes was unaffected by the cold shock. Data points that fall above the upper line correspond to genes for which the cold shock enhanced the level of expression. About two hundred genes were induced upon the cold shock in wild-type cells. In *desA⁻/desD⁻* cells, the cold shock induced the expression of the same set of genes, and, in addition, it induced the expression of several genes which are not cold-inducible in wild-type cells.

Cold-inducible genes can be divided into three groups according to the effects of the membrane rigidification caused by the *desA⁻/desD⁻* mutation: The first group consists of 15 genes whose cold inducibility was highly enhanced by the rigidification. This includes certain heat-shock genes, such as *hspA, clpB* and *dnaK2*, the *sbpA, cysA* and *cysT* genes for the sulfate transport system, the *sigB* gene for an RNA polymerase sigma factor, the *hik34* gene for a histidine kinase, and the *pbp* gene for a homologue of the penicillin-binding protein. The second group consists of 17 genes whose cold inducibility was moderately enhanced by the rigidification, most of which are genes for proteins of unknown function. The third group consists of 25 genes whose cold inducibility was unaffected by the rigidification. This includes genes which are known as cold-shock genes, such as *crhL* for an RNA helicase (Chamot et al. 1999) and *rbpA* for an RNA binding protein (Sato 1995; Maruyama et al. 1999). The third group also include several genes for ribosomal proteins, the *ndhD2* gene for NADH dehydrogenase subunit 4, the *ccmK, ccmN* and *ccmM* genes for the CO_2 concentration mechanism, the *secE* gene for the preprotein translocase subunit, the *rlpA* gene for a rear lipoprotein A, the *nrtA* gene for the nitrate/nitrite transport system substrate-binding protein, and the *hik31* gene for a histidine kinase.

These results indicate that the rigidity (or fluidity) of membranes plays a role in the regulation of gene expression upon cold shock. This suggests that the mechanism of cold perception might involve the rigidification of membrane lipids. Our earlier work demonstrated that the perception by *Synechocystis* of low temperature, with subsequent enhancement of the expression of genes for fatty acid desaturases, is caused by the rigidification of membrane lipids (Vigh et al. 1993). Our present findings suggest that the cold induction of the genes which were converted from non cold-inducible to cold-inducible by the membrane rigidification might require a higher degree of membrane rigidification than the cold induction of the genes for fatty acid desaturases. The rigidification did not affect the cold inducibility of cold-shock genes. This may suggest that the membrane rigidity in wild-type cells is sufficiently high for the maximum induction of these genes. We have identified a membrane-bound histidine kinase, Hik33, as a cold sensor in *Synechocystis* (Suzuki et al. 2000; Suzuki et al. 2001). A molecular structure predicted from the amino acid sequence indicates that Hik33 may span the plasma membrane twice and form a dimer. The dimeric structure could be influenced by physical properties of the membrane, such as rigidity, which are regulated by temperature and the extent of unsaturation of fatty acids. When the temperature is decreased or the fatty acids are more saturated, the histidine residue in the kinase domain of Hik33 might be phosphorylated. The phosphate group is then transferred to its cognate response regulator to transduce the cold signal.

3.4. *Northern blotting analysis of the expression of cold-regulated genes*

Northern blotting analysis revealed the cold-induced expression of *hspA* in *desA⁻/desD⁻* cells, with maximum induction at 24°C, which was hardly observed in wild-type cells. These results are consistent with the results of microarray analysis. However, the difference between the two types of cells was more apparent in the Northern blotting analysis than in the microarray analysis. Northern blotting analysis of the expression of the *dnaK*2 gene yielded a result similar to that obtained for the *hspA* gene. By contrast, Northern blotting analysis indicated that the cold inducibility of the *rbpA* and *crhL* genes was unaffected by the membrane rigidification, in agreement with the results of the microarray analysis.

5.   References

Chamot, D., Magee, W. C., Yu, E. and Owttrim, G. W. (1999) A cold shock-induced cyanobacterial RNA helicase. J. Bacteriol., 181, 1728-1732.

Kanesaki, Y., Suzuki, I., Allakhverdiev, S. I., Mikami, K. and N. Murata. (2002) Salt stress and hyperosmotic stress regulate the expression of different sets of genes in Synechocystis sp. PCC 6803. Biochem. Biophys. Res. Commun., 290, 339-348.

Maruyama, K., Sato, N. and Ohta, N. (1999) Conservation of structure and cold-regulation of RNA-binding proteins in cyanobacteria: probable convergent evolution with eukaryotic glycine-rich RNA-binding proteins. Nucleic Acids Res., 27, 2029-2036.

Murata, N. and Omata, T. (1988) Isolation of cyanobacterial plasma membranes. In (Packer, L. and Glazer, A. N., eds), Methods in Enzymology, Academic press, San Diego CA vol. 167, pp. 245-251.

Ono, T. and Murata, N. (1981) Chilling susceptibility of the blue-green alga Anacystis nidulans. I. Effect of growth temperature. Plant Physiol., 67, 176-181.

Sato, N. (1995) A family of cold-regulated RNA-binding protein genes in the cyanobacterium Anabaena variabilis M3. Nucleic Acids Res., 23, 2161-2167.

Suzuki, I., Kanesaki, Y., Mikami, K., Kanehisa, M. and Murata, N. (2001) Cold-regulated genes under control of the cold sensor Hik33 in Synechocystis. Mol. Microbiol., 40, 235-244.

Suzuki, I., Los, D. A., Kanesaki, Y., Mikami, K. and Murata, N. (2000) The pathway for perception and transduction of low-temperature signals in Synechocystis. EMBO J., 19, 1327-1334.

Szalontai, B., Nishiyama, Y., Gombos, Z. and Murata, N. (2000) Membrane dynamics as seen by Fourier transform infrared spectroscopy in a cyanobacterium, Synechocystis PCC 6803. The effects of lipid unsaturation and the protein-to-lipid ratio. Biochim. Biophys. Acta, 1509, 409-419.

Tasaka, Y., Gombos, Z., Nishiyama, Y., Mohanty, P., Ohba, T., Ohki, K. and Murata, N. (1996) Targeted mutagenesis of acyl-lipid desaturases in Synechocystis: evidence for the important roles of polyunsaturated membrane lipids in growth, respiration and photosynthesis. EMBO J., 15, 6416-6425.

Vigh, L., Los, D. A., Horváth, I. and Murata, N. (1993) The primary signal in the biological perception of temperature: Pd-catalyzed hydrogenation of membrane lipids stimulated the expression of the desA gene in Synechocystis PCC6803. Proc. Natl. Acad. Sci. USA., 90, 9090-9094.

MONITORING OF 12-oxo-PHYTODIENOIC ACID (OPDA)-INDUCED EXPRESSION CHANGES IN ARABIDOPSIS BY cDNA MACROARRAY

[1]Y. SASAKI-SEKIMOTO, [1]T. OBAYASHI, [1]M. MATSUUMI, [1]Y. KOBAYASHI, [2]E. ASAMIZU, [2]D. SHIBATA, [2]Y. NAKAMURA, [1]T. MASUDA, [1]H. SHIMADA, [1]K. TAKAMIYA, [2]S. TABATA and [1]H. OHTA.

[1]*Grad. Sch. Biosci. & Biotech., Tokyo Inst. Tech., Yokohama 226-8501, Japan*
[2]*Kazusa DNA Research Institute, Kisarazu 292-0812, Japan*

1. Introduction

Jasmonates are known as growth regulators, which have cyclopentanone or cyclopentenone ring, and synthesized through lipoxygenase pathway. Jasmonates are widely distributed in the plant kingdom and modulates wounding responses, disease responses, and anther development. Some kinds of jasmonates were shown to have specific effects on plants. 12-oxo-phytodienoic acid (OPDA), a precursor for jasmonic acid (JA) biosynthesis, promotes tendril coiling, and this effect is stronger than MeJA (Falkenstein E., et al., 1991). However, except for the fact, little is known about the OPDA specific function in various physiological events.

We identified methyl jasmonate (MeJA)-responsive genes whose mRNA levels were changed more than 3-fold (Sasaki et al., 2001). These included JA biosynthesis genes such as *LOX2*, *AOS*, *AOC*, *OPR1*, and *OPR3*, which showed changes of the mRNA levels mainly after 3 to 6 h of MeJA treatment. However, it remains to be clarified why jasmonate biosynthesis genes are induced by MeJA treatment. Recently, it was reported that *OPR3* deficient plants treated with exogenous OPDA powerfully up-regulated several jasmonate-responsive genes (Stintzi et al., 2001) suggesting OPDA itself functions as a signaling molecule. In this study, we adopted cDNA macroarray technique to screen OPDA-responsive genes and examine the effect of OPDA treatment on the expression of JA-responsive genes, particularly for JA-biosynthesis. Comparison between MeJA- and OPDA-responsive genes revealed 11 transcripts whose expression profiles were differently induced by each treatment. These results suggest the existence

335

N. Murata et al. (eds.), Advanced Research on Plant Lipids, 335–338.
© 2003 *Kluwer Academic Publishers. Printed in the Netherlands.*

of an OPDA-dependent signaling pathway in Arabidopsis.

2. Experimental

2.1 *Plant material*

MeJA was purchased from Wako and OPDA was synthesized as described in Kobayashi and Matsuumi, 2002. Seeds of *Arabidopsis thaliana* (Colombia) were grown in Murashige and Skoog medium, containing 1% (w/v) of sucrose under continuous light condition at 22°C. After 10 days, plants were treated with 30μM MeJA or OPDA.

2.2 *Hybridization of cDNA macroarray filters*

We basically used experimental conditions previously described (Sasaki et al., 2001), which have been applied for screening of MeJA-responsive genes. To study OPDA-responsive genes, we made new series of cDNA macroarray filters, which were spotted on 4512 ESTs generated from aboveground organs and flower buds (Asamizu et al., 2000). The new cDNA macroarray filters include the same 2880 ESTs, which were used previously (Sasaki et al., 2001). Total RNAs were isolated from OPDA-treated and untreated plants and purified using Rneasy Plant Mini kit (Qiagen). $[\alpha\text{-}^{33}P]$ dCTP labeled single-strand DNAs were synthesized from 10μg of total RNA using an oligo dT primer and RNA PCR kit Ver. 2.1 (Takara). Hybridization to the cDNA macroarray filters was carried out at 65°C for 16h, with Church phosphate buffer ($0.5M$ Na_2HPO_4, 1mM EDTA, 7% SDS) and the filters were washed twice with $0.2 \times SSC$, 0.1% SDS for 20 min at 65°C. The filters were exposed to an imaging plate (Fuji film).

2.3 *Identification of MeJA- and OPDA-responsive genes*

Radioactive images from two data sets (MeJA and OPDA treatment) were obtained with a high-resolution scanner (Storm, Amersham Pharmacia Biotech), and quantification of the signal intensity was carried out using an Array Vision (Amersham Pharmacia Biotech).

To prevent pseudo-positive, hypothetical membrane background was defined as the value of 0.8-fold of the minimum for the intensity of EST spots on each filter. Global normalization was adopted for normalizing the experimental conditions of the each filter. The relative signal intensity was calculated as the ratio of each signal to the median for the intensity of all the spots on the filter. The value thus estimated is called 'expression ratio'. Data were obtained from at least 4 independent experiments for MeJA and OPDA treatment. Genes that responded to MeJA or OPDA with reproducible results were selected by one-way repeated-measures analysis of variance ($P < 0.05$).

3. Results and Discussion

We screened MeJA- and OPDA-responsive genes using cDNA macroarray covering

2880 or 4512 ESTs respectively. Plants were treated with 30μM MeJA or OPDA. Total RNA was isolated from untreated controls and MeJA- or OPDA-treated plants at 15, 30, 60, 180, and 360min after application. Recently we have isolated 41 MeJA-responsive genes (Sasaki et al., 2001) using cDNA macroarray. Here, we identified 70 OPDA-responsive genes in the ESTs, which showed expression changes more than 3-fold. A comparison of expression profiles from both treatments revealed that MeJA and OPDA coinduced a series of genes, however, the induction profiles by the two molecules were different. If OPDA itself functions as a signaling molecule, there thought to be OPDA-responsive genes which show earlier induction than MeJA treatment. On the other hand, if exogenous OPDA is metabolized to MeJA in plants, it is likely that the induction of the genes are almost the same or later than MeJA treatment.

We selected a subset of genes, which was induced more than 3-fold at 60min after OPDA application but less than 2-fold by MeJA (TABLE 1). Eleven transcripts (e.g., glutathione-S- transferase 8 (GST8), and 5'-adenylylsulfate reductase) belonged to this group. In these genes, 12-oxo-phytodienoic acid reductase 1 (OPR1), a gene for JA-biosynthesis, was included. The expression profiles of OPR1, anthocyanin acyltransferase-like protein, 5'-adenylylsulfate reductase, GST8, a gene of unknown function, OPR3, and AOS were shown in Fig. 1. These transcripts are coinduced by both treatments. Among these genes, 5 genes including OPR1, and GST8 showed earlier induction by OPDA treatment than that by MeJA, whereas OPR3 and AOS showed the same profiles by both treatments.

We confirmed the response of these genes to MeJA and OPDA by Northern blot analysis. The expression profiles of the genes (shown in Fig.1) showed agreement with the result of cDNA macroarray. Above results suggest that these genes are likely to be regulated by OPDA-dependent signaling pathway in Arabidopsis.

TABLE 1 Relative Transcripts Abundance after 60min MeJA or OPDA treatment.

No.	gene function	fold	
		OPDA	MeJA
1	12-oxophytodienoate reductase (OPR1)	8.9	1.2
2	anthocyanin acyltransferase-like protein	3.1	0.9
3	5'-adenylylsulfate reductase	6.9	1.7
4	glutathione transferase 8 (GST8)	7.6	1.5
5	unknown 1	6.8	1.0
6	ABC transporter, putative	3.0	1.6
7	putative cytochrome p450	3.8	1.7
8	NADP dependent malic enzyme-like protein	3.4	1.1
9	heat shock protein 70	3.4	1.2
10	unknown 2	3.8	1.3
11	unknown 3	4.0	1.1

338

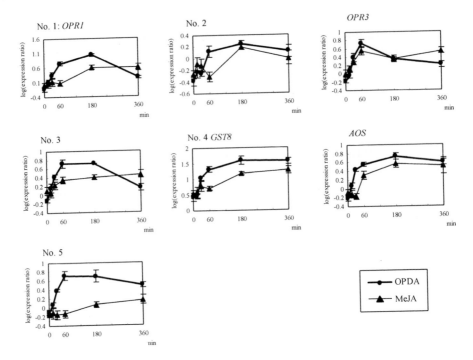

Fig. 1 Comparison between the expression profiles of OPDA and MeJA treatment. The y-axis shows the expression ratio of each transcript. Numbers of each graph is consistent with TABLE 1. Bars indicate the standard error of independent experiments. The numbers of independent experiments were as follows: OPDA-treatment 0, 15, 30, 60, 180 and 360 min, 4; MeJA-treatment 0, 60, and 180 min, 8; 15, 30, and 360 min, 4.

4. References

Asamizu, E., Nakamura, Y., Sato, S., Tabata, S. (2000) A large scale analysis of cDNA in Arabidopsis thaliana: generation of 12,028 non-redundant expressed sequence tags from normalized and size-selected cDNA libraries. DNA Res. 7, 175-180.

Falkenstein, E., Groth, B., Mithofer, A., and Weiler, E. W. (1991) Methyljasmonate and a-linolenic acid are potent inducer of tendril coiling. Planta. 185, 316-322.

Kobayashi, Y., and Matsuumi, M. (2002) Controlled syntheses of 12-oxo-PDA and its 13-epimer. Tetrahedron Lett. in press.

Sasaki, Y., Asamizu, E., Shibata, D., Nakamura, Y., Kaneko, T., Awai, K., Amagai, M., Kuwata, C., Tsugane, T., Masuda, T., Shimada, H., Takamiya, K., Ohta, H., and Tabata, S. (2001) Monitoring of methyl jasmonate-responsive genes in Arabidopsis by cDNA macroarray: self-activation of jasmonic acid biosynthesis and crosstalk with other phytohormone signaling pathways. DNA Res. 8, 153-161.

Stintzi, A., Weber H., Reymond, P., Browse, J., and Farmer, E. E. (2001) Plant defense in the absence of jasmonic acid: the role of cyclopentenones. Proc Natl Acad Sci U. S. A. 98, 12837-12842.

CHARACTERIZATION OF LIGHT-RESPONDING ACTIVATION OF PHOSPHOLIPASE D IN SUSPENSION-CULTURED PHOTOAUTOTROPHIC SOYBEAN (SB-P) CELLS

K. NISHIMURA, Y. SAKAE, D. EGUCHI, T. MUTA, M. JISAKA, T. NAGAYA, and K. YOKOTA

Department of Life Science and Biotechnology, Shimane University, Nishikawatsu-cho, Matsue, Shimane 690-8504, Japan

1. Introduction

Plant phospholipase D (PLD) has been implicated in many cellular processed such as lipid deterioration under senescence and stress, lipid turnover during normal growth and development, and lipid-derived signaling pathways [Wang, 2000]. Plant PLD activity is also known to be stimulated during germination, but the molecular mechanism for stimulation of PLD activity remains unclear. PLD pathway is considered to be linked to PLC pathway because these metabolites are enzymatically interconverted and phospholipase C (PLC) increases intracellular levels of Ca^{2+} to induce activation of Ca^{2+}-dependent enzyme such as PLD. As light stimulated turnover of phosphatidyl inositol catalyzed by PLC in various types of plants [Morse, et al., 1987], light might be a trigger for stimulation of PLD activity during germination

In the present study, we investigated effect of light on PLD activity in suspension-cultured photoautotrophic soybean (SB-P) cells in order to understand the involvement of PLD during photomorphogenesis.

2. Materials and Methods

2.1. *Cell culture*

Suspension-cultured photoautotrophic soybean (SB-P) cells were kindly gifted from Widholm (Ohio States University). Cultured SB-P cells were grown in KN1 medium containing Murashige-Skoog salts supplemented with 0.2 mg of kinetin per liter, 0.1mg of thiamine per liter, 1 mg of α-naphthaleneacetic acid per liter, and 3 g of sucrose per liter (pH 5.7-5.9) in 100-ml flasks. For experiments of light response,

N. Murata et al. (eds.), Advanced Research on Plant Lipids, 339–342.

SB-P cells were shaken at 107 rpm under continuous white fluorescent at 25℃ and subcultured every 14 days. In experiments of light response, the cells were grown in the light for 6 days after inoculation. The flasks of cell cultures were covered with two layers of aluminum foil, and shaken for additional 3 days to be adapted to darkness. After dark adaptation, the cells were re-illuminated and cultured up to 24 h to examine PLD activity, amounts of PA and chlorophyll, and expression of photosynthetic genes.

2.2. PLD assay

PLD activity was assayed according to the procedure of Ishimura and Horiuti (1978), which was based on the estimation of choline by using choline oxidase and peroxidase.

2.3. Determination of chlorophyll contents

Intracellular chlorophyll was extracted from cells with 80% (v/v) acetone. Contents of chlorophyll ($a+b$) were quantified by measuring an absorbance maximum at 590 nm and 635 nm with a spectrometer.

3. Results and Discussion

3.1. Activation of light-responding PLD activity in cultured SB-P cells

We examined the activity of PLD in response to light. As shown in Fig. 1, PLD activity was transiently stimulate to reach the maximum levels at 3 min. Approximately 20 min later, PLD activity was elevated to reach the maximum levels, followed by decreasing to maintain the basal levels. From these results, PLD activity in SB-P cells was twice stimulated after illumination.

3.2. Effect of 1-butanol, an attenuator of PLD signaling, on light-responding formation of chlorophyll in cultured SB-P cells

In order to examine the involvement of PLD in the signal transduction pathway responsible for light, we investigated the effects of 1-butanol (1-BuOH), a specific attenuator of PLD signaling, on light responding formation of chlorophyll in cultured SB-P cells. As shown in Fig. 2, amounts of chlorophyll in SB-P cells shifted from darkness to the light in the presence of 1-BuOH were reduced to the similar levels in the cell grown in darkness in the absence of 1-BuOH. On the contrary, 3-BuOH, an inactive attenuator of PLD signaling, was not effective. Therefore PLD was suggested to be involved in the light-responding formation of chlorophyll.

3.3. Effect of 1-BuOH on light-responding expression of photosynthetic genes in cultured SB-P cells

We examined the effect of 1-BuOH on light-responding expression of photosynthetic

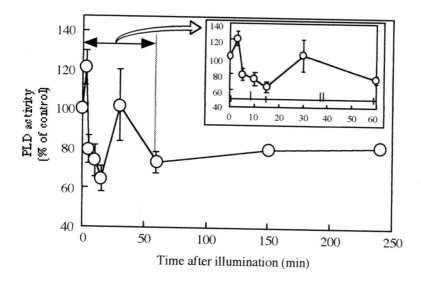

Fig. 1. Time-course profile of PLD activity in SB-P cells shifted from darkness to the light. SB-P cells were grown in the light for 6 days, and adapted to darkness for 3 days. The cells were shifted to the light to be cultured up to 4 h. The levels of PLD activity were normalized by protein contents, and were represented as the relative values to those at 0 min, which was indicated as 100%.

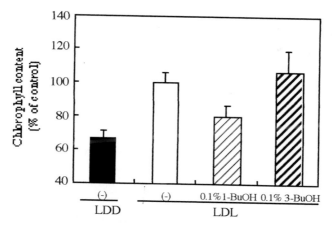

Fig. 2. Effect of 1-BuOH on light-responding formation of chlorophyll in SB-P cells. SB-P cells were grown in the light for 6 days, and then adapted to darkness for 3 days. 1- BuOH or 3-BuOH was added to the culture in darkness 1-h before the cells were shifted to the light to be cultured for 24 h. The levels of chlorophyll were normalized by protein contents, and were represented as the relative values to the cells re-illuminated without 1-BuOH (LDL (-)), which were indicated as 100%.

genes in SB-P cells. As shown in Fig. 3, the mRNA levels of *cab* encoding chlorophyll *a/b* binding protein and *chlI* encoding subunit I of magnesium-chelatase

342

in the cells shifted from darkness to the light in the presence of 1-BuOH were reduced to the similar levels in the cells grown in darkness without 1-BuOH. On the contrary, 3-BuOH was not effective. PLD was suggested to be involved in the light-responding expression of photosynthetic genes, concomitant with chlorophyll accumulation. PA, a metabolite converted from phosphatidyl choline (PC) by PLD, was suggested to be a lipid-derived mediator responsible for light signal transduction as well as wounding, protection of pathogen, and senescence.

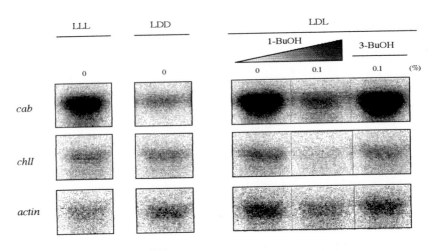

Fig. 3. Effect of 1-BuOH on light-responding expression of photosynthetic genes in cultured SB-P cells. 1-BuOH was added to the dark-adapted cells 1-h before the cells were shifted to the light to be cultured for 4 h. Thirty μg of total RNA was used for RNA gel blot analysis. A cDNA probe specific for soybean actin gene was used as an internal control. The radioactivity was detected with BAS 1500 (Fujix, Tokyo).

4. References

Chomczynski, P. and Sacchi, N. (1987) Single-step method of RNA isolation by acid guanidium thiocyanate-phenol-chloroform extraction. Anal. Biochem. 162, 156-159.

Ishimura, S. and Horiuti, Y. (1978) Enzymatic determination of phospholipase D activity with choline oxidase. J. Biochem. 83, 677-680.

Morse, M.J., Crain, R.C. and Satter, R.L. (1987) Light-stimulated inositol phospholipid turnover in *Samanea saman* leaf pulvini. Proc. Natl. Acad. Sci. USA 84, 7075-7078.

Munnik, T. (2001) Phospatidic acid: an emerging plant lipid second messenger. Trends Plant Sci. 6, 227-233.

Wang, W. (2000) Multiple forms of phospholipase D in plants: the gene family, catalytic and regulatory properties, and cellular functions. Prog. Lipid Res. 9, 109-149.

SALICYLIC ACID AND AROMATIC MONOAMINES INDUCE GENERATION OF ACTIVE OXYGEN SPECIES LEADING TO INCREASE IN CYTOSOLIC Ca^{2+} CONCENTRATION IN TOBACCO SUSPENSION CULTURE

SHOSHI MUTO AND TOMONORI KAWANO[1]

Nagoya University Bioscience Center, Chikusa-ku, Nagoya 464- 8601, Japan
[1]*Present address: Graduate School of Science, Hiroshima University, Higashi-Hiroshima 739-8526, Japan*

1. Introduction

Until late '90s, only four enzymatic pathways have been considered to be responsible for production of reactive oxygen species (ROS) during the plant defense mechanism, namely NADPH oxidase, pH-dependent cell wall peroxidase, germin-like oxalate oxidase and amine oxidases. Among four enzymes, NADPH oxidase and the pH-dependent cell wall bound peroxidase are considered to be the main sources of ROS in plants (Bolwell, 1995; Bolwell and Wojtaszek, 1997).

H$_2$O$_2$ is an electron-accepting substrate for peroxidase-dependent reactions, thus peroxidases are likely considered merely as ROS detoxifying enzymes. However, it actually produces ROS at certain occasions for defenses against infection with pathogens. ROS generation by extracellularly secreted peroxidase in elicitor-treated plants has been reported (Kiba et al. 1996), although the electron-donating substrates are obscure. Bolwell (1995) proposed the generation of H$_2$O$_2$ by a pH-dependent cell wall peroxidase, in which an elicitor is recognized by a putative receptor and ion channels are activated. The movements of H$^+$ and other ions may contribute to a transient alkalinization of extracellular matrix, and activate the pH-dependent peroxidase. However, this reaction solely depends on pH changes, requiring no specific substrates, thus the salicylic acid (SA)- and aromatic monoamine (AMA)-dependent ROS generation requires alternative mechanisms.

Plant peroxidases in the apoplastic space are considered to catalyze the generation of aromatic oxyl radicals from several aromatic compounds (Takahama and Yoshitama 1998). Production of organic radicals by peroxidase reactions often result in generation of ROS (Kagan et al. 1990). In our recent works, involvement of peroxidase in ROS generation was repeatedly examined, confirming its role in oxidative signaling induced by AMAs and SA.

AMAs in plants ----- From a phytocentoric point of view, active defenses not only against microbial invasions but also against herbivores are needed. AMAs act act as deterrents to insect

N. Murata et al. (eds.), Advanced Research on Plant Lipids, 343–346.
© 2003 *Kluwer Academic Publishers. Printed in the Netherlands.*

predator and foraging animals, for examples, N-dimethyltyramine is a feeding repellent for grasshoppers, and the presence of AMAs in herbage causes locomotor ataxia in sheep (Smith 1997). On the other hand, the resistance of sugarbeet to fungus *Cercospora beticola* is related to the presence of dopamine (Smith 1997). Thus AMAs may fulfill the criteria for chemical defenses against both classes of enemies. Then, we have hypothesized that AMAs protect the plants by stimulating the production of ROS.

SA in defense induction in plants ----- Any hint of the mode of SA action in induction of plant defense responses, has not been obtained until the first report on a SA-binding protein (identified as catalase) appeared in 1991 (Chen and Klessig). Since then researchers started to study the actual mode of SA action leading to defense response in relation to metabolism of ROS. However, the catalase model merely explains the inhibition of ROS breakdown by SA in plants, and the mechanism for SA-stimulated *de novo* generation of ROS has not been elucidated.

2. Detection of ROS Production Induced by AMAs and SA

We used three different methods to detect the ROS generated in tobacco cell suspension culture, in response to addition of SA and AMAs. In chemiluminescence method, a *Cypridina* luciferin analog, 2-methyl-6-phenyl-3,7-dihydroimidazo [*1,2-a*] pyrazin-3-one, was used as an O_2^--specific probe. The chemiluminescence studies revealed that both SA and some AMAs induce rapid production of O_2^- (Kawano et al., 1998, 2000a). Among AMAs, PEA and benzylamine were very active in induction of O_2^- generation, but the AMAs with -OH group(s) at the aromatic rings were shown to be less active.

Microscopic fluorometry was adopted to detect AMA-induced H_2O_2 production. A membrane-permeable fluorescent probe, dichlorofluorescin diacetate was applied to the tobacco cells. Since the endogenous esterase in the cells hydrolyzes the membrane-permeable modification of the probe and the membrane-impermeable dye is released only within cytosolic space, the ROS production (especially H_2O_2 derived from O_2^-) can be visualized using a laser scanning microscope (Kawano et al., 2000a).

The third method, electron spin resonance (ESR) spectroscopy has revealed that active AMAs such as PEA induce the generation of HO' (Kawano et al., 2000a). However, in the SA-treated tobacco cell culture, no HO' production was detected since SA act as a strong HO' scavenger (Kawano and Muto, 2000).

3. SA and AMAs as ROS-generating Peroxidase Substrates

The SA-dependent O_2^- generating activity in tobacco cell suspension culture was shown to be a proteinaceous factors highly localized in the extracellular space, and the activities are sensitive to depletion of dissolved oxygen and treatment with a peroxidase inhibitor (Kawano et al., 1998). The AMA-dependent ROS producing activity is also highly localized extracellularly and sensitive to the peroxidase inhibitor (Kawano and Muto, 2000). Enzyme activities with similar properties were also found in epidermal cell layers of *Vicia fava* (Mori et al., 2001).

To confirm that the SA- and AMA-dependent ROS production is peroxidase-dependent reactions, we established a model enzyme system using the commercial horseradish peroxidase (HRP). In the presence of dissolved oxygen and trace of H_2O_2 added to the system, HRP catalyzed the rapid production of $O_2^{\cdot-}$ following the addition of SA (Kawano and Muto, 2000) or AMAs (Kawano et al., 2000b).

4. ROS Triggers the Influx of Ca^{2+} in Tobacco Cells

Using the transgenic tobacco cells expressing aequorin, it has been revealed that the rapid and transient bursts of $O_2^{\cdot-}$ and H_2O_2 generation induced by SA and AMAs, are followed by increases in cytosolic Ca^{2+} concentration ($[Ca^{2+}]_c$). When the ROS production is lowered by treatment with several ROS scavengers and a peroxidase inhibitor, the $[Ca^{2+}]_c$ increase was also lowered (Kawano et al., 1998). Addition of HRP to the cell culture enhanced the SA-induced $[Ca^{2+}]_c$ increase, suggesting that HRP enhances the $O_2^{\cdot-}$ generation and stimulates the $[Ca^{2+}]_c$ increase (Kawano and Muto, 2000). The results show that SA and AMAs trigger the $[Ca^{2+}]_c$ increase as a consequence of the peroxidase inhibitor-sensitive production of ROS. It is conclusive that the SA- and AMA-induced extracellular redox signals are converted to the intracellular Ca^{2+} signal, probably leading to activation of defense mechanism (figure 1a).

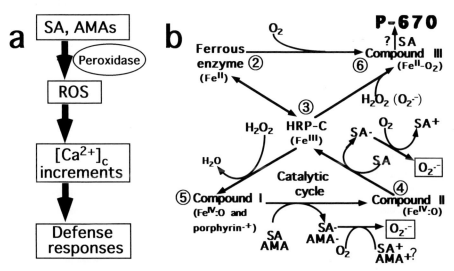

Figure 1. Model mechanisms for the actions of SA and AMAs. **a**, The signal transduction pathways common for both SA and AMAs in plants. Extracellular peroxidase-catalyzed oxidation of SA and AMAs lead to production of ROS and increase in $[Ca^{2+}]_c$. Ca^{2+} signaling may lead to the defense responses. **b**, Involvement of SA and AMAs in peroxidase reactions. Interrelationship among five redox states of peroxidase and P-670 are illustrated. Numbers (2–6) indicate the formal oxidation states. Reaction paths 3→5→4→3 indicate one catalytic cycle of the enzyme. AMA⁺ and SA⁺, an intermediate product of SA and AMA oxidation; SA˙ and AMA˙, SA- and AMA-free radicals.

5. Analysis of Enzyme Oxidation States

Since our previous works have revealed that both SA and AMAs are good substrates for peroxidase-catalyzed ROS production, we further examined the effects of SA and AMAs on the oxidation states of peroxidase (Kawano et al., 2002a). For this purpose, HRP was used as a highly purified model enzyme for spectroscopic monitoring of enzyme oxidation states during SA and AMA oxidation. We have obtained the data suggesting that the Compound II form of HRP-C does not utilize PEA as substrate. In contrast, addition of SA to Compound II resulted in rapid conversion of Compound II to the native form. Addition of H_2O_2 in combination with either of SA or PEA to HRP reaction mixture resulted in formation of Compound II. As illustrated in figure 1b, we proposed that SA is a good substrate for ROS-generating reaction catalyzed by both the Compound I and II, while AMAs (PEA) are utilized only in the Compound I-catalyzed reaction.

Compared to the action of AMAs, the behavior of SA in peroxidase reaction is more complex. Interestingly SA acts as an inactivator of peroxidase in the specific occasions where H_2O_2 presents in excess (Kawano et al., 2002b). We obtained a spectroscopic evidence in support of SA-dependent inactivation of HRP. Addition of SA to HRP arrested at a temporal inactive state (Compound III) in the presence of excess H_2O_2, resulted in rapid and irreversible inactivation of the enzyme yielding verdohemoproteins (P-670). The overall roles for SA and AMAs in peroxidase-catalyzed reactions are summarized in figure 1b.

This work was partly supported by Grant-in-Aid for Special Research on Priority Areas (no. 13039008) from Ministry of Education, Science, Sports, Culture and Technology of Japan.

References

Bolwell, G. P. (1995) *Free Rad. Res.* 23: 517-532.

Bolwell, G. P. and Wojtaszek, P. (1997) *Physiol. Mol. Plant Pathol.* 51: 347-366.

Chen, Z., Silva, H. and Klessig, D. F. (1993) *Science* 262: 1883-1886.

Kagan, V. E., Serbinova, E., A and Packer, L . (1990) *Arch. Biochem. Biophys.* 280: 33-39.

Kawano, T. and Muto, S. (2000) *J. Exper. Bot.* 51: 685-693.

Kawano, T., Muto, S., Adachi, M., Hosoya, H. and Lapeyrie, F. (2002a) *Biosci. Biotech. Biochem.* 66: 646-650.

Kawano, T., Muto, S., Adachi, M., Hosoya, H. and Lapeyrie, F. (2002b) *Biosci. Biotech. Biochem.* 66: 651-654.

Kawano, T., Pinontoan, P., Uozumi, N., Miyake, C., Asada, K., Kolattukudy, P. E. and Muto, S. (2000a) *Plant Cell Physiol.* 41: 1251-1258.

Kawano, T., Pinontoan, P., Uozumi, N., Morimitsu, Y., Miyake, C., Asada, K. and Muto, S. (2000b) *Plant Cell Physiol.* 41: 1259-1266.

Kawano, T., Sahashi, N., Takahashi, K., Uozumi, N. and Muto, S. (1998) *Plant Cell Physiol.* 39: 721-730.

Kawano, T., Sahashi, T., Uozumi, N. and Muto, S. (2000c) *Plant Peroxid. Newslett.* 14: 117-124.

Kiba, A., Toyota, K., Ichinose, Y., Yamada, T. and Shiraishi, T. (1996) *Annu. Phytopathol. Soc. Jpn.* 62: 508-512.

Mori, I. C., Pinontoan, R., Kawano, T., and Muto, S. (2001) *Plant Cell Physiol.* 42: 1383-1388.

Smith, T. A. (1977) *Phytochemistry* 16: 9-18.

Takahama, U. and Yoshitama, K. (1998) *J. Plant Res.* 111: 97-100.

Chapter 9:

Lipids and Function

A NOVEL GLYCOLIPID IN *SYNECHOCYSTIS* SP. PCC 6803 CELLS: STRUCTURAL CHARACTERISTICS

Z.N. WANG, Y.N. XU, Z.L. YANG, H.T. HOU, G.Z. JIANG, T.Y. KUANG
Key Laboratory of Photosynthesis and Environmental Molecular Physiology, Institute of Botany, Chinese Academy of Sciences
Nanxin Cun 20, Xiangshan, Beijing 100093, P. R. China

1. Introduction

Differences in fatty acid composition contribute to different membrane fluidity, which is very important for cells to survive unfavorable growth conditions (Nakashima et al., 1999). The demonstration of the roles of lipids is mostly based on differences in the composition of polar head groups. In fact, some fatty acids show characteristic roles in constructing membrane protein complexes. For example, trans-3-palmitoleic acid, is only found in PG of higher plants, and its percentage is related to the level of trimeric LHCII (Siegenthaler and Trémoli res, 1998). So it is very necessary to determine the fatty acid composition for a novel lipid in order to demonstrate its roles. This paper reported the fatty acid composition in a novel glycolipid in *Synechocystis* sp. PCC 6803 cells grown in BG-11 medium with glucose applied and the polar head group of the glycolipid was also investigated.

2. Materials and methods

2.1. Isolation of glycolipids

Glycolipids in the lipid extracts of *Synechocystis* sp PCC 6803 cells grown in BG-11 medium with 20 mM glucose applied were separated by silica gel G thin-layer chromatography (TLC) using $(CH_3)_2CO/C_6H_6/H_2O$ (91:30:8, v/v/v) as developing agent as described in Sato and Murata (1988) and identified as described in Su et al. (1980). The glycolipid, between MGDG and DGDG in TLC plate, was designated glycolipid-x.

2.2. Fatty acid composition assay

Fatty acid compositions in glycolipids were analyzed by gas chromatography (GC) as described previously (Sato and Murata, 1988) and heptadecanoic acid (C17:0) was

N. Murata et al. (eds.), Advanced Research on Plant Lipids, 349–352.

added as an internal standard. GC was carried out on an HP 6890 gas chromatograph. Standard fatty acid methyl ester mixture, LA-98232, purchased from *Supelco*, was employed to determine the retention times of different fatty acid methyl esters. The new peaks in chromatographs were further determined by GC-MS on a SHIMADZU QP5050A GC-MS spectrometer equipped with a DB-5 column (J&W Scientific).

2.3. FAB-MS of lipids

FAB-MS spectrum of glycolipids, dissolved in glycerol, was obtained on an APEX II FT-ICR spectrometer.

3. Results and Discussion

3.1. Fatty acid composition in glycolipid-x

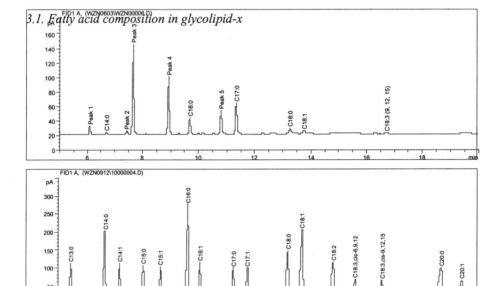

Figure 1. Gas chromatograms of fatty acid methyl esters of glycolipid-x (a, top) and the standard fatty acid methyl ester mixture (b, bottom).

Glycolipid-x is composed of fatty acid acyl chains, the methyl esters of which can not be marked with the given standard fatty acid methyl esters (Fig. 1a). With the aid of GC-MS, the unknown peaks can be determined. The fatty acid methyl esters corresponding to peak 1 and peak 2 are supposed to be $C_{11}H_{21}NO_2$ and $C_{12}H_{23}NO_2$, and the structural formulae are given in Fig. 2a, b. The molecular formulae corresponding to the other unknown peaks in gas chromatograms (Fig. 1a) based on the molecular ion peaks in their respective mass spectrums (data not presented), can be determined as follows: peak 3, $C_{16}H_{32}O_2$; peak 4, $C_{17}H_{34}O_2$; peak 5, $C_{18}H_{36}O_2$. The molecular formulae are the same as those of pentadecanoic acid (C15:0), palmitic acid

(C16:0) and heptadecanoic acid (C17:0) methyl esters, respectively. They should be the corresponding fatty acid methyl esters with branched chains.

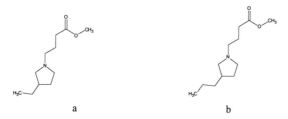

a   b

Figure 2. Structural formulae (a, b) corresponding to peak 1 and peak 2 in Fig. 1a, respectively.

3.2. Molecular formula determination

Glycolipid-x is seldom observed in cells grown in BG-11 medium containing both glucose and sodium thiosulfate. In the presence of 0.3% sodium thiosulfate, glycolipid-x can only be detected in cells grown with 100 mM glucose applied to BG-11 medium. The fatty acid composition was analyzed by gas chromatography, and it was found that branched C15:0 amounts up to 64.1% of the total fatty acids. From FAB-MS spectrum of glycolipid-x (Fig. 3). There are three 162-spaced ions at m/z 641.3636, 479.3184 and 317.2691, suggesting that one molecule of glycolipid contains two molecules of hexose residue. The ions at m/z 641.3636 and 663.3546 correspond to the formulae, $C_{30}H_{56}O_{14}H^+$ and $C_{30}H_{56}O_{14}Na^+$, respectively. The former is a proton-adducted molecular ion $[M+H^+]$ and the latter is $[M+Na^+]$. The molecular formula of glycolipid-x in this case can be written as $C_{30}H_{56}O_{14}$. There is a glycerol backbone left after two molecules of hexose residues and one molecule of branched pentadecanoic acyl chain are subtracted from the structural formula. Therefore, glycolipid-x contains only one fatty acid acyl chain.

There are several ions with a mass difference of 92 in the FAB-MS spectrum, which can be divided into two groups, 277.1492, 369.1968, 461.2454, 553.2908; 299.1310, 391.1789, 483.2260, 575. They should be proton-adducted and sodium-adducted ions of glycerol conjugate with different molecules.

As to whether the hexose composing glycolipid-x is galactose or glucose or else needs further investigating.

352

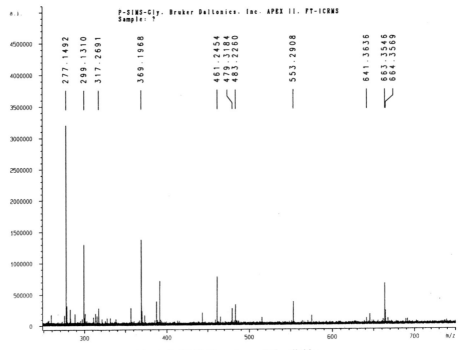

Figure 3. FAB-MS spectrum of glycolipid-x.

Acknowledgements:

Acknowledgements are given to National Natural Science Foundation of China (39890390) and the organizing committee of the 15th International Symposium on Plant Lipids.

References

Nakashima, S., Wang, S., Hisamoto, N., Sakai, H., Andoh, M., Matsumoto, K. and Nozawa, Y. (1999) Molecular Cloning and Expression of a Stress-responsive Mitogen-activated Protein Kinase-related Kinase from *Tetrahymena* Cells. J. Biol. Chem. 274, 9976-9983.

Sato, N. and Murata, N. (1988) Membrane lipids. Meth. Enzymol. 167, 251-259.

Siegenthaler, P. A. and Trémoli res, A. (1998) Role of acyl lipids in the function of photosynthetic membrane in higher plants. In P.A. Siegenthaler and N. Murata(eds.) Lipids in Photosynthesis: Structure, Function and Genetics. Kluwer Academic Publishers, Dordrecht, pp.145-173.

Su, W. A., Wang, W. Y. and Li, J. S. (1980) Assay techniques of plant lipoids and fatty acids. Plant Physiol Commun. 3, 54-60. (in Chinese)

THE MECHANISM OF BIOSYNTHESIS AND THE PHYSIOLOGICAL FUNCTION OF PHOSPHATIDYLCHOLINE IN *ARABIDOPSIS* DURING COLD ACCLIMATION

I. NISHIDA, R. INATSUGI AND M. NAKAMURA

Department of Biological Sciences, Graduate School of Science
The University of Tokyo
7-3-1 Hongo, Bunkyo-ku, Tokyo, 113-0033 Japan

1. Introduction

Cold-hardy plant species including *Arabidopsis thaliana* have the capacity to increase the freezing tolerance when exposed to low but non-freezing temperatures (Gilmour et al., 1988). This process, known as cold acclimation, is accompanied by various biochemical changes in the compositions of lipids, proteins, sugars and amino acids, and the physiological relevance of these biochemical changes to the survivability of plants at low temperatures remains to be evaluated.

Phosphatidylcholine (PC) is a major phospholipid in various plant membranes except thylakoid membranes. This lipid also serves as a metabolic precursor to major glycolipids in plastid membranes, such as monogalactosyldiacylglycerol (MGDG), digalactosyldiacylglycerol (DGDG) and sulfoquinovosyldiacylglycerol (SQDG) (for review, see Ohlrogge and Browse 1995). Thus, the efficient synthesis of PC is potentially important in the development and maintenance of plant cells at low temperatures. Besides these structural and metabolic roles, PC serves as a substrate for acyl-lipid desaturases and hence is potentially important for the regulation of membrane fluidity in response to low temperatures.

Cold-induced accumulation of PC was observed in a number of herbaceous and woody plant species (Yoshida, 1984). The role of di-unsaturated molecular species of PC in the development of the freezing tolerance has been hypothesized, although direct proof of this hypothesis remains to be established (Uemura et al., 1995; Uemura and Steponkus, 1994). In some cold-hardy woody plant species, cold-induced formation of vesicles beneath the plasma membrane was observed (Niki and Sakai, 1981), although physiological significance of the vesicle formation and its relevance to PC biosynthesis remain to be evaluated. To obtain a definite answer to the physiological function of PC at low temperatures, it is essential to know how the biosynthesis of PC is regulated at low temperatures and to obtain mutants defective in the biosynthesis of PC at low

N. Murata et al. (eds.), Advanced Research on Plant Lipids, 353–356.

temperatures.

PC is synthesized largely via the nucleotide pathway (or the CDP-choline pathway) in plants (Ohlrogge and Browse, 1995), which includes three steps of sequential reactions catalyzed by choline kinase (EC 2.7.1.32; CK), CTP:phosphorylcholine cytidylyltransferase (EC 2.7.7.15; CCT) and CDP-choline:diacylglycerol cholinephosphotransferase (EC 2.7.8.2; CPT). Among these enzymes, CCT is generally accepted as a rate-limiting enzyme in the nucleotide pathway. CCT activity was reported to increase in rye roots treated at 5 °C (Kinney et al., 1987). However, it remains unknown whether the increased CCT activity is caused by enhanced enzyme synthesis, activation of the enzyme, or by expression of a new isozyme.

The Arabidopsis genome initiative (2000) predicts that *A. thaliana* contained two genes for CCT in chromosome 2 (*At2g32260*) and chromosome 4 (*At4g15130*), which we designated *AtCCT1* and *AtCCT2*, respectively. In the present study, we investigated the expression of *CCT* isogenes in the rosette leaves of *Arabidopsis* subjected to chilling treatment. Our evidence showed that *AtCCT2*, but not *AtCCT1*, exhibits enhanced expression at low temperature and that enhanced CCT activity at low temperature could be sufficient to account for lipid changes induced by low temperature. We therefore conclude that the expression of CCT isozymes in *Arabidopsis* is differentially regulated in the biosynthesis and/or turnover of PC during cold acclimation.

2. Results and Discussion

2.1 *Expression of* AtCCT2 *but Not That of* AtCCT1 *Is Enhanced in the Rosette Leaves of Arabidopsis at Low Temperatures*

We isolated a cDNA for AtCCT2 by PCR from cDNAs derived from mRNA from cold-acclimated *Arabidopsis* plants. RNA gel blot analysis with specific probes revealed that the level of *AtCCT2* transcripts, but not that of *AtCCT1* transcripts, was enhanced at 2°C.

To examine the protein levels of AtCCT1 and AtCCT2 in the rosette leaves of *Arabidposis* at low temperature, we prepared antisera that specifically cross-react with AtCCT proteins: an antiserum A that was raised against a 20-amino-acid peptide in a carboxy-terminal region of AtCCT1 cross-reacted with AtCCT1 but not with AtCCT2 and an antiserum B that was raised against a recombinant AtCCT2 protein expressed in *E. coli* cross-reacted with both AtCCT1 and AtCCT2. In control plants that were grown at 23°C for 18 d under continuous illumination, both AtCCT1 (44 kDa) and AtCCT2 (38.5 kDa) proteins were detected with an antiserum B, suggesting that these CCT isozymes are expressed constitutively under ordinary growth conditions. By contrast, when *Arabidopsis* plants were grown for another 2 d at 2°C, a significant increase in the level of AtCCT2 was detected by immunoblot analysis. To obtain a quantitative result for the changes of AtCCT levels in the rosette leaves of *Arabidopsis* at low temperatures, we quantified the intensity of each band detected on immunoblots by the NIH image software 1.62, using a recombinant AtCCT2 protein as an internal standard. The result showed that the level of AtCCT2 increased twofold after 2 d at 2°C, and that no significant increase was observed for the levels of AtCCT1 at 2°C. From these results, we concluded that expression of AtCCT2 was selectively enhanced in the rosette leaves of *Arabidopsis* at low temperatures. These results also suggest that *CCT* isogenes in *Arabidposis* have differential roles in the biosynthesis and/or turnover of PC in *Arabidopsis* under cold acclimation conditions.

2.2 *The Absence of* AtCCT2 *Has No Deleterious Effect on the Growth of Arabidopsis at Low Temperatures*

To investigate the physiological significance of the expression of *AtCCT2* in *Arabidopsis* at low temperatures, we isolated a T-DNA-tagged gene disruptant for *AtCCT2* (see also an accompanying paper by Inatsugi et al., in this volume). The disruptant, designated *cct2*, was found to have a T-DNA insertion in the 2nd exon of *AtCCT2*. In *cct2* plants, no band corresponding to AtCCT2 protein (a 38.5-kDa band) was detected in rosette leaves developed at 23°C by immunoblot analysis, and no further increase of the 38.5-kDa band was detected in rosette leaves after cold treatment. This result offers another evidence for our above finding that expression of *AtCCT2* is enhanced in the rosette leaves of *Arabidopsis* at low temperatures. At moment, we could not find any difference in the rates of growth and increases in fresh and dry weight between the wild type and *cct2* plants either after 15 d at 23°C or after another 7 d at 5°C. Lipid analysis revealed that there was no change in the compositions of lipids and fatty acids between the wild type and *cct2* plants after 15 d at 23°C or after another 7 d at 5°C. These results suggest that the absence of *AtCCT2* has no deleterious effect on the growth of *Arabidopsis* at low temperatures. However, we still do not exclude the possibility that there might be physiological differences between the wild type and *cct2* plants under some growth conditions in which the rapid synthesis of PC might be required.

2.3 *Multiple Mechanisms May Regulate the Synthesis of PC in* Arabidopsis *at Low Temperatures*

As described above, the expression of AtCCT2 is up-regulated in the rosette leaves of *Arabidopsis* at low temperatures, but the growth profiles of *cct2* plants in comparison with those of the wild type suggest that the *AtCCT2* gene itself is not essential to the growth of *Arabidopsis* at low temperatures. How can we explain this discrepancy? Our results show that lipid compositions changed similarly between the wild type and *cct2* plants in response to low temperatures, which still holds for the hypothesis that alteration of lipid composition may be important for cold acclimation of *Arabidopsis*. Then, the question arises how the biosynthesis of PC at low temperatures is complemented in *cct2* plants. In *cct2* plants, we found that the level of AtCCT1 protein was up-regulated by 20% over the wild type and that the total CCT activity in the homogenates of the rosette leaves became comparable to that in the wild type at 23°C. Therefore, we suspect that the activity of AtCCT1 was up-regulated by some mechanisms in *cct2* plants, which might support the synthesis of PC at low temperatures.

To obtain a definite answer to the physiological role of PC at low temperatures, our efforts are going on to isolate a double disruptant of *AtCCT1* and *AtCCT2* genes.

3. References

Gilmour, S. J., Hajela, R. K. and Thomashow, M. F. (1988) Cold acclimation in *Arabidopsis thaliana*. Plant Physiol. 87, 745–750.

Kinney, A. J., Clarkson, D. T. and Loughman, B. C. (1987) The regulation of phosphatidylcholine biosynthesis in rye (*Secale cereale*) roots. Biochem. J. 242, 755–759.

Niki, T. and Sakai, A. (1981) Ultrastructural changes related to frost hardiness in the cortical parenchyma cells from mulberry twigs. Plant Cell Physiol. 22, 171–183.

Ohlrogge, J. and Browse, J. (1995) Lipid biosynthesis. Plant Cell 7, 957–970.

The Arabidopsis Genome Initiative (2000) Analysis of the genome sequence of the flowering plant *Arabidopsis thaliana*. Nature 408, 796–815.

Uemura, M. and Steponkus, P. L. (1994) A contrast of the plasma membrane lipid composition of oat and rye leaves in relation to freezing tolerance. Plant Physiol. 104, 479–496.

Uemura, M., Joseph, R. A. and Steponkus, P. L. (1995) Cold acclimation of *Arabidopsis thaliana*. Plant Physiol. 109, 15–30.

Yoshida, S. (1984) Chemical and biophysical changes in the plasma membrane during cold acclimation of Mulberry bark cells (*Morus bombycis* Koidz. cv Goroji). Plant Physiol. 76, 257–265.

Acknowledgement: This work was supported in parts by the Program for Promotion of Basic Research Activities for Innovative Biosciences (PROBRAIN) and by Grants-in-Aid for Scientific Research from the Japanese Society for the Promotion of Science.

INVOLVEMENT OF MEMBRANE LIPID FLUIDITY IN REGULATION OF MOLECULAR MECHANISM OF THE XANTHOPHYLL CYCLE

K. STRZAŁKA[1], D. LATOWSKI[1], K. BURDA[2], A. KOSTECKA-GUGAŁA[1], J. KRUK[1]

[1]*Department of Plant Physiology and Biochemistry, The Jan Zurzycki Institute of Molecular Biology and Biotechnology, Jagiellonian University, ul. Gronostajowa 7, 30-387 Kraków, Poland*
[2] *H. Niewodniczański Institute of Nuclear Physics, Radzikowskiego 152, 31-342 Kraków, Poland*

1. Abstract

Temperature dependence of violaxanthin de-epoxidase was measured using the system of isolated violaxanthin de-epoxidase enzyme (VDE) and unilamellar vesicles composed of various proportions of monogalactosyldiacylglycerol (MGDG) and egg yolk phosphatidylcholine (PC). It has been shown that at the constant ratio of VDE and MGDG in the assay system, the kinetics of de-epoxidation depended greatly on proportion of MGDG to PC as well as on temperature. A mathematical model of violaxanthin de-epoxidation was constructed and applied to calculate the probability of violaxanthin to antheraxanthin and to zeaxanthin conversion at different phases of de-epoxidation process. Measuremants of de-epoxidation rate and spin label mobility at different temperatures revealed that membrane fluidity is an important factor controlling conversion of violaxanthin to antheraxanthin. A model of the xanthophyll cycle is proposed in which VDE binding and the de-epoxidation reactions occur in MGDG rich domains. Availability of violaxanthin for de-epoxidation is a lateral diffusion-dependent process controlled by membrane fluidity. Significance of the presented results for understanding the mechanism of violaxanthin de-epoxidation in native thylakoid membranes is discussed.

N. Murata et al. (eds.), Advanced Research on Plant Lipids, 357–360.

2. Materials and Methods

2.1. *Materials*

MGDG and DGDG was purchased from Lipids Products. Egg yolk phosphatidylcholine (PC) and 5-doxyl-stearic acid (5-SASL) were obtained from Sigma (P2772). Violaxanthin was isolated from daffodil petals as described in Havir et al., (1997) and saponificated (Davies, 1976). VDE was isolated and purified from 7-day-old wheat leaves according to the method described by Hager and Holocher (1994).

2.2. *Methods*

The liposomes preparation was the same as described in Latowski et al., (2000).

De-epoxidation of violaxanthin was performed at 4, 12 and 25°C in liposomes composed of PC, MGDG and violaxanthin. The de-epoxidation reaction was initiated by addition of VDE with activity corresponded to 4 nmol de-epoxidated violaxanthin $min^{-1} ml^{-1}$. The pigments were extracted by mixing 750 µl of the reaction medium with 750 µl of the extraction solution containing chloroform : methanol : ammonia (1:2:0.004 v/v/v). Xanthophyll pigments were separated by HPLC, according to Gilmore and Yamamoto (1991)

Temperature dependent changes of the lipid fatty acyl chains order parameter in liposome membrane were recorded by EPR spin label measurements using a 5-SASL was (10^{-4} M) and Bruker ESP-300E spectrometer fitted with TM_{110} cavity. The final concentration of lipids was 10^{-2} M.

Mathematical model of violaxanthin de-epoxidation was described in Latowski et al. (2000) .

3. Results and Discussion

The rate of violaxanthin to antheraxanthin conversion depends on MGDG/PC ratio in the liposome membrane even if the absolute amount of MGDG in the reaction mixture and its proportion to violaxanthin and VDE remains constant (Tab. 1).

TABLE 1. Kinetic parameters of a de-epoxidation reaction calculated for the for violaxanthin de-epoxidation in liposomes, where MGDG and violaxanthin amounts were constant and PC concentration was changeed to obtain 5mol%, 15 mol% and 30 mol% of MGDG, by means of the mathematical model, where:

VA_0 - probability of violaxanthin to antheraxanthin conversion;

AZ_0- probability of antheraxanthin to zeaxanthin conversion;

MGDG [mol %]	VA_0	AZ_0
5	0.006	0.548
15	0.045	0.600
30	0.260	0.810

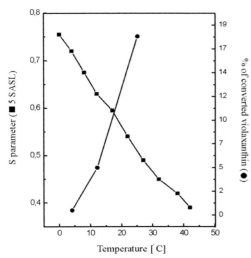

Figure 1. PC-MGDG liposome membrane fluidity and percent of violaxanthin converted after 1 min de-epoxidation reaction at different temperatures ● - % of converted violaxanthin; ■ - value of S parameter for 5-SASL.

The above results suggest that VDE binds only to certain membrane domains which are rich in MGDG and the de-epoxidation reactions take place in these domains. Violaxanthin has to enter the MGDG-enriched domains by lateral diffusion to be converted to antheraxanthin. The higher MGDG/PC ratio the higher the amount of such domains in the liposomal membrane, which shortens the diffusion path of violaxanthin molecules to these domains and results in higher rate of violaxanthin de-epoxidation (see the values of VA_0 in Tab. 1). The second reaction of de-epoxidation, i.e., conversion of antheraxanthin to zeaxanthin, also occurs in the MGDG rich domains, and is greatly facilitated because antheraxanthin, formed in such domains, has an immediate access to the VDE enzyme. Therefore, and in contrast to the de-epoxidation of violaxanthin to antheraxanthin, the conversion of antheraxanthin to zeaxanthin seems to be not limited by diffusion process. The conclusion that the conversion of violaxanthin to antheraxanthin is more sensitive to MGDG concentration than the conversion of antheraxanthin to zeaxanthin is supported by relatively high value of the AZ_0 parameter even in conditions of very low VA_0 (e.g. at 5 mol% of MGDG, Tab. 1). The lateral diffusion of violaxanthin in the membrane becomes faster at increased membrane fluidity. The molecules of violaxanthin may reach sooner the MGDG rich domains where they are de-epoxidated (Fig. 1). This model can also explain a temperature effect on the level and the time of antheraxanthin appearance in liposome system. The conversion of antheraxanthin to zeaxanthin is less dependent on the changes in membrane physical properties for the reasons already discussed. Application of the proposed model to the obtained results permits to understand why the conversion of violaxanthin to antheraxanthin is much slower and more sensitive to temperature than transition from the antheraxanthin to zeaxanthin. On the basis of our results and literature data (Arvidsson, 1997, Latowski et al. 2000) we postulate that changes in membrane fluidity may play an important role in regulation of the violaxanthin de-epoxidation rate in photosynthetic membranes.

360

4. Acknowledgement

This work was supported by grant No. 6PO4A 03817 from the committee for Scientific Research of Poland (KBN).

5. References

Arvidsson, P.-O., Carlsson, H., Stefansson, H., Albertsson, P.-A. & Åkerlund, H.-E. (1997) Violaxanthin accessibility and temperature dependency for de-epoxidation in spinach thylakoid membranes. *Photosynth. Res.* **52**, 39-48.

Davies, B.H. (1976) Carotenoids. In *Chemistry and Biochemistry of Plant Pigments* (Goodwin, T.W., ed.), Academic Press London New York San Francisco, pp.65-66.

Gilmore, A.M. & Yamamoto H.Y.(1991) Resolution of lutein and zeaxanthin using a non-endcappted, lightly carbon-loaded C_{18} high-performance liquid chromatographic column, *J. Chrom.*, 1991, vol. **543**, pp. 137-145.

Hager, A. and Holocher, K. (1994) Localization of the xanthophyll cycle enzyme violaxanthin de-epoxidase within the thylakoid lumen and abolition of its mobility by a (light-dependent) pH decrease. *Planta,* **192**, 581-589.

Havir, E.A., Tausta, L.S. and Peterson, R.B. (1997) Purification and properties of violaxanthin de-epoxidase from spinach. *Plant Science* **123**, 57-66.

Latowski, D., Burda, K. and Strzałka, K. (2000) A mathematical model describing kinetics of conversion of violaxanthin to zeaxanthin via intermediate antheraxanthin by the xanthophyll cycle enzyme violaxanthin de-epoxidase *J. Theor. Biol.*, **206**, 507-514.

Latowski, D., Kostecka, A. and Strzałka, K. (2000) Effect of monogalactosyldiacylglycerol and other thylakoid lipids on violaxanthin de-epoxidation in liposomes. *Bioch. Soc. Trans.* Vol. **28** Part. **6**, 810-812.

SULPHOQUINOVOSYLDIACYLGLYCEROL AND ADAPTATION SYNDROME

A.OKANENKO, N. TARAN, O.KOSYK

Plant Physiology & Ecology Department, Kyiv National Taras Shevchenko University, Volodymyrska, 64, UA-01033, Kyiv, Ukraine,

Abstract

The effect of drought factor induced the SQDG accumulation in leaves of drought resistant wheat plants and the drastic decrease of their content in sensitive plants. Heat and water stress combined induced SQDG accumulation in resistant plants, while it decreased in sensitive variety more drastically if compared with single factor effect. SQDG accumulation in wheat plants infected by *Puccinia graminis* and in kidney bean plants infected by tobacco mosaic and potato *x* viruses was observed as well. Lead stress changed drastically the glycolipid composition in laboratory experiments. Significantly lower SQDG content observed could be connected with possible competitive sulphur usage for sulphur containing polypeptide and protein synthesis judging by increasing protein content, or perhaps creating complexes with Pb. Thus we assume that SQDG could play special role in adaptation reactions connected mainly with stabilising the photosynthesis processes and perhaps heavy metal binding.

1. Drought adaptation

Lipids in interaction with sterols, carotenoids and proteins are substances determining molecular structure, physical and chemical properties, and functional activity of membranes. Studying the effect of the drought factors upon the lipid composition of different plants attracted our attention to sulpholipid behaviour. Therefore we consider expedient to investigate SQDG content changes in wheat leaves under the effect of various stresses. Our data showed the high temperature action induced SQDG accumulation in leaves and chloroplasts of drought resistant winter wheat plants while in sensitive plants the decrease of the content was observed when expressed in terms of chlorophyll or of mol% glycolipid (Fig. 1). Water deficit (WD) effect induced SQDG increase in resistant variety plants while the insignificant decrease of the content in sensitive plants took place (Fig.2). Fatty acid (FA) composition in resistant plant SQDG was characterised by relative stable saturation, whereas saturation of SQDG FA in soft plants increased from 54 up to 90% because of palmitic and stearic residue sharp increase and linolenic drastic fall. Combined action

N. Murata et al. (eds.), Advanced Research on Plant Lipids, 361–364.

362

sensitive variety more drastically, if compared with heat or WD influence (data not shown).

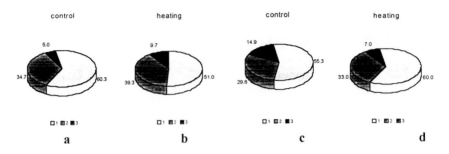

Figure 1. Glycolipid content (mol%) in wheat leaves treated with heat (45°Cx3h).
1 – MGDG, 2 – DGDG, 3 – SQDG; a, b – resistant variety; c, d – susceptible variety.

Thus the shifts observed under heat are similar to those under WD effect and their combined effect induced the SQDG accumulation sometimes much more significant than single factor effect The data presented in literature confirm that super optimal temperature effect induced the increase of SQDG content in *Atriplex lentiformis* plants grown in coastal (by 260%) and desert (Death Valley) habitat (by 64%). Both types of plants - from hot and temperate regions - accumulated SQDG when grown in regime 43/30° C (compared to 23/18° C regime). But plants from the hot region accumulated SQDG more readily. FA composition of the lipid was characterised by the increase of the palmitic and oleic residue content in plants from hot region and by the decrease of the linolenic acid content being

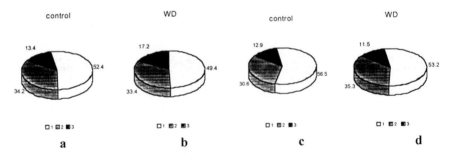

Figure 2. Glycolipid content in wheat leaves treated by WD.
1 – MGDG, 2 – DGDG, 3 – SQDG; a, b – resistant variety; c, d – susceptible variety.

more significant also in these plants in 43/30°C regime. Thus, total saturation was higher in plants from the hot region (Pearcy, 1978). The major changes in molecular species were an increase in dipalmitoyl- SQDG proportion (from 12 up to 20%) and a decrease in linolenoyl-palmitoyl SQDG proportion (from 40 to 30%) (Orr et al.,1987). The exposure of rye to drought (Pancratova et al., 1984) led to the increase of SQDG (by about 26%) in leaves on dry land plots in comparison with watered ones. Other investigations (Quartacci et al., 1995) showed that SQDG (especially) and PG amounts enlarged in drought tolerant and fell down in sensitive varieties. While WD, SQDG unsaturation increased in soft wheat plants, whereas in tolerant plants unsaturation remained unchangeable. These results differed from ours perhaps of the different WD conditions.

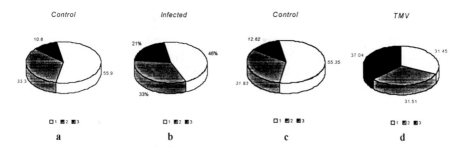

Figure 3. Glycolipid content in leaves of infected plants.
1 – MGDG, 2 – DGDG, 3 – SQDG; a - wheat control; b - wheat infected by *Puccinia graminis*; c - kidney bean plants infected by tobacco mosaic viruse.

Besides, we observed the accumulation of SQDG in wheat plants infected by *Puccinia graminis* (Fig. 3) (Taran et al.,2000) and in kidney bean plants infected by potato x and tobacco mosaic viruses (Senchugova et al., 1999). In the latter case SQDG functions, perhaps, as an virus development inhibitor taking into account the information about eukaryotic DNA polymerases and reverse transcriptase inhibiting activity (Ohta et al, 1999). Considering the growing lead pollution in our country, we investigated lead action upon wheat plants grown in the field and seedlings grown on distilled water. Various concentrations of Pb solutions caused SQDG decrease in

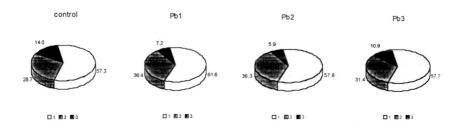

Figure 4. Glycolipid content in wheat seedling leaves treated with lead.
1 – MGDG, 2 – DGDG, 3 – SQDG; Pb1 – lead concentration 0.0048 µM Pb,
Pb2 – lead concentration 0.48 µM Pb   and Pb3 - lead concentration 48 µM Pb.

wheat leaves in laboratory experiments and in plants grown in field.   It was accompanied by protein content increase. However the data expressed in terms of mol% of total glycolipid quantity showed SQDG increase (Fig. 4) in plants supplied with toxic lead concentration comparing to variants with less lead concentration in solution.

Concerning the phenomenon observed we could suppose that in laboratory experiments SQDG decrease could be connected with possible competitive usage of sulphur (because of the sulphur limiting conditions of water culture) for sulphur containing polypeptide and protein synthesis judging by increasing protein content or perhaps creation of complexes with lead.

One could see that SQDG content increases both under the WD and/or temperature conditions in the resistant variety plants give good grounds for conclusion about special role of these compounds in adaptive reaction to extreme factor actions. As to the role of SQDG in the processes of adaptation it is likely to assume availability of the protective function for CF_1, F_1 and ATPases and stabilising function for D1/D2 dimer and LHC II. SQDG localised at the surface of the native D1/D2 heterodimer (Vijayan et al., 1998) might hold the dimer together (Kruijff et al., 1998). However it should not be excluded the using of water binding SQDG ability (Shipley et al., 1973). Thus, accumulation of definite SQDG molecular species is likely to prevent PSII degradation and perhaps perform heavy metal binding.

Abbreviations: PG - phosphatidyl glycerol; MG - monogalactosyl diacylglycerol; DG - digalactosyl diacylglycerol; SQDG - sulphoquinovosyl diacylglycerol; PS II- photo system II; RC - reaction centre; WD - water deficit.

Acknowledgements: We are grateful to the OC15-ISPL for financial support.

References

Kruijff, B. de, Pilon, R., Hof, R. van't, and Demel, R. (1998) Lipid-protein interactions in chloroplast protein import. in P.-A.Siegenthaler and N.Murata (eds.), Lipids in photosynthesis: structure, function and genetics. Advances in photosynthesis 6. Kluwer Academic Publishers, Dordrecht, pp. 191-208.

Ohta, K., Hanashima, S., Mizushina, Y., Yamazaki, T., Saneyoshi, M., Sugawara, F. and Sakaguchi, K. (2000) Studies on a novel DNA polymerase inhibitor group, synthetic sulfoquinovosylacylglycerols: inhibitory action on cell proliferation. Mutat. Res. 467, 139-152

Orr, G. and Raison, J. (1987) Compositional and thermal properties of thylakoid polar lipids of *Nerium oleander* L. in relation to chilling sensitivity. Plant Physiol. 84, 88-92.

Pancratova, S. and Karimova, F. (1984) Water supply influence upon winter rye water exchange and lipid composition. Depon. in VINITY, N 5537-84 Dep. (in Russian).

Pearcy, R. (1978) . Effect of growth temperature on the fatty acid composition of the lipids in *Atriplex lentiformis* (Torr) Wats. Plant Physiol. 61,484-486.

Quartacci, M.F., Pinzino, C., Sgherri, C. and Navari-Izzo, F. (1995) . Lipid composition and protein dynamics in thylakoids of two wheat cultivars differently sensitive to drought. Plant Physiol. 108, 191-197.

Senchugova, N., Taran, N. and Okanenko, A. (1999) Virus impact upon bean photo-synthesising tissue lipid composition. Arch Phytopath Pflanz. 32, 471-477

Shipley, G., Green, J. and Nichols, B. (1973) The phase behaviour of monogalactosyl, digalactosyl and sulphoquinovosyl diglycerides. Biochim Biophys Acta. 311, 531-544.

Taran, N., Okanenko, A. and Musienko, N. (2000) Sulpholipid reflects plant resistance to stress-factor action. Biochem. Soc. Trans.28, 922-924

Vijayan, P., Routaboul, J.-M. and Browse, J. (1998) A genetic approach to investigating membrane lipid structure and photosynthetic function. in P.-A.Siegenthaler and N.Murata (eds.), Lipids in photosynthesis: structure, function and genetics. Advances in photosynthesis 6. Kluwer Academic Publishers, Dordrecht, . pp. 263-285.

PLASTOQUINOL AND OTHER NATURAL MEMBRANE PRENYLLIPIDS MAY FORM PSEUDOCYCLIC ELECTRON TRANSPORT BY SCAVENGING SUPEROXIDE GENERATED IN PHOTOSYSTEM I

J. KRUK, M. JEMIOŁA-RZEMIŃSKA
Department of Plant Physiology and Biochemistry, The Jan Zurzycki Institute of Molecular Biology and Biotechnology, Jagiellonian University
Gronostajowa 7, 30-387 Kraków, Poland

1. Introduction

Among thylakoid membrane prenyllipids, plastoquinone-9 (PQ-9) and its reduced form plastoquinol (PQH_2-9) are the most abundant. Apart from the function in the electron transport these lipids are potent antioxidants, similar to α-tocopherol (α-Toc). The main source of active oxygen species produced in thylakoids is superoxide anion radical (O_2^-) which is generated when an electron is transferred onto an oxygen molecule:

$$O_2 + e^- \rightarrow O_2^-$$

It can be formed as a side-product in enzymatic reactions or due to autoxidation of low-potential (strongly-reducing) redox centers, for example vitamin K_1 and Fe-S centers (X and A/B) in photosystem I (PS I), or ferredoxin (Fd). The overall production rate of superoxide in chloroplasts may reach up to 30% of the total electron transport rate under stress conditions (Asada, 1999). Superoxide generated in the membrane interior after diffusion to the membrane surface undergoes dismutation to H_2O_2. If there were no O_2^- and H_2O_2 scavenging systems in stroma of chloroplasts, production rates of both these toxic oxygen derivatives would be high. When superoxide is formed in the aprotic interior of the membrane, its lifetime is extended and it can directly induce membrane lipid peroxidation or destruction of membrane proteins. In the water phase, after dismutation to H_2O_2, it is especially toxic to Calvin cycle enzymes and also to the oxygen evolving complex (OEC). Therefore, there are protecting systems preventing stroma and membrane components against toxic action of superoxide. In stroma, such an enzymatic system (Asada, 1999) is called water-water, Halliwell-Asada or ascorbate cycle. However, within thylakoid membranes, where O_2^- lifetime is extended, superoxide and H_2O_2 scavenging enzymes cannot be active. This function could fulfil

N. Murata et al. (eds.), Advanced Research on Plant Lipids, 365–368.
© 2003 *Kluwer Academic Publishers. Printed in the Netherlands.*

membrane prenyllipids such as α-Toc, PQH$_2$-9 or α-TQH$_2$ which are well known antioxidants (Kruk et al., 1997).

2. Materials and Methods

Spinach thylakoids were isolated according to the method described in Robinson and Yocum (1980). The oxygen consumption measurements were performed on dark adapted thylakoids (150 μg/ml chl) in 50 mM Hepes buffer, pH 7.5 containing 10 mM NaCl and 5 mM MgCl$_2$ using a three electrode system in the presence of 50 μM DCMU and 10 mM hydroquinone as an electron donor. Saturating light flashes of 5 μs (full width at half-maximum) were provided by a xenon lamp (Stroboscope 1539A from General Radio). The samples were illuminated by 15 flashes spaced 300 ms apart. Hexane extraction of thylakoids was performed by 3-fold extraction of lyophylized thylakoids (1 mg chl) with 1 ml hexane for 20 min, each extraction. After centrifugation and removing the supernatant, thylakoids were dried in a rotatory evaporator and suspended in the buffer giving final concentration of 1 mg/ml chl and used as a stock solution. Vitamin K$_1$ (1 μM) was added to all the extracted samples.

3. Results and Conclusions

In our study we examined if membrane prenyllipids are scavengers of O$_2^-$ in thylakoid membranes by measuring oxygen consumption with the fast oxygen electrode. For the measurements of O$_2^-$ generation, (which manifests in oxygen uptake) in thylakoid membranes under light illumination, it is necessary to inhibit photosynthetic oxygen evolution by DCMU and to use an artificial electron donor, such as hydroquinone (HQ) (Fig. 1).

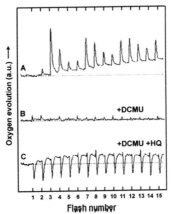

Fig. 1. (A) Oxygen signal of spinach thylakoids under short saturating flashes, (B) same as (A) after addition of 50 μM DCMU, (C) same as (B) after addition of 10 mM HQ.

We have found that cytochrome b₆-f inhibitors (DNP-INT, DBMIB) were not inhibitory for the HQ electron donation while plastocyanin removal or plastocyanin inhibitors (KCN) inhibited considerably this reaction. This means that electron donation site of HQ is as shown in Fig. 4. For the measurements of inhibition of superoxide generation by prenyllipids we used two thylakoid preparations, one hexane-extracted to remove native prenyllipids and the other one without the extraction.

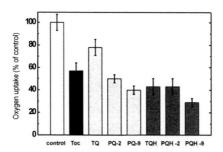

Fig. 2. Inhibition of oxygen uptake by different prenyllipids (prenyllipid:chl ratio = 1:5) in flash-illuminated hexane- extracted spinach thylakoids in the presence of DCMU and HQ.

The results shown in Fig. 2 show that all the investigated prenyllipids inhibited oxygen uptake by PS I. The well-known antioxidant, α-Toc was relatively poorly active in the measured reaction. The highest activity of the PQ-9/PQH₂-9 couple is probably due to the closest location of these molecules to the site of superoxide generation, i.e. at the membrane interior. The inhibition pattern measured on the non-extracted thylakoids (Fig. 3) shows that the inhibitory action of both PQ-2 forms, α-TQH₂ and PQ-9 is lower than that in the case of the extracted thylakoids even at higher prenyllipids concentration used for these samples.

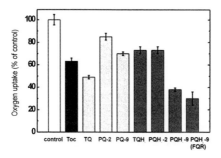

Fig. 3. Inhibition of oxygen uptake by different prenyllipids (prenyllipid:chl ratio = 1:2.5) in flash-illuminated spinach thylakoids in the presence of DCMU and HQ.

368

These differences may be caused by poor incorporation of the added prenyllipids into the lipid bilayer. Interestingly, PQH_2-9 formed in thylakoids by enzymatic reduction of PQ-9 with ferredoxin-PQ reductase was even more active in the inhibition of oxygen consumption than the externally added PQH_2-9.

Our results show that: (1) O_2^- production in thylakoids can be reliably measured using fast oxygen electrode in the presence of oxygen evolution inhibitors (DCMU) and a suitable electron donor (HQ), (2) membrane prenyllipids, such as PQ-9/PQH_2 couple, α-Toc and α-TQH_2 inhibit light-induced O_2 uptake in thylakoids by scavenging O_2^- radical, (3) the reactions of PQ-9/PQH_2 couple with O_2^- contribute to a native pseudocyclic electron transport around PS I, (4) scavenging of O_2^- by prenyllipids within thylakoid membranes protects both the membrane components against toxic action of O_2^-, as well as reduces the level of O_2^- diffusing towards membrane surface and inhibits formation of toxic H_2O_2 in chloroplasts.

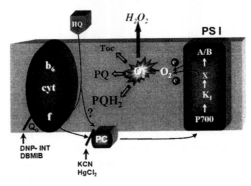

Fig. 4. The model showing generation of superoxide radical in PS I and its scavenging by membrane prenyllipids. The size of the prenyllipid symbol reflects their reactivity with superoxide.

The model presented in Fig. 4 shows the protective function of the investigated prenyllipids against superoxide generated by PS I in thylakoid membranes.

4. Acknowledgements

This work was supported by KBN grant No. 6P04A 031 20.

5. References

Asada, K. (1999) The water-water cycle in chloroplasts: Scavenging of active oxygens and dissipation of excess photons. Ann. Rev. Plant Mol. Biol. 50, 601-639.

Kruk, J., Jemioła-Rzemińska, M. and Strzałka, K. (1997) Plastoquinol and α-tocopherol quinol are more active than ubiquinol and α-tocopherol in inhibition of lipid peroxidation. Chem. Phys. Lipids 87, 73-80.

Robinson, H.H. and Yocum, C.F. (1980) Cyclic photophosphorylation reactions catalyzed by ferredoxin, methyl viologen and anthraquinone sulfonate. Biochim. Biophys. Acta 590, 97-106.

FLUORESCENCE ANISOTROPY OF PLANT MEMBRANE PRENYLLIPIDS AS A TOOL IN STUDIES OF THEIR LOCALIZATION AND MOBILITY IN MODEL MEMBRANES.

M. JEMIOŁA-RZEMIŃSKA, J. KRUK, K. STRZAŁKA

Department of Plant Physiology and Biochemistry, The Jan Zurzycki Institute of Molecular Biology and Biotechnology, Jagiellonian University

Gronostajowa 7, 30-387 Kraków, Poland

1. Introduction

Recent studies show that apart from the relatively well-characterised role as redox active compounds in energy transducing membranes, other functions of prenylquinones have come to the fore. One of the most important for the membrane defence system appears to be antioxidant activity exhibited by reduced forms of plant prenylquinones (QH$_2$) similar to that of α-tocopherol (α-Toc) (Munne-Bosch et al.2002). Located within the lipid bilayer, close to the site where free radicals and active oxygen species are generated, prenylquinones exhibit the high scavenging efficiency protecting fatty acids, pigments and other membrane-localized compounds from oxidative damage. Since the efficiency of this protection depends, among others, on bimolecular collision rate between a prenyllipid and the oxidant therefore, the detailed knowledge of localization, orientation and mobility of QH$_2$ and α-Toc in their lipid environment seem to be crucial for determining their antioxidant activity.

Bearing in mind that extensively used in membrane studies the fluorescent probe approach has also some shortcomings, we have exploited the intrinsic fluorescence of natural plant prenyllipids as a very convenient method of directly observing the molecule. In this report we describe the fluorescence anisotropy measurements of plastoquinols, as well as α-tocopherol incorporated in liposomes composed of egg yolk lecithin and synthetic dipalmitoylphosphatidylcholine.

2. Materials and Methods

All chemical and solvents were of highest available purity. Prenylquinones were reduced following the procedure of Kruk et al., (1992).

N. Murata et al. (eds.), Advanced Research on Plant Lipids, 369–372.

Small unilamellar vesicles (SUV) of dipalmitoylphosphatidylcholine (DPPC) and egg yolk lecithin (EYPC) containing PQH$_2$-2, PQH$_2$-9 and α-Toc were prepared by the injection method described in Jemioła-Rzemińska et al., (1996). The final phospholipid concentration was 0.5 mM and prenyllipid content in the samples was in the range of 1 to 10 mol%.

Solid poly(vinyl alcohol) (PVA) was dissolved in hot dimethylsulphoxide and added to prenyllipid in order to obtain the ratio of QH$_2$:PVA = 1:250 (w/w). The sample was cast on the quartz plate and the solvent was then allowed to evaporate under vacuum.

All fluorescence anisotropy measurements were performed using the Perkin-Elmer LS-50B fluorometer. An excitation wavelength of 290 nm was used and the fluorescence emission of prenyllipids in liposomes was detected at a wavelength of 330 nm. Unless otherwise indicated, measurements were made at 21°C. A Schott k-V 320 cut-off filter was used for emission of prenyllipids embedded in PVA films. An optical contact between the sample casted on a quartz plate and the propylene glycol was kept in order to reduce light scattering at the surface.

The contribution of scattered light to the fluorescence emission (I_{VV}^S and I_{VH}^S) was determined independently for reference solution without quinols. The steady-state fluorescence emission anisotropy was calculated as:

$$A = \frac{\left(I_{VV} - I_{VV}^S\right) - G\left(I_{VH} - I_{VH}^S\right)}{\left(I_{VV} - I_{VV}^S\right) + 2G\left(I_{VH} - I_{VH}^S\right)} \qquad G = \frac{I_{HV}}{I_{HH}}$$

where I_{VV} and I_{VH} are emission intensities of vertically polarized light measured parallel (I_{VV}) and perpendicular (I_{VH}) to the plane of excitation light, G is the instrument grating correction factor. The anisotropy values represent the averaged values from several independently prepared samples.

3. Results and Conclusions

In accordance with expectation, our results indicate that fluorescence anisotropy (FA) of plastoquinols varies depending upon the structure of the surrounding lipids (Figs. 1A, 1B). In an unsaturated lipid such as EYPC the FA values are considerably lower in contrast to fully saturated DPPC liposomes. At lower concentrations, both investigated plastoquinol homologues display similar FA values, regarding the same kind of liposomes. The explanation could be offered by the fact that PQH$_2$-9 fluorescence efficiency in hydrophobic environment is considerably lower than that in polar medium. Thus, we measure mainly the FA of PQH$_2$-9 molecules close to the membrane surface. The significantly lower anisotropy of PQH$_2$-9 than that of PQH$_2$-2 at higher (6-10%) prenyllipid content could indicate disordering of PQH$_2$-9 molecules at their high concentrations.

Throughout the whole range of the investigated concentrations, fluorescence anisotropy of both plastoquinols is higher than that of α-Toc regardless the type of phospholipid used for liposome preparation. However, the difference observed between these two kinds of prenyllipids is higher in case of the lipid bilayer composed of DPPC.

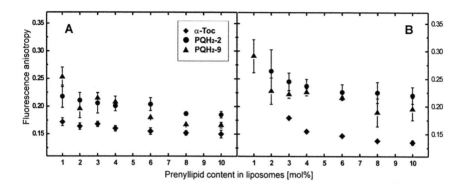

Figure 1. Fluorescence anisotropy of α-tocopherol (α-Toc), plastoquinol-2 (PQH$_2$-2) and plastoquinol-9 (PQH$_2$-9) in EYPC (A) and DPPC (B) liposomes at different prenyllipid content measured at 21°C.

In order to get the information about the relative mobility of α-Toc and PQH$_2$ in liposomes based on the anisotropy results, the fundamental anisotropy (A$_0$) of these molecules was measured in parallel experiments (Table 1). The obtained values correspond to the total immobilisation of the investigated fluorophores in liposomes. As it can be clearly seen, the anisotropy values of α-Toc incorporated into EYPC and DPPC liposomes (Figs. 1A, 1B) are close to its maximal anisotropy value determined in immobilised matrices.

TABLE 1. Values of fundamental anisotropy (A$_0$) and the angular displacement of transition moments (β) determined for different prenyllipids in PVA films and propylene glycol.

Prenyllipid	PVA (T = 21°C)			propylene glycol (T = -68°C)		
	λ_{max} (nm)	A$_0$	β (deg)	λ_{max} (nm)	A$_0$	β (deg)
α-Toc	333	0.167±0.017	38.6	330	0.160±0.005	39.2
PQH$_2$-9	337	0.302±0.005	23.8	324	0.283±0.008	26.2

The ratio of (A$_0$-A) to A$_0$ expressed as percent value was calculated for different prenyllipids, representing their rotational freedom (Figs. 2A, 2B). The motion of α-Toc seems to be severely restricted in the lipid bilayer, whereas plastoquinol molecules display significant mobility of in DPPC membranes and even higher in EYPC

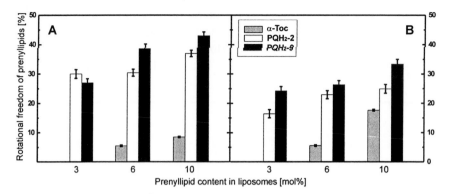

Figure 2. Rotational freedom of α-tocopherol (α-Toc), plastoquinol-2 (PQH$_2$-2) and plastoquinol-9 (PQH$_2$-9) incorporated into EYPC (A) and DPPC (B) liposomes, expressed as [(A$_0$-A)/A$_0$] x 100%.

In the light of our finding it seems to be likely that α-Toc acts mainly at the membrane surface while PQH$_2$ showing high rotational motion may penetrate the membrane to a considerable depth, therefore having opportunity to react with the lipid peroxyl radicals at the site of their generation.

4. Acknowledgements

We are indebted to Dr H. Koike (Himeji Institute of Technology, Hyogo, Japan) and Hoffmann-La Roche (Switzerland) for kind gifts of quinones.
This work was supported by the grant No. 6 P04A 028 19 from Committee for Scientific Research (KBN) of Poland.

5. References

Munne-Bosh, S., Alegre, L. (2002) The function of tocopherols and tocotrienols in plants. Crit. Rev. Plant Sci., 21, 31-57.

Kruk, J. Strzałka, K., Leblanc, R.M. (1992) Monolayer study of plastoquinones, α-tocopherol quinone, their hydroquinone forms and their interaction with monogalactodiacylglycerol. Charge-transfer complexes in a mixed monolayer, Biochim. Biophys. Acta 1112, 19-26.

Jemioła-Rzemińska, M., Kruk, J., Skowronek, M., Strzałka, K. (1996) Location of ubiquinone homoloques in liposome membranes studied by fluorescence anisotropy of diphenyl-hexatriene and trimethylammonium-diphenyl-hexatriene, Chem. Phys. Lipids 79, 55-63.

THE ROLE OF BETAINE LIPIDS IN ADAPTATION OF LICHEN PELTIGERA APHTHOSA TO LONG-TERM DEHYDRATION

E.R. KOTLOVA[1], N.F. SINYUTINA[2]

[1]Dept. Lichenology & Bryology, Komarov Botanical Institute
Prof. Popov Street 2, St. Petersburg 197376, Russia
[2]Dept. Plant Physiology & Biochemistry, St. Petersburg State University
Universitetskaya Emb. 7/9, St. Petersburg 199034, Russia

1. Introduction

Diacylglyceryl-N,N,N-trimethylhomoserine (DGTS) is a common lipid component of a number of green algae and cryptogamic plants (Sato, 1992; Kato et al., 1996) as well as some fungi (Künzler, Eichenberger, 1997). The cellular localization and site of biosynthesis of DGTS are associated with non-plastidial structures (Künzler et al., 1997; Moore et al., 2001). As regards the biosynthesis of DGTS, S-adenosyl-L-Met has been demonstrated to be a precursor for both the homo-Ser moiety of the headgroup and the methyl units (Moore et al., 2001). In spite of the progress in our knowledge about DGTS distribution and biosynthesis, there is an obvious lack of understanding of its physiological role. According to our previous results obtained on the lichen material the amount of DGTS remains unchangeable in stress conditions whereas the concentration of phospholipids decreases (Bychek-Guschina et al., 1999). This phenomenon may be simply explained by the greater resistance of DGTS molecules because the ether bond in the sn-3 position of glycerol moiety of DGTS is much more stable to hydrolysis than ester bond of phospholipids. However, since in some stresses the DGTS amount increases (Kotlova, 2000), it has been suggested that not only specificity of structure but also biosynthesis are responsible for the relative constancy of DGTS concentration, which seems to provide lichens with survival in different stress conditions. The present work describes experiments designed to test this proposal.

2. Materials and methods

The lichen *Peltigera aphthosa* containing green alga *Coccomyxa* sp. and ascomycete fungus as main symbionts was used throughout this work. Lichen thalli were collected, dried in air and stored in paper bags at room temperature in the dark for 1 month

N. Murata et al. (eds.), Advanced Research on Plant Lipids, 373–376.
© 2003 *Kluwer Academic Publishers. Printed in the Netherlands.*

(control), 6 and 18 months. In the preliminary experiment in order to reveal distribution of polar lipids among every partner of lichen association only freshly collected lichen thalli were used. The homogenate of lichen thalli was separated on algal and fungal component by centrifugation with sucrose density gradient (Kotlova, 2000).

Before labelling experiments 0.5 g of air-dried samples were activated by floating on 20 ml of distilled water at 18°C in the light and applied after either 0.4 h pre-incubation (partial recovery) or 20 h (full recovery). Then full or partly recovered lichen thalli were incubated on 30 ml H_2O with 5 µCi [2-^{14}C]acetate in the light at 18°C for 3 h.

The lipids were extracted, separated and quantified as previously described (Bychek-Guschina et al., 1999). The results are expressed as means ± SE (n=3).

3. Results and discussion

The photo- and mycobiont of the lichen *Peltigera aphthosa* are characterized by lipids which are quite typical of green algae and ascomycete fungi. Among glycerolipids of photobiont *Coccomyxa* sp., monogalactosyldiacylglycerol is the most abundant component, followed by digalactosyldiacylglycerol, sulfoquinovosyldiacylglycerol and phosphatidylglycerol which are typical of plastidial structures (data are not shown), as well as phosphatidylholine (PC), phosphatidylethanolamine (PE), phosphatidylinositol (PI). The membrane lipids of mycobiont consist of PE, PC as major components and PI, diphosphatidylglycerol (DPG), phosphatidylserine (PS) as minor ones. The distribution of polar glycerolipids among lichen symbionts is shown in Table 1.

TABLE 1. The distribution of polar glycerolipids among symbionts of the lichen *Peltigera aphthosa* (% of total polar lipids)*

	Lipid classes					
	DGTS	PC	PE	PI	DPG	PS
Photobiont	38.8	29.2	18.5	13.5	–	–
Mycobiont	10.1	30.7	34.9	13.7	8.3	2.3

* – for the photobiont only extra-plastidial polar lipids are considered.

DGTS was found to be accumulated not only in photobiont cells, where it is the most abundant extra-plastidial polar lipid, but also in mycobiont cells. The latter was unexpected because cultured lichen fungi do not produce betaine lipids (Bychek-Gushchina, 1997). Since the directed flow of different metabolites from the photobiont to the mycobiont is well documented, the accumulation of DGTS in fungal cells can be explained by the transport of either some precursor and/or co-factor(s) required for betaine lipid synthesis.

The various environmental stresses, including water stress, can significantly affect lipid composition of lichens. In spite of the fact that poikilohydric lichens are quite tolerant to long-term dehydration and their membrane structures can be reconstructed even after more than a year of water lack, dehydration causes a considerable decrease of polar lipid content. However, this decrease was first due to changes in PC and PE (Figure 1). At the same time there were no changes in DGTS amount after 6 months of dehydration and so drastic reduction after 18 months. In that way in the course of dehydration the rate of DGTS gradually rises from 16% of the total cytoplasmic polar lipids in control samples to 45% in 18 month stressed ones.

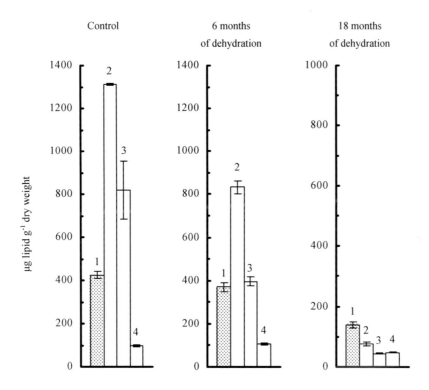

Figure 1. The effects of long-term dehydration on the content of extra-plastidial polar lipids in the lichen *Peltigera aphthosa*. 1 – DGTS, 2 – PC, 3 – PE, 4 – PI.

In order to clarify whether the increased proportion of DGTS is the result of nonequal breakdown of phospho- and betaine lipids or is also connected with the features of betaine lipid synthesis the experiments with [2-^{14}C] acetate incorporation were undertaken (Table 2).

The main results could be summarized as follows: 1) the specific radioactivity of partly recovered lichen samples (0.4 h pre-incubation with water) was always less than that of full recovered samples; 2) after both short-term dehydration (control) and long-term dehydration the biosynthesis of DGTS was recovered much faster than that of phospholipids; 3) the specific radioactivity of PC and PI in 20 h pre-incubated lichen samples dramatically decreased with the time of dehydration, while that of DGTS and PE was high enough by the end of the experiment. Taken together the results obtained may indicate an important role of betaine lipids in adaptation of lichens to long-term dehydration.

TABLE 2. The specific radioactivity (dpm μg^{-1} lipid) of polar extra-plastidial lipids from the lichen *Peltigera aphthosa* submitted to long-term dehydration

Time of dehydration	Pre-incubation time	Lipid classes			
		DGTS	PC	PE	PI
Control					
	0.4 h	3.4±0.3	1.4±0.3	3.3±0.4	6.9±0.7
	20 h	5.1±0.6	50.2±3.2	60.8±5.4	76.7±5.9
6 months					
	0.4 h	2.2±0.1	3.6±0.2	8.3±0.5	3.3±0.3
	20 h	5.3±0.8	26.9±1.0	22.9±2.5	28.1±0.8
18 months					
	0.4 h	3.3±0.1	2.7±0.1	5.8±0.4	1.4±0.2
	20 h	4.8±0.4	10.4±1.0	76.0±2.4	13.7±1.1

This work has been supported by RFBR (grant № 02-04-49600).

4. References

Bychek-Gushchina, I.A. (1997) Study of biochemical aspects of lichen symbiosis. I. Lipids and fatty acids in cultured lichen symbionts. Biochemistry (Moscow) 62, 574-580.

Bychek-Guschina, I.A., Kotlova, E.R. and Heipieper H. (1999) Effects of sulfur dioxide on lichen lipids and fatty acids. Biochemistry (Moscow) 64, 61-65.

Hofmann, M. and Eichenberger, W. (1996) Biosynthesis of diacylglyceryl-*N,N,N*-trimethylhomoserine in *Rhodobacter sphaeroides* and evidence for lipid-linked *N* methylation. J. Bacteriol. 178, 6140-6144.

Kato, M., Sakai, M., Adachi, K., Ikemoto, H. and Sano, H. (1996) Distribution of betaine lipids in marine algae. Phytochemistry 42, 1341-1345.

Kotlova, E.R. (2000) Antioxidative Systems of Lichens. Ph.D. thesis. St. Petersburg. [In Russian].

Künzler, K., Eichenberger, W. and Radunz, A. (1997) Intracellular localization of two betaine lipids by cell fractionation and immunomicroscopy. Z. Naturf. 52b, 487-495.

Künzler K. and Eichenberger W. (1997) Betaine lipids and zwitterionic phospholipids in plants and fungi. Phytochemistry 46, 883-892.

Moore, T.S., Du, Z. and Chen, Z. (2001) Membrane lipid biosynthesis in *Chlamydomonas reinhardtii*. In vitro biosynthesis of diacylglyceryl trimethylhomoserine. Plant Physiol. 125, 423-429.

Sato, N. (1992) Betaine lipids. Bot. Mag. 105, 185-197.

A STUDY OF THE PHYSIOLOGICAL FUNCTION OF PHOSPHATIDYLETHANOLAMINE IN *ARABIDOPSIS*

J. MIZOI, M. NAKAMURA, I. NISHIDA
Department of Biological Sciences, Graduate School of Science,
The University of Tokyo
7-3-1 Hongo, Bunkyo-ku, Tokyo, 113-0033 Japan

1. Introduction

Phosphatidylethanolamine (PE) is a major non-bilayer lipid that is widely distributed in extra-plastid membranes of plant cells. PE is synthesized via three metabolic pathways, *i.e.* the nucleotide pathway, the phosphatidylserine decarboxylation pathway and the base exchange pathway in biological systems, and the nucleotide pathway is thought to be the main pathway for the *de novo* synthesis of PE in plants (Wang and Moore, 1991). Among three enzymes involved in the nucleotide pathway, CTP:phosphorylethanol-amine cytidylyltransferase (ECT; EC 2.7.7.14) is thought to be a rate-limiting enzyme (Wang and Moore, 1991). PE is involved in membrane fusion in bacteria and fungi, serves as a substrate for the synthesis of GPI anchors in eucaryotes (Menon and Stevens, 1992) and *N*-acylphosphatidylethanolamine in animal and plant tissues (Chapman, 2000), and has a crucial function in cell division of mammalian cells (Emoto et al., 2000). However, very little information has been obtained about the physiological function of PE in plants. In this work, we introduce our studies to elucidate the physiological function of PE using transgenic *Arabidopsis* plants.

2. Experimental Procedures

The open reading frames for *AtECT1* (At2g38670), *AtCCT1* (At2g32260) and *AtCCT2* (At4g15130) were amplified from an *Arabidopsis* cDNA library and sub-cloned into the *E. coli* expression vector pQE30 (QIAGEN, Germany). Recombinant proteins were expressed in *E. coli* M15[pREP4] cells, and the crude cell lysates prepared by sonication were used for the measurement of ECT activity. The ECT activity was measured according to a method modified from a CCT assay method described in Nishida et al. (1996), except that phosphorylethanolamine was used instead of phosphorylcholine as a substrate. A cDNA for *AtECT1* was inserted in the antisense

N. Murata et al. (eds.), Advanced Research on Plant Lipids, 377–380.

orientation between the cauliflower mosaic virus 35S promoter and the nopaline synthase terminator from pBI221 (Clontech USA), and the resultant fragment was sub-cloned into the binary vector pPZP221 (Hajdukiewicz et al., 1994). *Arabidopsis* Col-0 plants were transformed by a vacuum infiltration method. For all experiments, independent homozygous T_3 lines of transformants with a single T-DNA insertion were used.

3. Results and Discussion

To construct transgenic plants with altered PE metabolism, we first characterized the ECT activity in *Arabidopsis*. The *Arabidopsis* genome initiative (2000) predicted three putative cytidylyltransferase genes: one codes for a protein that was very similar to known ECTs, and the other two encode isozymes of CTP:phosphorylcholine cytidylyl-transferases (CCT; EC 2.7.7.15) that were designated AtCCT1 and AtCCT2 (Nishida et al., in this volume). A cDNA for putative ECT was amplified by PCR and the cDNA was expressed in *E. coli* M15[pREP4] cells. The ECT activity was detected in the cell lysates of *E. coli* expressing the recombinant protein, verifying the identity of the cDNA. We therefore named the gene *AtECT1*.

We also examined whether AtCCT1 and AtCCT2 could exhibit ECT activity. Under our assay conditions, both recombinant proteins for AtCCT1 and AtCCT2 expressed in *E. coli* exhibited the ECT activity, although the activity was almost 10 fold smaller than that measured for CCT (Figure 1). This finding is quite interesting because ECT and CCT have been thought to be distinct enzymes in biological systems including plants : purified CCT from caster bean endosperm did not utilize phosphorylethanol-amine as a subatrate (Wang and Moore, 1990); CCT purified from rat liver exhibited no detectable ECT activity (Vermeulen et al., 1993); and the yeast double mutant *pss1ect1* was unable to grow on an ethanolamine-supplemented medium (Min-Seok et al., 1996). Therefore, it is interesting to know if AtCCT1 and AtCCT2 exhibit ECT activity *in vivo*. Because we could not detect the CCT activity for the recombinant AtECT1 protein under our assay conditions, we concluded that AtECT1 is the sole phosphorylethanol-amine-specific cytidylyltransferase in *Arabidopsis*.

AtECT1 was predicted to have a transmembrane domain at the N-terminal region of the protein, and this feature was not seen in other ECTs from yeast and mammals. Differences in the structure and substrate specificity of cytidylyltransferases between *Arabidopsis* and other organisms suggest that regulatory systems for the synthesis of PE and PC might differ between plant and other organisms.

To study the physiological function of PE, we transformed *Arabidopsis* plants with an antisense-*AtECT1* cDNA under the control of the cauliflower mosaic virus 35S promoter, and several lines of transformants that accumulated high levels of antisense-*AtECT1* transcripts were established. In these transgenic lines, however, the ECT activity in the rosette leaves was not significantly reduced when compared with

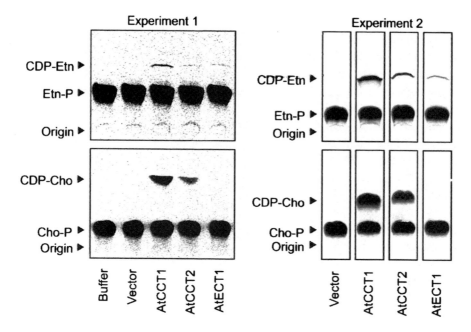

Figure 1. ECT and CCT activity in lysates of E. coli cells that are expressing recombinant *Arabidopsis* cytidylyltransferases. In Experiment 1, the values of the ECT activity of CCT isozymes that were measured under kinetic conditions were 10-fold smaller than those of CCT activity of AtCCT1 and AtCCT2. In Experiment 2, the ECT activity of CCT isozymes were measured with excess amounts of enzymes. In both experiments, however, no CCT activity was detected for AtECT1.

control lines that were transformed with a control vector. We do not have any data to conclude if the ECT activity in the rosette leaves of the antisense-*AtECT1* plants should be ascribed to the residual ECT protein or to the ECT activity of AtCCT isozymes, the accumulation of endogenous levels of *AtECT1* transcripts in the rosette leaves of the antisense-*AtECT1* plants suggested that AtECT1 proteins could be still translated. However, a significant proportion of antisense-*AtECT1* plants exhibited abnormal morphological features, such as very short or zigzag-shaped flower stalks in most cases and even shrunk rosette leaves in the severest case. The proportion of abnormal plants varied depending on the growth conditions (data not shown). Although a genetic link between these abnormal features and the expression of antisense-*AtECT1* should be established in the future, several lines of circumstantial evidence suggest that abnormal features and the expression of antisense-*AtECT1* were closely correlated. Firstly, similar abnormal features were observed among independent T-DNA insertion lines from independent transformation events. Secondly, every transgenic line that accumulated antisense-*AtECT1* transcripts produced plants with abnormal phenotypes, and finally the abnormal phenotypes were also observed in back crossed lines (data not shown).

In the wild type of *Arabidopsis*, *AtECT1* transcripts accumulated in almost all organs examined. However, high levels of accumulation were observed in elongating stems, shoot apices and flowering buds from young bolting plants, and in flowers, shoot apices and flowering buds from mature bolted plants. It is interesting to note that abnormal shapes of antisense-*AtECT1* plants were observed in stems where relatively high levels of *AtECT1* transcripts accumulated. Thus, we expect that the expression of the antisense *AtECT1* cDNA caused abnormal features only in tissues or cells that accumulated high levels of *AtECT1* transcripts, which might be prerequisite for the induction of RNA interference and hence the inhibition of PE metabolism.

4. References

Chapman, K.D. (2000) Emerging physiological roles for *N*-acylphosphatidylethanolamine metabolism in plants: signal transduction and membrane protection. Chem. Phys. Lipids. 108, 221-230.

Emoto, K. and Umeda, M. (2000) An essential role for membrane lipid in cytokinesis: regulation of contractile ring disassembly by redistribution of phosphatidylethanolamine. J. Cell Biol. 149, 1215-1224.

Hajdukiewicz, P., Svab, Z. and Maliga, P. (1994) The small, versatile *pPZP* family of *Agrobacterium* binary vectors for plant transformation. Plant Mol Biol. 25, 989-994.

Menon, A.K. and Stevens, V.L. (1992) Phosphatidylethanolamine is the donor of the ethanolamine residue linking a glycosylphosphatidylinositol anchor to protein. J. Biol. Chem. 267, 15277-15280.

Min-Seok R., Kawamata, Y., Nakamura, H., Ohta, A. and Takagi, M. (1996) Isolation and characterization of *ECT1* gene encoding CTP:phosphoethanolamine cytidylyltransferase of *Saccharomyces cerevisiae*. J. Biochem. 120, 1040-1047.

Nishida, I., Swinhoe, R., Slabas, A.R. and Murata, N. (1996) Cloning of *Brassica napus* CTP:phosphocholine cytidylyltransferase cDNAs by complementation in a yeast *cct* mutant. Plant Mol. Biol. 31, 205-211.

The Arabidopsis Genome Initiative (2000) Analysis of the genome sequence of the flowering plant *Arabidopsis thaliana*. Nature. 408, 796-815.

Vermeulen, P.S., Tijburg, L.B.M., Geelen, M.J.H. and Van Golde, L.M.G. (1993) Immunological characterization, lipid dependence, and subcellular localization of CTP:phosphoethanolamine cytidylyltransferase purified from rat liver. Comparison with CTP:phosphocholine cytidylyltransferase. J. Biol. Chem. 268, 7458-7464.

Wang, X. and Moore, T.S. Jr. (1990) Phosphatidylcholine biosynthesis in castor bean endosperm. Plant Physiol. 93, 250-255.

Wang, X. and Moore, T.S. Jr. (1991) Phosphatidylethanolamine synthesis by castor bean endosperm. J. Biol. Chem. 266, 19981-19987.

GLYCOLIPID FATTY ACIDS WHILE DROUGHT ACTION

N. SVYETLOVA, A. OKANENKO, N. TARAN, M. MUSIENKO
Plant Physiology & Ecology Department, Kyiv National Taras Shevchenko University, Volodymyrska, 64, UA-01033, Kyiv, Ukraine,

Abstract

The results of soil water deficit action on the fatty acid glycolipid composition of different wheat variety chloroplasts (*Triticum aestivum* L.) are presented. It is shown that the decrease of glycolipid fatty acid unsaturation and activation of peroxidation processes in chloroplasts of sensitive plants grown under stress condition occur. The chloroplasts of drought resistant variety was characterised both by the accumulation of unsaturated fatty acid residues in digalactosyl diacylglycerol and by the absence of changes in sulphoquinovosyl diacylglycerol without increasing peroxidation. The role of transformations in the composition of chloroplast glycolipid fatty acids is discussed.

Introduction

The conception of regulating plant adaptive reaction in environmental stress assumes that membrane lipid components could be considered as a substrate for stress hormone (jasmonic acid) production and the formation of oxylipin signal intermediates (Mulligan et al., 1997; Crellman and Mullet, 1995).

Methods

Winter wheat (*Triticum aestivum* L.) 10-day seedlings with different resistance to drought grown in sand at 70% humidity of its FM – control and at 40% humidity FM – experiment; at 25^0C / 22^0C (day/night) and 16 h photoperiod with 400 W/m^2 illumination at a plant level and air relative humidity 65-70% were used: *Albatros odesky* variety is drought resistant; *Myronivska 808* variety is drought sensitive. The amount of fatty acids was determined by the method of gas chromatography (Goldbert and Vigdergaus, 1990).

Results

The FA residue analysis of wheat chloroplast membrane glycolipid showed that among unsaturated FA residues in MGDG and DGDG fractions $C_{18:3}$ prevailed, among saturated FA $C_{16:0}$ prevailed (Figures 1,2,3). Besides, the level of saturating FA residues

381

N. Murata et al. (eds.), Advanced Research on Plant Lipids, 381–384.

382

in the lipids studied can be arranged in the following sequence: MGDG < DGDG< SQDG. The general tendency of changes in MGDG FA residues in plants of both varieties treated with soil WD was an increase in $C_{16:0}$ content and a decrease in $C_{18:1}$ residues to be more appreciable in chloroplasts of Myronivska 808 sensitive variety.

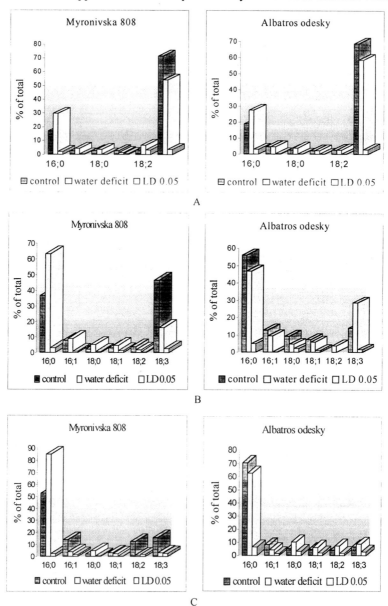

Figure 1. The content of MGDG (A), DGDG (B), SQDG (C) fatty acids in chloroplasts of *Triticum aestivum* L. varieties differing upon water deficit action (Myronivska 808 – as sensitive to drought; Albatros odesky – as resistant to drought; control – while soil humidity 70%; stress – while soil humidity 40%).

An essential drop in $C_{18:3}$ and $C_{18:1}$ acid residues in sensitive variety chloroplasts took place. The amount of $C_{16:1}$, $C_{18:1}$ and $C_{18:2}$ residues remained practically in unchangeable steady variety chloroplasts. Changes in MGDG FA residues caused a two-fold decrease in the ratio of unsaturated / saturated FA in Myronivska 808 variety chloroplasts and a 1.4-fold in Albatros odesky variety chloroplasts.

In DGDG fraction of sensitive variety chloroplasts the accumulation of $C_{16:0}$ and $C_{18:0}$ residues was accompanied by a decrease in $C_{18:3}$ and $C_{18:2}$ residues whereas the saturation of FA residue content fell down in steady variety plants. Steady plants were characterised by a two-fold increase in $C_{18:3}$ residue and $C_{18:2}$ appearance (3.9 against traces). It could argue about the formation of more unsaturated FA, particularly by $C_{18:1}$ desaturation.

In SQDG fraction the general tendency of both varieties was an increase in $C_{18:0}$ content (more essential in sensitive plants). Besides, the soft variety plants were characterised by an increase in $C_{16:0}$ content accompanied by $C_{18:2}$ (10-fold) and $C_{18:1}$ (5-fold) content drop. Thus, one could see stable SQDG saturation in steady variety and its increase in soft one.

Changes in the composition of glycolipid FA residues are considered to reflect to a certain degree the intensity of LP. Our results showed the increase in LP level of drought sensitive variety chloroplasts (Myronivska 808) and a relative stability of this process occurring in drought resistant variety (Albatros odesky) while WD action (Figure 4).

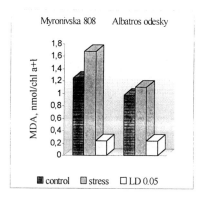

Figure 4. Influence of soil water deficit on lipid peroxidation in chloroplasts of different drought resistance winter wheat varieties (Myronivska 808 – as sensitive to drought; Albatros odesky – as resistant to drought).

The acceleration of free radical oxidation processes in drought steady variety chloroplasts was not observed and stable MDA content while specific increase in DGDG unsaturation level took place. Besides, C_{16} and C_{18} fatty acids being as oxylipin precursors may act also as inductors of metabolic transformations of lipoxygenase signal system in components in plants during formation of general adaptive responses to drought.

Acknowledgements. We are grateful to the OC 15-ISPL for financial support.

Abbreviations: FA – fatty acids; FM – full moisture; MGDG – monogalactosyl diacylglycerol; DGDG – digalactosyl diacylglycerol; SQDG – sulphoquinovosyl diacylglycerol; $C_{16:0}$ – palmitic acid; $C_{16:1}$ – palmitoleic acid; $C_{18:0}$ – stearic acid; $C_{18:1}$ – oleic acid; $C_{18:2}$ – linoleic acid; $C_{18:3}$ – linolenic acid; LP – lipid peroxidation process; MDA – malone dialdehyde; WD – water deficit.

References

Mulligan, R.M., Chory, J. and Echer, J.R. (1997) Signalling in plants. Proc. Natl. Acad. Sci. USA. 94, 2793-2795.

Creelman, R.A. and Mullet, J.E. (1995) Jasmonic acid distribution and activity in plants: regulation during development and response to biotic and abiotic stress. Proc. Natl. Acad. Sci. USA. 92, 4114-4119.

Goldbert, K.A. and Vigdergaus, M.S. (1990) Introduction to gas chromatography. Chemistry, Moscow.

SEASONAL CHANGES IN THERMOTROPIC BEHAVIOR OF PHOSPHO- AND GLYCOLIPIDS FROM *LAMINARIA JAPONICA*

Adjustments of lipid physical state

N.M. SANINA, S.N. GONCHAROVA, E.Y. KOSTESKY
Far Eastern National University
Sukhanov St., 8. Vladivostok 690600, Russia

1. Introduction

Marine macrophytes belong to poikilotherms, which are the most temperature sensitive organisms. Their compensatory mechanism is directed to maintain the liquid-crystalline state of biomembrane lipid matrix, and, hence, it underlies the thermal adaptation of these organisms (Hazel, 1995). During the last few years, extensive evidence has shown that a change in membrane fluidity appears to be the primary event for the perception and transduction of temperature signal (Vigh et al., 1993, Los et al., 1997). In this connection, thermal behavior of major glyco- and phospholipids of brown algae *Laminaria japonica,* harvested in winter, were studied by differential scanning calorimetry (DSC) and polarizing microscopy. Fatty acid composition of lipid samples was analyzed to interpret the thermal transitions of lipids. Results were compared with earlier received data on this macrophitic algae collected in summer (Sanina et al., 2000).

2. Materials and Methods

L. japonica, Aresch (Phaeophyta) was harvested in Possiet Bay (the Sea of Japan) in winter and summer from seawater of 3°C and 20°C, respectively. Freshly collected algae were thoroughly cleaned to remove epiphytes, small invertebrates and sand particles and then heated for 2 min in boiling water to inactivate enzymes. Total lipid extracts from about 10 kg of algae were obtained according to the method of Folch et al. (1957). Crude glyco- and phospholipids were isolated from total lipid extract by column chromatography on silica gel by elution with acetone, acetone/benzene/acetic acid/water (200:30:3:10, by vol.) and a gradient of chloroform and methanol. Then, lipids were purified by preparative silica TLC using chloroform/methanol/water (65:25:4, by vol.). Their purity was checked by two-dimensional

N. Murata et al. (eds.), Advanced Research on Plant Lipids, 385–388.

silica TLC. Chromatographically pure phospholipids were solubilized in chloroform and introduced into standard aluminium pans. Vacuum-dried samples of approx. 10 mg were sealed into pans and placed in a DSM-2M differential scanning calorimeter (Puschino, Russia). Samples were either heated or cooled at 16 °C/min in a temperature range between −100 °C and 80°C at a sensitivity of 5 mW. The peak in the plot of heat capacity versus temperature was recorded as the phase-transition temperature, T_{max}. The temperature range was calibrated by using naphthalene, mercury and indium. Temperature ranges of isotropic transitions of the same phospholipids were defined by means of POLAM-P-312 polarizing microscopy (Russia) with a heated stage, at a magnification of x100. Analysis of acyl chains linked to phospho- and glyco-lipids was carried out by GLC as described by Khotimchenko (1993).

3. Results and Discussion

As shown in Figure 1, thermograms of phospho– and glycolipids from *L. japonica,* collected in winter, were characterized by the complex profile and sited in the wide temperature range from -82 to -48°C up to 44 to 78°C, that is close to the respective values in summer. Endothermic peaks were revealed both in the low- and high-temperature ranges on the heating curves of phospholipids (Fig. 1 A-C). Peak maximum temperatures (T_{max}) of phosphatidylcholine (PC) and phosphatidylethanolamine (PE) were defined at -19°C and -10°C, that was lower by 10-12°C and 32-36°C than in summer, respectively (Sanina et al., 2000). As known, PC and PE are mainly located in the chloroplast envelops and other non-photosynthesizing membranes (Quinn and Williams, 1978). Their decreased T_{max} appears to be the result of the low-temperature induced synthesis of desaturases (Suzuki et al., 2000), which take part in the biosynthesis of both phospho- and glycolipids. Then, the latter are transferred to the chloroplast thylakoids (Morre et al., 1991). Indeed, unsaturation index (UI) of phospho- and glycolipids, excepting monogalactosyldiacyglycerol (MGDG), increased at the lower ambient temperature (Table 1).   Since ambient temperatures influence the fluidity of membrane phospholipids, it is possible that the change in physical state makes phospholipids available as substrates for cellular phospholipases, resulting in production of second messengers in response to a change in temperature. Released second messengers then may be involved in the process of thermal signal transduction.

The contrast seasonal change was observed in glycolopids: MGDG, digalactosyldiacyglycerols (DGDG) and sulphoquinovosyldiacyglycerol (SQDG). Despite the higher values of the ratio of unsaturated to saturated fatty acids, T_{max} of glycolipids increased by 24-52°C in winter in comparison with summer. In contrast with PC and PE, glycolipids and phosphatidylglycerol (PG) are the main constituents of photosynthesizing membranes. Thus, more rigid glycolipids, surrounding photosynthetic complexes in thylakoid membranes, is probably necessary to inhibit the photosynthetic activity of algae in winter (Gombos et al., 1994). As a rule, UI of phospho- and glycolipids increased due to the percentage elevation of eicosapentaenoate (20:5n3). Additionally, in all glycolipids, the content of 18:4n3 is also higher in winter than in summer. In contrast with polyunsaturated fatty acids (PUFA), the sums of saturated and monounsaturated fatty acids are reduced. The higher percentage of PUFA seems to be important in the protection of the photosynthetic machinery from low-temperature photoinhibition (Gombos et al., 1994). Especially sharp seasonal reorganization of fatty acid composition of DGDG appears to result in very unusual DSC behavior of this glycolipid of algae collected in winter. There were the powerful exothermic and endothermic peaks at –27°C and 67°C, respectively, on thermogram of DGDG. Temperature range of the later peak correlated with temperature of isotropic melting observed under polarizing microscope. It seems to be due to the especially strong seasonal reorganization of fatty acid composition.

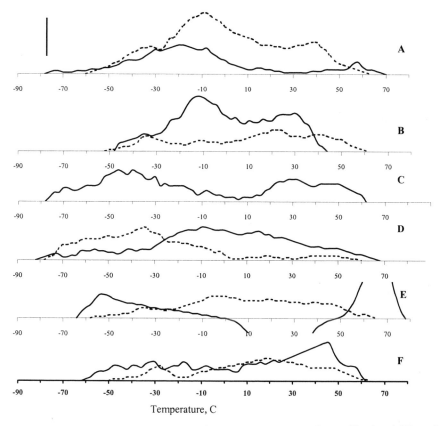

Figure 1. DSC thermograms of major lipids isolated from *Laminaria japonica*, harvested in winter (solid curve) and summer (dotted curve). PC (A), PE (B), PG (C), MGDG (D), DGDG (E) and SQDG (F). Vertical bar represents 0.5 mW. Scanning rate, 16 °C/min. Sample weight, 10 mg

High-temperature peaks, revealed on the thermograms of other lipids, seem to be arisen from their isotropic melting too. Temperature of the beginning of the isotropic melting decreased at the lower ambient temperature. Hence, it may be resulted in enhanced vulnerability of *L. japonica* to the higher ambient temperatures during the low-temperature acclimatization.

TABLE 1. Fatty acid composition of glyco- and phospholipids from *L. japonica* harvested in winter (W) and summer (S) (% of the total fatty acids)

SFA, MUFA, PUFA, saturated, monounsaturated and polyunsaturated fatty acids, respectively; Fatty acids with contents of less than 2% are excepted.

Fatty acid	MGDG		DGDG		SQDG		PC		PE		PG
	W	S	W	S	W	S	W	S	W	S	W
14:0	5.8	5.0	3.4	9.0	1.1	3.6	18.5	12.7	3.0	4.4	1.5
16:0	2.9	5.5	1.9	20.0	25.5	45.2	12.4	12.4	9.5	29.3	27.0
16:1	1.1	4.0	7.1	20.5	2.6	4.3	2.6	4.4	4.6	6.1	12.1
18:0	0.9	0.5	0.6	3.3	1.5	3.6	0.4	0.7	1.9	3.8	0.3
18:1n9	4.2	9.6	1.2	14.1	17.0	21.9	2.6	12.1	3.9	9.2	13.1
18:2n6	6.3	11.1	1.1	10.9	5.1	7.7	8.3	12.4	2.8	5.4	9.5
18:3n6	3.3	8.0	0.1	1.8	1.4	0.9	0.7	0.4	0.2	0.7	0.2
18:3n3	5.4	8.7	4.5	5.2	11.4	3.0	2.7	1.3	3.0	2.8	29.0
18:4n3	54.3	20.3	38.0	3.2	9.3	0.9	0.8	0.4	1.6	0.8	0.9
20:4n6	1.5	9..9	2.6	2.1	2.0	3.0	25.0	29.1	44.8	26.9	3.2
20:5n3	10.9	15.9	33.2	3.4	9.9	1.4	17.7	7.5	18.5	2.3	1.4
SFA	9.8	11.5	5.9	35.0	28.1	53.2	32.0	26.6	15.5	40.0	29.5
MUFA	5.9	14.0	12.2	37.7	19.6	30.0	5.2	17.6	9.0	19.9	25.4
PUFA	83.1	74.5	81.5	27.3	46.6	16.8	62.0	54.0	73.5	39.2	44.7
UI	324	387	504	120	182	79	247	214	318	169	157

Acknowledgements. The research was supported in part by NATO Collaboration Linkage Grant (CLG №978844) and Award REC-003 of the U.S. Civilian Research & Development Foundation for the Independent States of the Former Soviet Union (CRDF).

References

Hazel, J.R. (1995) Thermal adaptation in biological membranes. Is homeoviscous adaptation the explanation? Annu Rev. Physiol. 57, 19-42.

Gombos, Z, Wada, H. and Murata, N. (1994) The recovery of photosynthesis from low-temperature photoinhibition is accelerated by the unsaturation of membrane lipids: a mechanism of chilling tolerance. Proc. Natl. Acad. Sci. USA. 91, 8787-8791.

Folch, J., Less, M. and Sloane-Stanley, G. H. (1957) Isolation and purification of total lipids from tissues. J. Biol. Chem. 226, 497-509

Khotimchenko, S. V. (1993) Phytochemistry. Fatty acids and polar lipids of seagrasses from the Sea of Japan. 33, 369-372

Morre, D. J., Morre, J.T., Morre, S.R., Sundqvist, C., Sandelius, A.S. (1991) Chloroplast biogenesis. Cell-free transfer of envelope monogalactosylglycerides to thylakoids. Biochim. Biophys. Acta. 1070, 437-45

Quinn, P.J., Williams, W.P. (1978) Plant lipids and their role in membrane function. Progr. Biophys. Molec. Biol. 34, 109-173

Sanina, N. M., Kostetsky, E. Y., Goncharova, S. N. (2000) Thermotropic behaviour of membrane lipids from brown marine alga *Laminaria japonica.* Biochem. Soc. Trans. 28, 894-897

Suzuki, I, Los, D. A. and Murata, N. (2000) Perception and transduction of low-temperature signals to induce desaturation of fatty acids. Biochem. Soc. Trans. 28, 628-630

Vigh, L., Los, D.A., Horvath, I. and Murata N. (1993) The primary signal in the biological perception of temperature: Pd-catalyzed hydrogenation of membrane lipids stimulated the expression of the desA gene in Synechocystis PCC6803. Proc. Natl. Acad. Sci. 90, 9090-9094

CYANOBACTERIAL LIPID COMPOSITION WITH REGARD TO THE REGULATORY ROLE OF GLUCOSE

N.F. MYKHAYLENKO, O.K. ZOLOTAREVA
M.G. Kholodny Botany Institute, Nat. Acad. Sci. Ukraine
2 Tereschenkivska Str., 01004, Kyiv, Ukraine

Introduction

Cyanobacteria, the group of organisms in which oxygenic photosynthesis and aerobic respiration are active in the same thylakoid membrane (Schmetterer, 1994), are similar by their lipid composition to the inner envelope membranes and thylakoid membranes of higher plant chloroplasts (Wada, Murata, 1998). It seems to be of great interest to scrutinize the effects of glucose, a key metabolite of a photosynthetic cell, on the cyanobacterial membrane lipid status.

Glucose inhibits photosynthetic processes not only via feedback regulation of enzyme activities, but also on the level of gene expression. Hexokinase (HK), the first enzyme in the hexose assimilation pathway, is one of the key sensors and signal transmitters of sugar repression (Jang et al., 1997, Smeekens, 2000). Sugars that are the substrates of HK cause repression of higher plant photosynthesis already at 1 to 10 mM concentrations (Jang, Sheen, 1994). Mannose, a glucose epimer at the second carbon atom, that can be phosphorylated by HK but doesn't undergo further metabolism, is capable of repressing promoters of genes for photosynthetic components at considerably lower concentrations than glucose itself (Jang, Sheen, 1994, Pego et al., 2000). Since glucose inhibition of photosynthesis in cyanobacteria is much less studied than that in higher plants, we studied chlorophyll and polar lipid content in glucose- and mannose-cultivated cyanobacteria to determine the role of HK-mediated photosynthetic repression in cyanobacterial mixotrophic growth.

Materials and Methods

Freshwater cyanobacterium *Spirulina platensis* was studied in comparison with two different

---

Abbreviations: DGDG, digalactosyl diacylglycerol; HK, hexokinase; MGDG, monogalactosyl diacylglycerol; PG, phosphatidylglycerol; SQDG, sulfoquinovosyl diacylglycerol.

N. Murata et al. (eds.), Advanced Research on Plant Lipids, 389–392.
© 2003 *Kluwer Academic Publishers. Printed in the Netherlands.*

strains of soil cyanobacterium *Nostoc linckia* gathered on the opposite slopes of 'Evolution Canyon' (Lower Nahal Oren, Mt. Carmel, Israel) (Krugman et al., 2001). The strain from north-facing slope was designated *N. linckia* 'N' and that from south-facing slope – *N. linckia* 'S'. Cyanobacteria were grown in sterile conditions at 28°C on liquid media described in (Zender, Gorham, 1960) for *N. linckia* and in (Pinevich et al., 1970) for *S. platensis*. The cultures were illuminated 12 hours per day with white light at an intensity of 2200 lux. On the 15th day of growth the cultivation conditions were changed by adding 5 mM glucose, 50 mM glucose or 5 mM mannose. Chlorophyll *a* and polar lipid content were determined after 4 days of growth with sugars.

Cyanobacterial cells were harvested by centrifugation at 3,000 g and washed thrice with distilled water. Chlorophyll *a* concentration was measured according to Wettstein (1957). Lipids were extracted and analysed as described previously (Mikhaylenko, Skorokhod, 1998). Results are the averages from three independent experiments, each analysed thrice. The standard deviation was always less than 5%.

Results and Discussion

Growth profiles of all cyanobacterial strains studied exhibited considerable similarity in response to sugar addition (Fig. 1) despite much higher growth rates of *Spirulina* as compared with *Nostoc*. 50 mM glucose evoked active growth of *S. platensis* and *N. linckia* 'N' almost at once after addition but in the case of *N. linckia* 'S' the effect was prominent only on the fourth day. In the presence of 5 mm glucose substantial biomass increase was observed on the first day of cultivation in *S. platensis*, on the fourth day – in *N. linckia* 'N' and only after a week of growth – in *N. linckia* 'S'. The retardation of growth in response to adding of 5 mM mannose in *Spirulina*, in contrast to *Nostoc*, was rather moderate (5-15%) and lowering with time.

The decrease of chlorophyll *a* content in the presence of mannose was appreciable in all strains (Table 1) while 50 mM glucose didn't affect this parameter significantly in *N. linckia* 'S' and 5 mM glucose – in *S. platensis* as well.

Table 1. Cyanobacterial chlorophyll *a* content after 4 days of growth with sugars, mg / g dry weight.

	Spirulina platensis	Nostoc linckia 'N'	Nostoc linckia 'S'
Control	23.10 ± 0.91	13.57 ± 0.70	12.39 ± 0.15
+ 5 mM glucose	22.39 ± 0.86	11.84 ± 0.37	12.72 ± 0.47
+ 50 mM glucose	10.99 ± 0.46	10.55 ± 0.21	12.23 ± 0.37
+ 5 mM mannose	16.69 ± 0.54	12.45 ± 0.30	11.39 ± 0.34

Polar lipid compositions of cyanobacteria are compared in Table 2. The general trends show the relative decrease of MGDG in 5mM glucose- and 5 mM mannose-grown organisms and increase of PG content under providing with both glucose and mannose (however the latter effect was much more intense in *S. platensis* than in *N. linckia*). DGDG part lowered in *Spirulina* cultivated on 50 mM glucose and 5 mM mannose but rose in mannose-grown

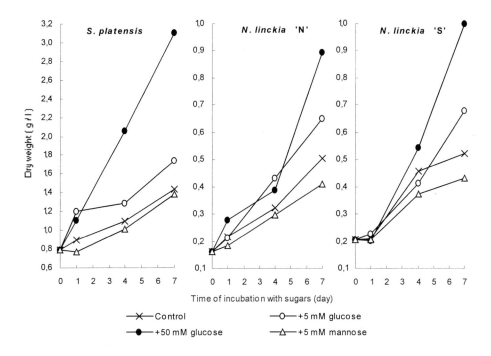

Figure 1. Growth of *Spirulina platensis* and *Nostoc linckia* in the presence of sugars.

Nostoc. The proportion of SQDG was nearly the same in all strains and didn't vary significantly under experiment conditions. *S. platensis* in comparison with *N. linckia* was characterised by somewhat lower parts of galactolipids but strikingly higher PG proportion.

The increase of PG content in the presence of glucose was shown in our laboratory earlier on another species of *Nostoc* (Mikhaylenko, Skorokhod, 1998), however increase of galactolipid content and disappearance of SQDG were observed at the same time.

Molar ratio of PG to glycolipids always rose under sugar supplementation and that increment was also much more prominent in *S. platensis*.

Summarising, it might be said that the regulatory effect of glucose on cyanobacterial cell metabolism lied not only in the inhibition of photosynthetic processes (probably via hexokinase-mediated mechanism) but also in the quantitative redistribution of polar lipids in favour of phosphatidylglycerol, the only cyanobacterial glycerolipid also inherent in extra-plastidial plant membranes. As compared with *Nostoc linckia*, *Spirulina platensis* had much higher adaptive potential resulted in the retaining of higher growth rates accompanied with significant rearrangements of pigment and polar lipid content under cultivation in the presence of sugars.

Table 2. Cyanobacterial polar lipid content after 4 days of growth with sugars.

	Polar lipids, mol%				Phospholipid / glycolipids, molar ratio
	MGDG	DGDG	SQDG	PG	
Spirulina platensis					
Control	45.8	12.3	23.1	18.8	0.23
+ 5 mM glucose	38.7	12.3	23.7	25.3	0.34
+ 50 mM glucose	41.1	8.9	21.4	28.6	0.40
+ 5 mM mannose	37.6	9.4	24.1	28.9	0.41
Nostoc linckia 'N'					
Control	50.5	14.4	27.6	7.5	0.08
+ 5 mM glucose	47.0	14.9	29.4	8.7	0.10
+ 50 mM glucose	50.3	14.5	26.1	9.1	0.10
+ 5 mM mannose	43.0	18.1	25.9	13.0	0.15
Nostoc linckia 'S'					
Control	51.8	16.1	25.2	6.9	0.07
+ 5 mM glucose	44.3	15.2	31.6	8.9	0.10
+ 50 mM glucose	51.0	15.9	26.0	7.1	0.08
+ 5 mM mannose	46.5	20.7	24.7	8.1	0.09

References

Jang, J.-C. and Sheen, J. (1994) Sugar sensing in higher plants. Plant Cell 6, 1665-1679.

Jang, J.-C., Leon, P., Zhou, L. and Sheen, J. (1997) Hexokinase as a sugar sensor in higher plants. Plant Cell 9, 5-19.

Krugman, T., Satish, N., Vinogradova, O.N, Beharav, A, Kashi, Y. and Nevo, E. (2001) Genome diversity in the cyanobacterium *Nostoc linckia* at 'Evolution Canyon', Israel, revealed by inter-HIP1 size polymorphisms. Evolutionary Ecology Research 3, 899-915.

Mikhaylenko, N.F., Skorokhod, T.Ph. (1998) Membrane lipid composition of the cyanobacteria *Nostoc punctiforme* and *Spirulina platensis* under various conditions of carbon nutrition. in G. Garab (ed.), Photosynthesis: Mechanisms and Effects, Vol. III. Kluwer Academic Publishers, Dordrecht, pp. 1831-1834.

Pego, J.V., Kortstee, A.J., Huijser, C. and Smeekens, S.C.M. (2000) Photosynthesis, sugars and the regulation of gene expression. J. Exp. Bot. 51, 407-416.

Pinevich, V.V., Verzilin, N.N. and Mikhajlov, A.A. (1970) Studies of *Spirulina platensis*, a novel object for high-intensive cultivation. Russ. Plant Physiol. 17, 1037-1046.

Schmetterer, G. (1994) Cyanobacterial respiration. in D.A. Bryant (ed.), The Molecular Biology of Cyanobacteria. Kluwer Academic Publishers, Dordrecht, pp. 409-435.

Smeekens, S. (2000) Sugar-induced signal transduction in plants. Annu. Rev. Plant Physiol. Plant Mol. Biol. 51, 49-81.

Wada, H. and Murata, N. (1998) Membrane lipids in cyanobacteria. in P.-A. Siegenthaler and N. Murata (eds.), Lipids in Photosynthesis: Structure, function and Genetics. Kluwer Academic Publishers, Dordrecht, pp. 65-81.

Wettstein, D. (1957) Experimental Cell Research, 12, 427.

Zender, A., Gorham, E. (1960) Factors influencing the growth of *Microcystis aeruginosa Kütz. emend. Elenk.* Can. J. Microbiol. 2, 195-200.

ARE GALACTOLIPIDS ENGAGED IN THE MOLECULAR MECHANISM OF THE XANTHOPHYLL CYCLE?

D. LATOWSKI[1,3], H.-E. ÅKERLUND[2], K. STRZAŁKA[1]

[1]*Department of Plant Physiology and Biochemistry, The Jan Zurzycki Institute of Molecular Biology and Biotechnology, Jagiellonian University, ul. Gronostajowa 7, 30-387 Kraków, Poland*
[2]*Department of Plant Biochemistry, University of Lund, Box117, S-221 00 Lund, Sweden*
[3]*Department of Chemistry, Pedagogical University, ul. Podchorążych 2 30-084 Kraków, Poland*

1. Introduction

Two major lipids present in the thylakoids membrane are monogalctosyldiacylglicerol (MGDG) and digalactosyldiacylglicerol (DGDG), constituting 50% and 25% of the total amount of thylakoid lipids, respectively (Yamamoto et al., 1974, Siefierman et al., 1987, Webb and Green, 1991). Although both MGDG and DGDG contain sugar residues in their molecules they create different structures in water. DGDG is characterised by high hydration level (about 50 water molecules per lipid) (Sen and Hui, 1988) and its critical packing parameter value is between 0.5 – 1 (Israelachvili and Mitchell, 1975). These features are responsible for liposomes formation in DGDG water solution. The DGDG values of the critical packing parameter and hydration level are similar to the another lipid, phosphatidylcholine (PC), which forms bilayer. On the other hand, MGDG in the water solution creates inverted hexagonal structures due to its much lower hydration level (only 5 water molecules per lipid) (Sen and Hui, 1988) and critical packing parameter value which is superior to one (Israelachvili and Mitchell, 1975). Another lipid which is characterised by similar features and thanks to them may form inverted hexagonal structures is phosphatidylethanoloamine (PE) (Sen and Hui, 1988). It is known that MGDG and PE form reversed hexagonal structures over a wide temperature range of -15°C to 80°C and this extent is highly dependent on the degree of unsaturation of their acyl chains.

MGDG was found as an indispensable lipid for get optimal activity of violaxanthin de-epoxidase (VDE) during *in vitro* studies (Yamamoto et al., 1974). VDE is one of two enzymes engaged in xanthophyll cycle, important protective mechanism existing in

N. Murata et al. (eds.), Advanced Research on Plant Lipids, 393–396.

thylakoid membranes of all higher plants, ferns, mosses and several algal groups (Stransky and Hager, 1970). That enzyme is situated in the lumen of thylakoids (Hager, 1969) and catalyses the de-epoxidation of violaxanthin to zeaxanthin *via* antheraxanthin (Yamamoto et al., 1962). VDE is activated when the thylakoids lumen pH is acidic and a suitable reducing agent (ascorbate) is available (Hager, 1969).

In present studies we have tested what features of lipids are important for VDE activity. We studied if presence of the sugar residues in lipid molecules or ability to form inverted hexagonal structures by these lipids, plays a key role in VDE activity.

2. Materials and Methods

2.1. *Chemicals*
MGDG and DGDG was purchased from Lipids Products. Egg yolk phosphatidylcholine (PC) was obtained from Sigma (P2772) and egg yolk phosphatidylethanoloamine (PE) from Fluka (60647). Violaxanthin was isolated from daffodils petals as described in (Havir et al., 1997) and saponificated (Davies, 1976) VDE isolated and purified from 7-day-old wheat leaves according to the method described by Hager and Holocher (1994).

2.2. *The liposomes preparation*
The liposomes preparation was the same as described in Latowski et al., (2000).

2.3. *^{31}P-NMR studies*
^{31}P-NMR spectra of liposomes suspended in Na-citrate buffer pH 5.1, containig 10% D_2O were recorded at 202.5 MHz using a Bruker AMX-500. Generally, a sweep width of 41.7 kHz and a repetition 2.6 s using 30° radio frequency pulses were used. The expotential multiplication of free induction decay resulted in a 100 Hz line broadenig. The number of scans was 28000. All spectra were recorded at 17° C.

2.4. *Study on VDE activity*
The enzyme activity was determined by dual-wavelength measurements (502-540 nm) using DW-2000 SLM Aminco spectrophotometer at 25°C. The reaction mixture contained 0.33 µM violaxanthin; 12.9 µM lipid in methanol (one of following: MGDG, DGDG, PC or PE), VDE and 30 mM ascorbic acid. All reagents were suspended in 0.1 M sodium citrate buffer, pH 5.1. The de-epoxidation reaction was initiated by addition of ascorbic acid.

3. Results and Discussion

A comparison of the results presented in Fig. 1 and 2 shows that violaxanthin de-epoxidation occurs both in assay mixture containing galactolipids and phospholipids. However, not all kinds of galacto- and phospholipids resulted in VDE activity. The violaxanthin conversion to zeaxanthin is observed for MGDG and PE and is not observed for DGDG and PC. These results show that it is not the sugar moiety in lipids molecules but rather sort of structures which are formed by these lipids which plays an important role in VDE activity.

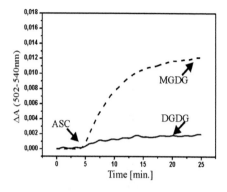

Figure 1. VDE activity when MGDG was present in assay mixture and when MGDG was replaced with DGDG

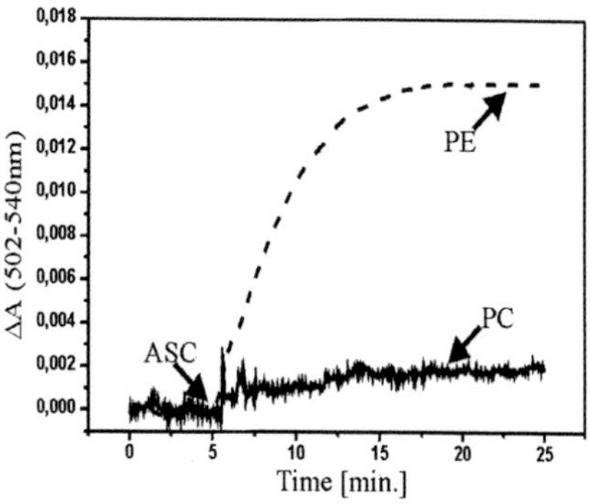

Figure 2. VDE activity when PE or PC were present in assay instead of MGDG

In this way MGDG might be necessary for VDE activity not as galactolipid but as non-bilayer prone lipid forming inverted micelles. In our opinion, inverted hexagonal structures which are created by both MGDG and PE are needed for VDE activity. The enzyme activity is higher when MGDG is replaced with PE in assay mixture. This seems to be a now finding because, up to now, MGDG was considered to be optimal lipid to obtain maximal activity of VDE. It is possible that physical or chemical parameters of PE inverted hexagonal structures are better for violaxanthin de-epoxidation than similar structures created by MGDG.

The violaxanthin conversion to zeaxanthin in PC liposomes was not observed until the enrichment of such liposomes by MGDG. The ^{31}P-NMR analysis showed appearance of the inverted hexagonal structures in PC liposomes enriched by MGDG (data not shown).

We postulate that violaxanthin de-epoxidation requires the presence of the inverted hexagonal structures which may exist in natural membranes (Quinn and Williams 1983).

4. Acknowledgement

This work was supported by grant No. 6PO4A 07321 from the committee for Scientific Research of Poland (KBN).

5. References

Davies, B.H. (1976) Carotenoids. In *Chemistry and Biochemistry of Plant Pigments* (Goodwin, T.W., ed.), Academic Press London New York San Francisco, pp.65-66.

Hager. A. (1969) Lichtbedingte pH-Erniedringung in einem Chloroplasten-Kompartiment als Ursache der Enzymatichen Violaxanthin-zu Zeaxanthin – Umwandlung Beziehungen zur Photophosphorylierung. *Planta* **89**, 224-243.

Hager, A. and Holocher, K. (1994) Localization of the xanthophyll cycle enzyme violaxanthin de-epoxidase within the thylakoid lumen and abolition of its mobility by a (light-dependent) pH decrease. *Planta,* **192**, 581-589.

Havir, E.A., Tausta, L.S. and Peterson, R.B. (1997) Purification and properties of violaxanthin de-epoxidase from spinach. *Plant Science* **123**, 57-66.

Israelachvili, J.N. and Mitchell, D.J. (1975) A model for the packing of lipids in bilayer membranes. *Biochim. Biophys. Acta.* **389**, 13-19.

Latowski, D., Kostecka, A. and Strzałka, K. (2000) Effect of monogalactosyldiacylglycerol and other thylakoid lipids on violaxanthin de-epoxidation in liposomes. *Bioch. Soc. Trans.* Vol. **28** Part. **6**, 810-812.

Sen, A. and Hui, S.-W. (1988) Direct measurement of headgroup hydration of polar lipids in inverted micelles. Chem. Phys. Lipids 49, 179-184.

Siefierman-Harms, D., Ninnemann, H. and Yamamoto, H.Y. (1987) Reassembly of solubilized chlorophyll-protein complexes in proteolipid particles-comparison of monogalactosyldiacylglycerol and two phospholipids. *Biochim. Biophys. Acta,* **892**, 303-313.

Stransky, H. and Hager, A. (1970) The carotenoid pattern and the occurence of the light induced xanthophyll cycle in various classes of algae part 6 chemo systematic study. *Arch. Microbiol.* **73**, 315 - 323.

Quinn, P.J. and Williams W.P. (1983) The structural role of lipids in photosynthetic membranes. *Biochim. Biophys. Acta.* **737**, 223-266.

Webb, M. and Green, B. (1991) Biochemical and biophysical properties of thylakoid acyl lipids. *Biochim. Biophys. Acta,* **1060**, 133-158.

Yamamoto, H.Y., Chenchin, E.E. and Yamada, D.K. (1974) Effect of chloroplast lipids on violaxanthin de-epoxidase activity. *Proc. 3rd Int. Cong. Photosyn., Elsevier, Amsterdam pp.1999-2006.*

Yamamoto, H.Y., Nakayama, T.O.H. & Chichester, C.O. (1962) Studies on the light and dark interconversions of leaf xanthophylls. *Arch. Biochem. Biophys.* **97**, 168-173.

Chapter 10:

Biotechnology

LIPID SYNTHESIS CATALYZED BY LIPASE IMMOBILIZED IN BIOCOMPATIBLE MICROEMULSION-BASED ORGANOGEL SYSTEMS

MASANAO IMAI
Department of Food Science and Technology,
College of Bioresource Sciences, Nihon University
1866 Kameino, Fujisawa, Kanagawa 252-8510, Japan
KAZUHITO NAGAYAMA and KANA FUKUYAMA
Department of Materials Science and Engineering,
Kochi National College of Technology
200-1 Monobe, Nankoku, Kochi 783-8508, Japan

1. Introduction

Microemulsion-based organogels (MBGs) formed by the addition of natural biopolymer become of interest as novel tools of enzyme immobilization for hydrophobic reactions. Because such gel matrices are effective in reactions involving interfacially active enzyme, they can be advantageously applied to lipid-transformations catalyzed by lipase (Jenta et al., 1997; Nagayama et al., 1998).

The activity of lipase has been reported to be influenced by the compositional changes of the MBG and to be stable over a long period (Jenta et al., 1997; Nagayama et al., 2002).

In developing MBG system catalytic processes, the use of phospholipids as amphiphile molecule is one of the main factors that should be considered. However, there have been only a few reports on the reactivity of enzyme immobilized in the MBGs based on phospholipid and natural biopolymer (Backlund et al., 1996; Stamatis and Xenakis, 1999).

In the present study, we demonstrate esterification in phospholipids microemulsion-based organogels containing lipase. Lipase activity is examined as a function of the water and gelatin concentrations in the MBG. The operational stability of lipase is also investigated in the repeated batch reactions.

2. Materials and Methods

2.1. Chemicals

N. Murata et al. (eds.), Advanced Research on Plant Lipids, 399–402.

Lecithin from soybean (EPIKURON 200) was supplied by Nihon Sieber Hegner (Tokyo) and Lucas Meyer (Hamburg). Lipase (from *Candida rugosa*, Type VII, 840 units/mg) and gelatin (from bovine skin, Type B, Bloom 225) were purchased from Sigma (St. Louis, MO). 2,2,4-Trimethylpentane (isooctane) was obtained by Wako Pure Chemical Industries (Osaka). Lauric acid and butyl alcohol were also from Wako. All other chemicals used were of analytical reagent grade.

2.2. *Preparation of microemulsion-based organogels (MBGs)*

The microemulsion phase was prepared by the addition of lipase solution to isooctane (2 cm^3) containing the desired lecithin concentration. The buffer consisted of 50 mM KH2PO4—50 mM NaOH (pH 7.0). Gelatin became swollen after the addition of buffer. The microemulsion and gelatin phases were individually heated at 328 K. These solutions were vigorously mixed until homogeneous dispersion was achieved, and the mixture was then cooled to room temperature (298±2 K). The MBGs thus obtained were stored at 253 K. The water content in the microemulsion phase was determined using a Karl Fisher titrator (MKS-1s; Kyoto Electronics, Kyoto).

2.3. *Reaction procedure*

The prepared MBGs (4 cm^3) were cut into cubes of approximately 2 mm by using a razor blade, and then immersed into 10 cm^3 of isooctane with magnetic stirring (4 s^{-1}) at 298 K. The reaction was initiated with the addition of 10 cm^3 isooctane containing the desired concentrations of substrates.

Lipase activity was determined from the region of the linear increase of butyl laurate concentration. Butyl laurate concentration was analyzed by gas chromatography (GC-8A; Shimadzu, Kyoto) using a glass column packed with FFAP/Uniport S (60/80 mesh, GL Sciences, Tokyo) and a TCD detector.

3. Results and discussion

3.1. *Effect of water content on reaction rate*

Figure 1 shows the effect of water content in the MBG on the reaction rate. The water content in the MBG was defined as a sum of water content in the microemulsion and in gelatin phase. When the water content was less than 50% v/v, the MBGs could not be formed; instead of dispersion, the gelatin separated into two phases.

The reaction rate increased with increasing water content, reaching a maximum at a water content of 70% v/v. The optimal water content in the lecithin-MBG system was about 2-fold higher than that in AOT-MBG system (Nagayama et al., 2002).

The structure of the lecithin MBG has proposed as a rigid gelatin network surrounded by water channels intertwined with oil channels separated by the lecithin film (Backlund et al., 1996). Although the relationship between the gel structure and the MBG composition has yet to be elucidated, the lipase activity seems to be affected by

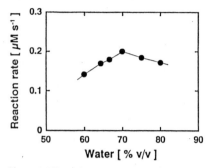

Figure 1 Effect of water content on reaction rate. Lauric acid = 100 mM; butyl alcohol = 100 mM; gelatin = 22.5% w/v; lecithin = 18 mM.

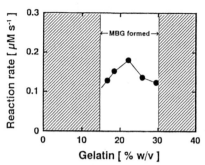

Figure 2 Effect of gelatin content on reaction rate. Lauric acid = 100 mM; butyl alcohol = 100 mM; water = 70% v/v; lecithin = 18 mM.

the change of gel structure depending on the water content.

3.2. Effect of gelatin content on reaction rate

Figure 2 shows the effect of gelatin content in the MBG on the reaction rate. For gelling of the lecithin microemulsion phase, the MBGs could be formed in the gelatin content of 15 to 30% w/v. The reaction rate increased with increasing gelatin content, reaching a maximum at a gelatin content of 22.5% w/v. The lipase activity was also considered to differ by the change of gel structure depending on the gelatin content.

3.3. Operational stability of immobilized lipase

The operational stability of the immobilized lipase during reuse in batch reactions is shown in Figure 3.

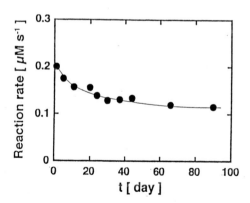

Figure 3 Reaction rate in reuse of the immobilized lipase in batch reactions. Lauric acid = 100 mM; butyl alcohol = 100 mM; water = 70% v/v; gelatin = 22.5% w/v; lecithin = 18 mM.

Each batch reactions were continued for 24 h. After each reaction, MBGs were washed five times with fresh isooctane (20 cm^3) and the reused.

As shown in the figure, the lipase activity was successfully maintained for 90 days. This result indicates that the lecithin MBG system is effective as immobilization carrier of lipase for the esterification reaction.

Water molecules, which are formed as a by-product in the reaction progresses, were mainly accumulated in the MBGs. Hydrophobic-hydrophilic balance in immobilization carrier plays a significant role on the enzyme reactivity. When the accumulation of water within the MBGs is gradually enhanced the hydrophilicity around lipase immobilized, the lipase activity decreased with repeating the batch reactions.

4. Conclusions

Soybean lecithin MBGs containing *Candida rugosa* lipase were used as a catalyst for the esterification of lauric acid with butyl alcohol. The reaction rate was found to be highest at a water content of 70% v/v and a gelatin content of 22.5% w/v. The lipase activity was successfully preserved for 90 days.

5. Acknowledgements

This work was supported by a Grant-in Aid for Encouragement of Young Scientists (No.13750720) from the Japan Society for the Promotion Science. The authors are grateful to Nihon Sieber Hegner K.K. and Lucas Meyer Co. for providing lecithin from soybean.

6. References

Backlund, S., Eriksson, F., Hedstrom, G., Laine, A. and Rantala M. (1996) Lipase-catalyzed enantioselective eserification using different microemulsion-based gels. Colloid Polym. Sci. 274, 540-547.

Jenta, T.R.J., Batts, G., Rees, G.D. and Robinson, B.H. (1997) Biocatalysis using gelatin microemulsion-based organogels containing immobilized *Chromobacterium viscosum* lipase. Biotechnol. Bioeng. 53, 121-131.

Nagayama, K., Karaiwa, K. and Imai, M. (1998) Esterification activity and stability of *Candida rugosa* lipase in AOT microemulsion-based organogels. Biochem. Eng. J. 2, 121-126.

Nagayama, K., Karaiwa, K., Ueta, A. and Imai, M. (2002) Reaction conditions and continuous mixed-flow esterification reactions by *Candida rugosa* lipase immobilized in AOT microemulsion-based organogels. Seibutsu-kogaku 80, 95-101. [in Japanese]

Stamatis, H. and Xenakis, A. (1999) Biocatalysis using microemulsion-based polymer gels containing lipase. J. Mol. Catal. B:Enzym. 6, 399-406.

PRODUCTION OF NUTRACEUTICAL FATTY ACIDS IN OILSEED CROPS

XIAO QIU[1*], HAIPING HONG[1], NAGAMANI DATLA[1], DARWIN W. REED[2], MARTIN TRUKSA[1], ZHIYUAN HU[1], PATRICK S.COVELLO[2], SAMUEL L. MACKENZIE[2]
[1] Research & Development, Bioriginal Food & Science Corporation, 102 Melville Street, Saskatoon, SK, Canada S7J 0R1
[2] National Research Council of Canada, Plant Biotechnology Institute, 110 Gymnasium Place, Saskatoon, SK, Canada S7N 0W9
[*] For Correspondence: Xiao.Qiu@nrc.ca; 306-975-9558

Nutraceutical fatty acids refer to those fatty acids that are not usually associated with food, but appropriate dietary supplements of those fatty acids can promote our health or provide protection against some diseases. Based on this tentative definition, long chain polyunsaturated fatty acids (LCPUFA) and some fatty acids with conjugated double bonds would stay in this category. Polyunsaturated fatty acids contain two or more double bonds and have long been recognized as essential nutrients for mammals (Gill & Valivety, 1997; Spector 1999). For instance, docosahexaenoic acid (DHA), rich in the cells of mammalian retina and brain tissues, was shown to be essential for the normal function of human eye and brain, as well as beneficial for patients who suffer from other diseases (Horrocks and Yeo, 1999; Uauy et al, 2001). Conjugated fatty acids are newly recognized nutraceutical compounds, within which, conjugated linoleic acid (CLA) is presently the most popular. Dietary supplements of this fatty acid was shown to inhibit chemical-induced skin and stomach cancers, reduce development of atherosclerosis and enhance immune function in mammals (Pariza, 1997).

There are hundreds of different fatty acids identified in nature, of which many have potentials for nutraceutical and pharmaceutical uses. However, these fatty acids mostly occur in wild plant species or oily microorganisms that are not readily cultivated or cultured. The oil content and yields are low. Comparatively, the cultivated oilseed crops have high oil content and seed yield, but only produce a limited set of fatty acids that usually contain less than two double bonds and any polyunsaturates are in the *cis* and methylene-interrupted configuration.

Transgenic plants are now considered an efficient way to produce biological compounds. The cost-efficiency and little side-effect have recently attracted much attention to production of various compounds in plants (Somerville and Bonetta, 2001). Utilisation of oilseed crops to produce the nutraceutical fatty acids is economically attractive (Abbadi et al, 2001).

N. Murata et al. (eds.), Advanced Research on Plant Lipids, 403–406.
© 2003 *Kluwer Academic Publishers. Printed in the Netherlands.*

Our goal is to produce nutraceutical fatty acids in oilseed crops especially in flax. Flax is one of the most important oilseeds worldwide and is one of the major oilseed crops in Canada. The traditional linseed oil, an excellent drying oil used in paints, vanishes, linoleum and printing, contains about 50% to 70 % of linolenic acid. A new type of flaxseed oil, solin oil, has recently been exploited as the edible oil, which contains approximately 70% linoleic acid. These two types of flaxseed oil provide the appropriate substrates for production of the two series of essential fatty acids.

As the first step towards our goal, we put a lot of time and energy into the establishment and refinement of the enabling technologies. This included improvement of the flax transformation system and creation of an efficient flax expression cassette. We then started to identify genes involved in the biosynthesis of nutraceutical fatty acids and transfered them into flax plants to produce the target fatty acids.

Gamma-linolenic acid (GLA) is one of essential fatty acids for mammals and has been widely used in the marketplace of nutraceutical and pharmaceutical industries (Gunstone, 1992; Huang and Ziboh, 2001). It is synthesized by a $\Delta 6$ desaturase, which converts linoleic acid (LA, 18:2-9,12) to GLA (18:3-6,9,12), as well as alpha-linolenic acid (ALA, 18:3-9,12,15) to stearidonic acid (SDA, 18:4-6,9,12,15). At present, the predominant sources of GLA are oils from plants such as evening primrose (*Oenothera biennis*), borage (*B. officinalis*) and black currant (*Ribes nigrum*) (Phillips and Huang, 1996). However, these GLA sources are not ideal for dietary supplementation because these plant species are difficult to grow and yields are low in Canadian prairie when compared to typical oil seed crops.

Production of GLA in flax was first attempted by using a borage $\Delta 6$ desaturase. We transformed two flax genotypes with the binary construct that contains the borage $\Delta 6$ desaturase cDNA under the control of a heterologous seed-specific promoter (*B. napus* napin storage protein promoter). Transgenics analysis showed that the GLA level is low, only about 0.1 to 2% of the total fatty acids in seeds. We believed that low level production of $\Delta 6$ desaturated fatty acids in transgenic flax might be either due to the low activity of the enzyme or due to the low level expression of the gene since *B. napus* napin promoter might not work optimally in flax (Qiu et al, 2002).

Then we isolated a cDNA encoding a $\Delta 6$ desaturase from *Pythium irregular* and introduced this gene into flax and *Brassica juncea*. In flax transgenics with a heterologous seed-specific promoter, of which the majority exhibited two new fatty acids, GLA and SDA, but levels are still low, constituting 0.1 to 4.3 % of total fatty acids in seeds. Whereas in transgenics with a flax native seed-specific promoter, the GLA content was much higher, representing up to 20 % of the total fatty acids in the transgenic seeds (Fig 1). Moreover, expression of the desaturase in *Brassica juncea* under the control of the napin promoter resulted in very high level production of GLA, accounting for up to 40 % of the total seed fatty acids. The GLA is almost exclusively incorporated into triacylglycerol with only trace amounts found in the other lipids. (Hong et al, 2002).

Eicosanpentaenoic acids (EPA) and arachidonic acids (AA) are both $\Delta 5$ essential fatty acids. Beneficial effects of these fatty acids on reduction of the risk of cardiovascular disease and arterial thrombosis have well been documented. In addition some other favorable effects on arthritis, renal disorders, diabetes and various cancers were also reported (Das 1990; Babcock et al, 2000).

We identified two of the Δ5 desaturases responsible for EPA and AA biosynthesis respectively from *Pythium* and *Thraustochytrium*. Expression of the *Pythium* desaturase under the control of native seed-specific promoter resulted in accumulating up to 10% of Δ5-unsaturated polymethylene-interrupted fatty acids (Δ5-UPIFAs) in seeds. Moreover, when the substrate 20:3(8,11,14) was exogenously supplied, transgenic plants could convert the substrate into AA. These data suggest the possibility that Δ5 desaturated fatty acids can be produced in oilseed crops on a large scale.

CLA is a newly recognized nutraceutical compound, which has recently drawn attention of both pharmaceutical and nutraceutical industries (Haumann, 1996; Pariza, 1997). With realization of the benefits of CLAs, market demand for the product is growing. Unfortunately, there is no rich natural source for CLAs. Linoleic acid can be converted to CLA by chemical methods (Chen et al, 1999), but CLA derived from the chemical process is a mixture of several regiospecific and stereospecific isomers.

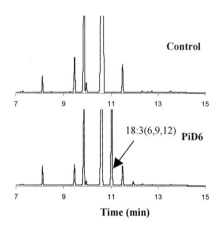

Figure 1: Gas chromatography analysis of fatty acid esters of transgenic flax expressing a Δ6 desaturase (PiD6) from *Pythium irregulare*.

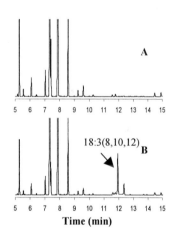

Figure 2: Gas chromatography analysis of fatty acid esters of transgenic *Brassica juncea* expressing a conjugase (CoFac2) from *Calendula officinalis*. A, control; B, CoFac2.

By using a PCR-based cloning strategy, we identified the cDNA, *CoFac2*, from *Calendula officinalis* developing seeds that encodes a fatty acid conjugase. When expressed in yeast, unexpectedly, we found it could use both mono and di-unsaturated fatty acids as substrates, resulting in conjugated di- and tri-enoic products, respectively. Because the only substrates available in yeast are Δ9 mono-unsaturates the CLA produced by CoFac2 is an unusual isomer with two conjugated double bonds at the 8 and 10 positions (Qiu et al, 2001). This finding suggests the possibility of production of this unusual fatty acid in plants. Introducing *CoFad2*, under the control of a seed specific promoter, into *Brassica juncea* was therefore attempted. The results indicated, however, that when *CoFad2* was provided the correct substrate in the transgenic seeds it only produced conjugated linolenic acid 18:3-8,10,12 (Fig 2) in seeds.

References:

Abbadi, A., Dormergue, F., Meyer, A., Riedel, K., Sperling, P., Zank T.K. and Heinz, E. (2001) Transgenic oilseeds as sustainable source of nutritionally relevant C20 and C22 polyunsaturated fatty acids? Eur. J. Sci. Technol. 103:106-113.

Babcock, T., Helton, W.S. and Espat, N.J. 2000. Eicosapentaenoic acid (EPA): an antiinflammatory omega-3 fat with potential clinical applications. Nutrition 16: 1116-1118.

Chen, C.A., Lu, W. and Sih, C.J. (1999) Synthesis of 9Z, 11E-octadecadienoic and 10E, 12Z-octadecadienoic acids, the major components of conjugated linoleic acid. Lipids 34: 879-884.

Das, U.N. (1990) Gamma-linolenic acid, arachidonic acid, and eicosapentaenoic acid as potential anticancer drugs. Nutrition 6: 429-434.

Gill, I. and Valivety, R. 1997. Polyunsaturated fatty acids, part 1: occurrence, biological activities and applications. TIBTECH 15: 401-409.

Haumann, B.F. (1996) Conjugated linoleic acid offers research promise. Inform 7: 152-159.

Hong, H.P., Datla, N., Reed, D.R., Covello, P.S., MacKenzie, S.L. and Qiu, X. (2002) High level production of γ-linolenic acid in Brassica juncea using a Δ6 desaturase from *Pythium irregulare*. Plant Physiol (in press).

Horrobin, D.F. (1992) Nutritional and medical importance of gamma-linolenic acid. Prog. Lipid Res. 31: 163-194.

Horrocks, L. A. & Yeo, Y. K. Health benefit of docosahexaenoic acid. *Pharmacol. Res.* 40, 211-215 (1999)

Huang, Y.S. and Milles, D.E. (1996) Gamma-linolenic acid: Metabolism and its roles in nutrition and medicine. AOCS Press, Champaign, Illinois.

Huang, Y.S. and Ziboh, A. (2001) Gamma-linolenic acid: recent advances in biotechnology and clinical applications. AOCS Press, Champaign, Illinois.

Pariza, M.W. (1997) Conjugated linoleic acid, a newly recognized nutrient. Chemistry & Industry 16: 464-466.

Qiu, X., Hong, H.P., Datla, N., MacKenzie, S.L., Tayler, C.D., Thomas, L.T. 2002. Expression of borage Δ-6 desaturase in *Saccharomyces cerevisiae* and oilseed crops. Can. J. Bot. 80:42-49.

Qiu, X., Reed, D.W., Hong, H.P., Mackenzie, S.L. and Covello, P.S. 2001. Identification and analysis of a gene from *Calendula officinalis* encoding a fatty acid conjugase. Plant Physiol. 125: 847-855.

Somerville, C.R. and Bonetta, D. (2001) Plants as factories for technical materials. Plant Physiol 125: 168-171.

Spector A.A. Essentiality of fatty acids. Lipids 34:S1-S3.

Uauy, R., Hoffman, D. R., Peirano, P., Birch, D.G. & Birch, E. E. Essential fatty acids in visual and brain development. *Lipids* 36, 885-95 (2001).

APPLICATION OF hpRNA-MEDIATED GENE SILENCING TECHNIQUES TO MODIFICATION OF FATTY ACID COMPOSITION

QING LIU, CLIVE HURLSTONE, SURINDER SINGH AND ALLAN GREEN
CSIRO Plant Industry, GPO Box 1600, ACT 2601, Australia

1. Introduction

The down-regulation of fatty acid desaturase enzyme activity by antisense or cosuppression approaches has been successful in genetic modification of fatty acid composition in a number of oilseed crops. Improvements in the efficacy of gene silencing have been made by using specifically designed genetic constructs containing inverted-repeats (IR) that encode RNAs having regions of self-complementarity. These constructs can reliably generate hairpin RNA (hpRNA) transcripts that invoke sequence-specific RNA degradation targeted to the double-stranded region of hpRNA and to homologous endogenous mRNA molecules. We have previously reported the use of this technology to genetically modify the fatty acid composition of cottonseed oil in order to improve its nutritional value and functional properties (Liu et al., 2000). As a result of transforming cotton with a hpRNA construct expressing inverted repeat of cotton microsomal $\Delta12$-desaturase (*ghFAD2-1*) the oleic acid level in cottonseed oil increased from normally 15% up to 78%, mainly at the expense of linoleic acid which is reduced down to as low as 4%. Likewise, the silencing of the cotton $\Delta9$-desaturase (*ghSAD-1*) by a similar approach, led to an increase in stearic acid level in cottonseed oil from normally 2% to as high as 40%. Here we report the further development of the homozygous high-oleic (HO), high-stearic (HS) lines and their hybrids. The inheritance of these novel traits has also been studied. A new generation of hpRNA constructs aimed at further improving the silencing efficacy and specificity, as well as simultaneously silencing multiple genes, are also described.

2. Results and Discussion

2.1 Inheritance of the HO and HS traits

T$_2$ seed population were examined from a range of independently derived primary transformants (T$_1$ plants) showing the HO or HS traits. Among the HO lines, two distinct segregation patterns were evident. In about three quarters of the HO lines, both homozygous and heterozygous seeds contained over 70% oleic acid, indicating the near-

N. Murata et al. (eds.), Advanced Research on Plant Lipids, 407–410.

maximum suppression of the *ghFAD2-1* gene. It is apparent that the silencing effect is inherited as a completely dominant trait in these lines. In the rest of the HO lines, the accumulation of oleic acid in the T_2 seeds revealed a continuous distribution between the wild type value and that of the maximum suppression of the target gene. Several independent transgenic HO lines of the first type were bred to homozygosity and the novel fatty acid composition was found to be stably inherited. In contrast, the segregation pattern of the high-stearic trait in the HS lines appeared to be more complex with a relatively wide variation among individual seeds. A similar range in stearic acid levels was observed between individual seeds in populations that were homozygous for the transgene, suggesting that the variation is non-genetic.

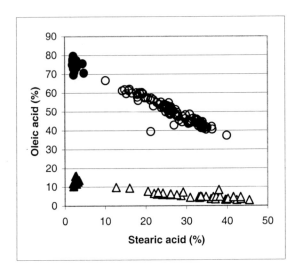

Fig. 1. Stearic and oleic acid accumulation in an F_2 seed population derived from a cross between the HO and HS lines. Null seeds (▲), seeds containing only the *ghSAD-1*-IR transgene (△), seeds containing only the *ghFAD2-1*-IR transgene (●) and seeds containing both *ghSAD-1*-IR and *ghFAD2-1*-IR transgenes (○).

By intercrossing the HS and HO genotypes it was possible to simultaneously down-regulate both *ghFAD2-1* and *ghSAD-1*, demonstrating that hpRNA induced PTGS in independent genes can be combined without any diminution in the degree of silencing. In such an F_2 population (Fig. 1), all the high-oleic seeds, both homozygous and heterozygous, contained similar fatty acid composition. High-stearic seeds, however, displayed greater variations of stearic acid level. The relatively high variation in fatty acid composition of these seeds are likely due to the high variation in stearic acid rather than the oleic acid. Nevertheless, the total amount of stearic acid and oleic acid was a constant figure, about 77%. This reflects the substrate and product relationship between these two fatty acids. The variation of fatty acid composition in these hybrids was persistent in subsequent generations where both transgenes were homozygous.

A common feature of various PTGS phenomena is a specific decrease in the level of mRNAs of both the introduced transgene and the homologous target gene. This is demonstrated in the Northern blot analysis (Fig. 2).

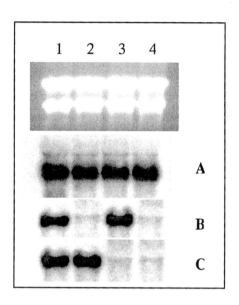

Fig. 2. Northern blot analysis of *ghFAD2-1* and *ghSAD-1* expression in middle maturity cotton embryos. Lane 1, WT; lane 2, *ghSAD-1*-IR; lane 3, *ghFAD2-1*-IR; lane 4, *ghSAD-1*-IR x *ghFAD2-1*-IR. A cotton KASII gene was used as a control, showing similar expression levels among the four genotypes (A). *ghSAD-1* mRNA (B) was drastically reduced in the *ghSAD-1*-IR line, and the *ghFAD2-1* mRNA (C) was similarly reduced in the *ghFAD2-1*-IR line. Similar reductions of both desaturase mRNAs (B&C) were demonstrated in the hybrid.

2.2 Increased Silencing Efficacy by Using an ihpRNA Construct

A modification of the hpRNA construct was made by replacing the spacer between the two inverted repeats with a spliceable intron. The inclusion of an intron in the transgene leads to enhanced expression and therefore increased efficacy of gene silencing (Smith et al., 2000). To differentiate from the hpRNA construct, we named this type of constructs ihpRNA constructs. We have also used gene-specific fragments, such as the 5' untranslated region as the inverted repeat. Such a design is aimed at increasing the specificity of silencing towards a targeted member of a multigene family, while not affecting the expression of other members of the same gene family.

Using an ihpRNA construct containing inverted repeats of a 98-bp 5' UTR from the *ghFAD2-1* gene separated by a 1.1-kb *ghFAD2-1* intron (Fig. 3), we have produced lines with oleic acid content as high as those achieved by hpRNA construct. The ihpRNA strategy, however, achieved a higher frequency of silencing and all the high-oleic lines appear to have the maximum suppression of the *ghFAD2-1* gene.

410

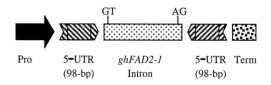

Pro   5′-UTR   *ghFAD2-1*   5′-UTR   Term
          (98-bp)   Intron   (98-bp)

Fig. 3. Schematic illustration of the ihpRNA construct consisting of a pair of 98-bp inverted repeats of the 5′ UTR of *ghFAD2-1* gene separated by the *ghFAD2-1* intron (~1,100-bp) that was used to construct the silencing chimeric gene.

2.3 Simultaneous Silencing of Multiple Genes with a Single ihpRNA Construct

The combination of transgenes silencing a number of different target genes can be achieved by sexual crossing, cotransformation or retransformation, but these are laborious processes. Furthermore, with these methods the transgenes are inserted at independent locations and the combination may breakdown by segregation in the progeny and in subsequent crosses. An alternative strategy is to use a single chimeric transgene to simultaneously suppress multiple genes by fusing partial sequences together. We are currently evaluating the effectiveness of this approach to down-regulate the cotton fatty acid thioesterase gene (*ghFatB-1*) in combination with *ghSAD-1* and *ghFAD2-1* silencing. Approximately 300-bp cDNA fragments from each of three genes, in the order of *ghSAD-1, ghFAD2-1* and *ghFatB-1*, were fused together as a whole unit. In the binary construct, this chimeric gene is separated from its inverted repeat by a spliceable intron derived from *ghFAD2-1*. This approach is aimed at further reducing the palmitic acid level in the high-oleic and high-stearic germplasm.

3. Conclusions

The stable inheritance of the HO and HS traits gained through the hpRNA mediated PTGS has been demonstrated in selfed and crossed progeny. The inclusion of an intron in the inverted repeat constructs (ihpRNA construct) targeted against the *ghFAD2-1* can dramatically increase the silencing efficacy. The use of short gene-specific UTR fragments of the target genes enhances the specificity of gene silencing and may make it possible to silence multiple genes simultaneously using a single chimeric ihpRNA construct.

4. References

Liu, Q., Singh, S. and Green, A. (2000) Genetic modification of cottonseed oil using inverted repeat gene silencing techniques. Biochemical Society Transactions 28, 927-929.

Smith, N., Singh, S. Wang, M., Stoutjesdijk, P., Green, A. and Waterhouse, P. (2000) Total silencing by intron-spliced RNAs. Nature 407: 319-320.

TARGETING GENE EXPRESSION IN MESOCARP AND KERNEL OF THE OIL PALM FRUITS

A. SITI NOR AKMAR, R. ZUBAIDAH,
M.A. MANAF and W. NURNIWALIS
*Advanced Biotechnology and Breeding Centre,
Malaysian Palm Oil Board, P.O. Box 10620,
50720 Kuala Lumpur, Malaysia*

Introduction

Production of specialty oils for industrial applications would be a very attractive proposition for the oil palm since it is the most productive oil crops (Sambanthamurthi et al., 2000). Palm fruits have two storage tissues; mesocarp and kernel that can be the target for producing genetically manipulated products. The mesocarp and kernel oil differ in fatty acid composition as well as the period at which the oil accumulates during oil palm fruit development. A study on the regulation of gene expression during period of oil synthesis in both tissues is of interest to serve as background information for genetic manipulation. Progress has also been made in the isolation of tissue-specific promoters for targeting gene expression to either of these tissues.

Expression Profile of Mesocarp and Kernel-specific cDNAs

Stearoyl-ACP desaturase (SAD), a fatty acid biosynthetic gene is use as a model in our study of gene expression profile in mesocarp and kernel tissues during period of oil synthesis. The regulation of SAD at both transcriptional and post-transcriptional levels at different stages of mesocarp and kernel development were analysed using Northern blot analysis with gene-specific probe and with Western blot analysis using antibodies raised against the oil palm SAD. It was found that the expression of SAD1 is induced in mesocarp and kernel tissues in phase with oil synthesis. This is followed by an increase in enzyme levels suggesting transcriptional control of gene expression plays an important role (Siti Nor Akmar et al., 1999). In the efforts to isolate promoter sequences for targeting gene expression to the mesocarp or kernel tissues, several cDNAs with selective

N. Murata et al. (eds.), Advanced Research on Plant Lipids, 411–414.
© 2003 *Kluwer Academic Publishers. Printed in the Netherlands.*

expression in these tissues have been isolated via various approaches including differential screening (Siti Nor Akmar et al., 1995), subtractive hybridization (Siti Nor Akmar et al., 2002), differential display and random sequencing in combination with transcript profiling of expressed sequence tag (ESTs). Some of these cDNAs showed strong correlation in expression pattern with SAD1 (Fig. 1).

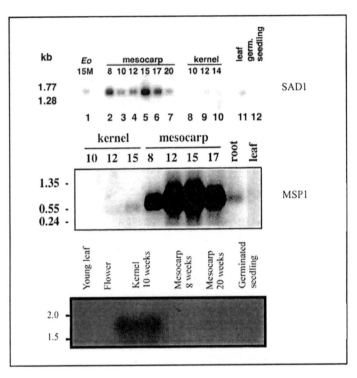

Figure 1. Expression profile of stearoyl-ACP desaturase (SAD1), MSP1 and glutelin genes in various tissues of the oil palm

Detailed studies at the cellular levels of the expression pattern of SAD and the mesocarp and kernel-specific genes have been carried out using RNA *in situ* hybridization. It was shown that the transcripts of the mesocarp specific cDNA, MSP1 are mainly localized in the parenchyma cells surrounding the vascular bundles and this paralleled that observed for the SAD transcripts. The transcripts for the kernel specific glutelin gene are uniformly distributed in the parenchyma cells of the endosperm tissue.

Transient Assay Analysis of Mesocarp-specific Promoter

The promoter sequences of a mesocarp-specific gene as well as a kernel-specific glutelin gene have been isolated and fully sequenced. Mesocarp-specific activity of the promoter corresponding to the mesocarp-specific gene, MSP1 was analysed in a transient assay system on oil palm tissue slices using β-glucuronidase (GUS) and green fluorescence protein (GFP) as reporter genes. The activity of MSP1 promoter was also compared with that of the constitutive cauliflower mosaic virus (CaMV) 35S promoter. It was shown that blue GUS spots can be detected on both mesocarp and leaf tissues bombarded with contruct containing CaMV 35S promoter, however when the CaMV 35S promoter was replaced with the MSP1 promoter, activity was only detected in the mesocarp tissues (Fig. 2). The result indicating mesocarp-specific activity of MSP1 promoter was further confirmed using GFP as the reporter gene.

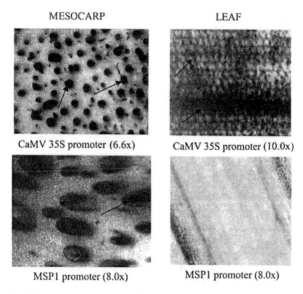

MESOCARP　　　　　　LEAF

CaMV 35S promoter (6.6x)　　　CaMV 35S promoter (10.0x)

MSP1 promoter (8.0x)　　　MSP1 promoter (8.0x)

Figure 2. GUS expression driven by MSP1 promoter versus CaMV 35S promoter

Transformation Vectors containing Mesocarp-specific Promoter

Backbone transformation vectors containing the mesocarp-specific promoter, nopaline synthase (NOS) terminator without (pMP1) and with (pMP2) plastid targeting sequence from oil palm SAD1 gene were produced (Fig. 3). These can be

414

used for insertion of gene of interest for transforming oil palm to alter palm oil composition. In this construct a rare cutter site (*Asc* I) was inserted to serve as the site for insertion of target gene. The construct containing plastid-targeting sequence is specially designed for introducing fatty acid biosynthetic without plastid targeting sequence or for inserting desired gene in an antisense orientation.

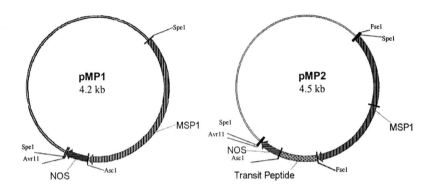

Figure 3. Backbone vectors pMP1 and pMP2 for oil palm transformation. pMP1 contains mesocarp-specific promoter (MSP1), NOS terminator and a rare cutter (*Asc* I) for insertion of target gene. pMP2 is similar to pMP1 with the addition of a plastid targeting sequence from oil palm stearoyl-ACP desaturase gene (SAD1).

Acknowledgement

We thank the Director General of MPOB for permission to publish this paper and the contributions of the Technical Assistances of Gene Expression Laboratory.

References

Sambanthamurthi, R., Parveez G.K.A and Cheah S.C. (2000). Genetic engineering of oil palm in Yusof Basiron, B.S. Jalani and K.W Chan (eds.), Advances in Oil Palm Research. Malaysian Palm Oil Board, pp. 284-331

Siti Nor Akmar, A., Cheah, S.C. and Murphy, D.J. (2002). Isolation and characterization of two divergent type 3 metallothioneins from oil palm, *Elaeis guineensis*. Plant Physiol. Biochem. 40, 255-263.

Siti Nor Akmar, A., Cheah, S.C., Aminah, S., Leslie, C.L.O, Sambanthamurthi, R. and Murphy, D.J. (1999). Characterization and regulation of the oil palm (*Elaeis guineensis*) stearoyl-ACP desaturase genes. J. of Oil palm Res. (Special Issue), 1-17.

Siti Nor Akmar, A., Shah, F.H. and Cheah, S.C. (1995). Construction of oil palm mesocarp cDNA library and the isolation of mesocarp-specific cDNA clones. Asia Pacific Journal of Molecular biology and Biotehnonology 3 No. 2, 106-111.

TRANSGENIC OIL PALM: WHERE ARE WE?

G.K.A. PARVEEZ[1], R. OMAR A.M.Y. MASANI, H.F. HALIZA, A.M. NA'IMATULAPIDAH, A.D. KUSHAIRI, A.H. TARMIZI AND I. ZAMZURI.
Advanced Biotechnology and Breeding Centre, Malaysian Palm Oil Board (MPOB), P.O. Box 10620, 50720 Kuala Lumpur, Malaysia. [1](corresponding author: parveez@mpob.gov.my)

1. Introduction

The challenge that the oil palm industry will face in the 21st century is the ability to maintain profitability in the face of labor shortage and limited land resources. At present, palm oil contributes to around 20% of world oils and fats production. It is envisaged that the demand for oil will grow faster then the rise in supply. By the year 2020, it is expected that nearly 26% of the world's oils and fats demand will be met by palm oil. It is anticipated that palm oil will capture approximately 50% of the world's oils and fats trade (Rajanaidu and Jalani, 1995). Due to this projected demand, it is important to increase the yield of oil palm as well as to improve the palm oil quality at a faster rate than has been achieved by conventional breeding (Parveez, 1998). MPOB has identified genetic engineering as a promising technology to overcome these limitations. Genetic engineering, with all the advantages over conventional breeding, could be used to produce transgenic oil palms with high-value added fatty acids and novel metabolites. The main goal of the MPOB genetic engineering programme is to change oil quality and in particular to increase oleic acid content (Cheah et al., 1995). Currently the *in vitro* transformation programme aims to develop transgenic oil palm with the following characteristics: high oleic acid, high stearic acid, high lycopene, and synthesizing biodegradable plastics (Parveez et al., 1999). Establishment of a reliable transformation and regeneration system is essential for genetic engineering. Production of transgenic oil palm has been developed successfully using the Biolistics method. The transgenic plants age more then 4 years old, still expressing the *gus* and *bar* genes. Leaf painting of the transgenic plants with Basta solution proved that the transgenic plants are resistant to the herbicide Basta. Production of callus from the above transgenic plants showed positive *gus* gene expression. Construction of transformation vectors for bombardment has been carried out extensively and a number of transformation vectors have been successfully produced. Transformation of the vectors into oil palm has been carried out. Latest achievements on genetic transformation, transformation vectors construction and transformation will be elaborated.

N. Murata et al. (eds.), Advanced Research on Plant Lipids, 415–418.

2. Production of Transgenic Oil Palm

2.1 *Biolistics Method*
First batch transgenic plants produced by microprojectile bombardment were confirmed by PCR, Southern hybridization and thin layer chromatography (Parveez, 2000). The palms are 4 years old now and had started flowering in the polybags. They should start producing fruit soon. However, approval from GMAC is still not obtained for field evaluation.

2.2 *Transformation using GFP as a selectable marker gene.*
Six constructs carrying *gfp* gene driven by different promoters (CaMV35S, Ubiquitin and HBT) were used to bombard oil palm embryogenic calli. Generally the number of spot and it brightness reduced drastically after 3 days of bombardment and totally faded out after 2 weeks. Longest GFP expression was observed up to 9 months, however, doubling of cells containing *gfp* (proliferation) was not observed so far. It might be due to GFP toxicity (Na'imatulapidah and Parveez, 2001). Experiments using co-bombardment with Basta resistant gene and using organelle targeted GFP were carried out to address the possible toxicity effect. The test is on going and data are being collected for the above. PCR and southern analysis to confirm stable integration of *gfp* gene is also on going.

3. Production of high oleic acid oil palm

Two approaches were considered for channeling palmitic acid to produce more oleic acid in oil palm: i) increase KAS II activity, and ii) reduce thioesterase activity towards palmitoyl ACP (Parveez et al., 2000). Construction of transformation vector carrying antisense copy of palmitoyl ACP thioesterase gene (Abrizah, 2001) driven by constitutive and mesocarp specific promoters has been achieved successfully. Construction of backbone plasmids carrying mesocarp specific promoter, with or without transit peptide have been successfully produced as well (Arif et al., 2001). Construction of transformation vectors for KAS II will be carried out soon. The constructs obtained to date have been used to transform oil palm. Resistant embryogenic calli is being obtained routinely and some of them have regenerated to produce small shoots. Regeneration for producing complete transgenic palms is on going.

4. Increasing Stearic Acid Content

Oil palm contains an active Δ9-stearoyl-ACP desaturase (*Δ9SAD*), which effectively desaturates stearoyl-ACP into oleoyl-ACP. Therefore, by introducing an antisense copy of the *Δ9SAD* gene, stearic acid could be accumulated in the oil palm (Parveez et al., 2000). The *Δ9SAD* gene has been isolated from oil palm (Siti Nor Akmar, 1997). Construction of transformation vectors carrying antisense copy of the gene driven by both the constitutive and mesocarp specific promoters have been successful carried out. Transformation of the constructs into oil palm callus has just been initiated and some resistant embryogenic calli obtained. Proliferation and regeneration of transgenic palms from the selected calli are progressing.

5. Production of Biodegradable Thermoplastics

In bacteria, PHB (polyhydroxybutyrate) is derived from acetyl-coenzyme A by a sequence of three enzymatic reactions. The first enzyme, *3-ketothiolase* catalyses the reversible condensation of two acetyl-CoA moieties to form acetoacetyl-CoA. *Acetoacetyl-CoA reductase* subsequently reduces acetoacetyl-CoA to D-(-)-3-hydroxybutyryl-CoA, which is then polymerized by the action of *PHB synthase* to form PHB (Anderson and Dawes 1990). Production of PHB in plants (Poirier et al., 1992; Nawrath et al, 1994; Mitsky et al., 1997) makes it possible to produce PHB in a large-scale. Introduction of these genes into oil palm may lead to the accumulation of PHB in oil palm tissues. Collaboration between Malaysian and MIT to introduce the three genes into oil palm was initiated in 1999 (Moffat, 1999).

Construction of transformation vectors for PHB and PHBV driven by different promoters for each gene have been successfully carried out (Masani et al., 2001). Construction of transformation vectors using mesocarp specific promoter is in progress. The constructs obtained to date have been used to transform oil palm. Resistant embryogenic calli is being obtained routinely and some of them have proliferated and regenerated to produce small shoots. Regeneration for producing complete transgenic palms is on going.

6. Production of lycopene (carotenoid)

Oil palm is one of the richest sources of α- and ß-carotene and are the major carotenoids of oil palm. An interesting observation was that oil palm mesocarp has very little or no lycopene (Kaur and Sambanthamurthi 2000). Lycopene is a more valuable nutraceutically than α- or ß-carotene. Among the health benefits of lycopene are antioxidants and anticancer properties (Canfield, 1995; Mayne, 1996). Since both α and ß-carotene are produced from lycopene by the action of α- and ß-lycopene cyclase, down-regulating these cyclase genes would result in accumulation of lycopene.

Using RT-PCR approach, partial length lycopene β-cyclase and lycopene ϵ-cyclase were isolated from oil palm. The results from sequence analysis showed that the clones are highly identical to other plant lycopene β-cyclase and lycopene ϵ-cyclase genes (about 80% identity). These clones are now being used for construction of transformation vectors using constitutive vectors.

7. Acknowledgement

The authors thank the Director General of MPOB for permission to present this paper. We would also like to acknowledge Drs. Siti Nor Akmar, Mohd Arif, Sh. Shahrul, Umi Salamah and Mrs. Abrizah for the genes and and technical help. Thanks are also due to Fatimah Tahir, Mohd Ali Abu Hanafiah, Nik Rafeah Nik, Suhaila Abdul Wahab, Siti Marlia Silong, Zurina Mohd Noor, Noorul Azwana Azizan and Norazlinda Yazid for their technical assistance.

8. References

Abrizah, O (2001) Isolation and characterization of acyl-acyl carrier protein (ACP) thioesterase gene from oil palm (*Elaeis guineensis*). *PhD Thesis*, University of Bristol, U.K.

Anderson, A.J. and Dawes, E.A. (1990) Occurrence, metabolism, metabolic role and industrial uses of bacterial polyhydroxylalkanoates Microbiological Reviews 54, 450-472

Arif, MAA, A.W. Nurniwalis and A. Siti Nor Akmar (2001) Construction of backbone transformation vectors for modifying palm oil composition. Proc. 13th Natl. Biotechnology Seminar, 10-13 November, Penang, pp653- 657.

Canfield, L. M. (1995). β–carotenoid metabolites: Potential importance to human health. *Malaysian Oil Science Tech.* 4: pp 43-46.

Cheah, S.C., Sambanthamurthi, R.,,Abdullah, S.N.A., Othman, A., Manaf, M.A.A., Ramli, U.S. and Parveez, G.K.A. (1995) in Plant Lipid Metabolism (Kader J.C. and Mazliak P., eds.) pp. 570-572, Kluwer Academic Publishers,Netherlands

Kaur J S and Sambanthamurthi R (2000). Preliminary studies on carotene profiles in the developing oil palm mesocarp. First MMBPP Symposium. Kuala Lumpur, 16-18 November, Kuala Lumpur.

Masani, A.M.Y., Na'imatulapidah A.M.; Ho C.L., Rasid, O and Parveez, G.K.A. (2001). Transfer of (PHB) genes into oil palm for the production of biodegradable plastics. *Second MMBPP Symposium*, 5-6 November, Kuala Lumpur, pp.20.

Mitsky, T.A., taylor, N.B., Gruys, K.G., Hao, M., La Vallee, B.J., Colburn, S.M., Slater, S.C., Rodriquez, D.J., tran, M., Thorn, G.M., Wu, Z., Shah, D.T., Valentin, H.E., Houmiel, K.L., Gunter, C.A., Mahadeo, D.A, Reiser, S.E., Kishore, G.M. and Padgette, S.R. (1997) Production of biodegradable plastics in plants. 5th *International Congress of Plant Molecular Biology*, Singapore, 21-27 September 1997. Kluwer Academic Publishers, USA.

Mayne, S. T. (1996). Beta-carotene, carotenoids and desease prevention in human. FASEB J. 10: pp 690-701

Na'imatulapidah A.M. and Parveez, G.K.A. (2001). Evaluation of GFP as selectable marker for oil palm. *Second MMBPP Symposium*, 5-6 November, Kuala Lumpur, pp.55.

Moffat, A.S. (1999) Richer oils from a palm. *Science*, 285, 370-371

Nawrath, C., Poirier, Y. and Sommerville, C. (1994) Targeting of the polyhydrxybutyrate biosynthetic pathway to the plastids of *Arabidopsis thaliana* results in high levels of polymer accumulations. Proc. Natl. Acad. Sci. USA 91, 12760-12764.

Parveez, G K A (1998) Optimization of parameters involved in the transformation of oil palm using the biolistic method *Ph.D. thesis*, Universiti Putra Malaysia

Parveez GKA, Ravigadevi, S., Abdullah, S.N.A., Othman, A., Ramli, U.S., Rasid, O., Masri M.M. and Cheah, S.C. (1999) Proceedings of the 1999 PORIM International Palm Oil Congress. Pp. 3-13, PORIM, Kuala Lumpur.

Parveez, GKA (2000) Production of transgenic oil palm (*Elaeis guineensis* Jacq.) using biolistic techniques. In *Molecular Biology of Woody Plants* Volume 2 (Eds: Jain, S. M and Minocha, S.C.). Kluwer Academic Publishers, Netherlands, Pp. 327-350.

Parveez, GKA; Rasid, O; Alizah, Z; Masri, MM; Majid, NA; Fadilah, HH; Yunus, AMM and Cheah, SC (2000). Transgenic Oil Palm: Production and projection. *Biochemical Society Transactions 28 (6):* 969-972.

Poirier, Y., Dennis, D.E., Klomparens, K. and Somerville, C. (1992) Polyhydroxybutyrate, a biodegradable thermoplastic, production in transgenic plants Science 256, 520-523.

Rajanaidu, N and Jalani, BS (1995). World-wide performance of DXP planting material and future prospects. *In* Proceedings of 1995 PORIM National Oil Palm Conference. - Technologies in Plantation, The Way Forward. 11-12 July 1995. Kuala Lumpur: Palm Oil Research Institute of Malaysia, pp.1-29.

Siti Nor Akmar A., Murphy, D.J. and Cheah, S.C. (1997) 5th International Congress of Plant Molecular Biology Singapore, 21-27 September 1997. No 696, Kluwer Academic Publishers, USA.

GENETIC MANIPULATION OF FATTY ACIDS IN OIL PALM USING BIOLISTICS: STRATEGIES TO DETERMINE THE MOST EFFICIENT TARGET TISSUES

SHAH, F.H.,TAN, C.L., CHA, T.S. and FATHURRAHMAN
School of Bioscience and Biotechnology,
Faculty of Science and Technology,
National University of Malaysia, 43000, Bangi, Malaysia

1. Abstract

Stearoyl-ACP desaturase and palmitoyl-ACP thioesterase are two important genes in the fatty acid pathway had been successfully isolated from oil palm and put in an antisense orientation into transformation vectors driven by mesocarp-specific promoter and constitutive promoter such as PSP'-AP-VF6, pAPT18 and pADST35.These vectors had been successfully transferred into various target tissues such as immature embryos, callus cultures and embryogenic cultures of *Elaeis guineensis*. Previous work in the laboratory using immature embryos had successfully detected the transgene using PCR. This work reports on the experiments and results obtained to determine which is the best target tissue for oil palm transformation. Results showed that some of the resistant cultures had also been successfully selected from the selection media and regenerated into plantlets. Based on our results, we found that embryogenic cultures responded better to the selection procedures compared to callus cultures because of their higher viability rate and regeneration ability. Immature embryos also responded well on the selection media but chimerism is the main problem. PCR experiments on regenerated plantlets had successfully detected the transgene in 10% of the total plantlets analysed in *E. guineensis*, Dura and 15.9% in *E guineensis*, Tenera. Currently more transformed plantlets are being analysed for presence of the transgene.

2. Introduction

Palm oil is Malaysia's most important commodity, contributing about 17% of total fat and oil in the world and used in both edible and non-edible sectors. There is interest in altering the metabolism of fatty acids in plant tissues because of the influence of fatty foods on food quality and their significance in biological processes. There has been considerable interest in reducing the saturated fatty acid content of foods. However, highly saturated fatty acids are needed for certain food and oleochemical industries. Therefore, our objective is to engineer a new population of oil palm plants that can produce specific fatty acids to cater to the demands of various industries. Many research has been carried out to manipulate the fatty acid biosynthetic pathway at the gene level to produce trangenic plants with desired fatty acid composition in various oil producing plants (Dehesh et al., 2001, Rangasamy and Ratledge, 2000). Though much work had been done to manipulate fatty acid in oil palm, there is no report yet reporting on success in fatty acid profile change. Various key genes in the fatty acid biosynthetic

419

N. Murata et al. (eds.), Advanced Research on Plant Lipids, 419–422.
© 2003 *Kluwer Academic Publishers. Printed in the Netherlands.*

pathway had been isolated and characterized from *Elaeis guineensis*, tenera and *E. oleifera* such as δ-9-stearoyl-ACP desaturase and two isoforms (Shah et al., 2000b), β-ketoacyl-ACP synthase II (Cha and Shah, 2000), β-ketoacyl-ACP synthase I, oleoyl thioesterase (Shah and Asemota, 2000a), palmitoyl-ACP thioesterase (Shah and Cha, 1999) and 3-ketoacyl-ACP synthase III (Shah and Shahabuddin, 2000c). Tissue specific clones from mesocarp and kernel have also been isolated and characterized. Previous research had successfully altered the fatty acid biosynthetic pathway via genetic transformation either using the antisense (Cartea et al., 1998) or overexpression (Dehesh et al., 2001) strategy in rapeseed, canola and tobacco (Dehesh et al., 2001). Research in our laboratory had shown successful integration of stearoyl-ACP desaturase in the antisense orientation and obtained a transformation rate of 2% (Shah et al., 2000d). In this research we had used different target tissues such as immature embryos, callus cultures and embryogenic cultures to determine the best target tissue used for oil palm transformation. Also, various selection strategies were tested on different stages of regeneration to select transgenic cell lines and were subsequently regenerate. These plantlets were analyzed for the transgene using the PCR. This communication reports on the progress of our work.

3. Materials and methods

Three transformation vectors used in this research were PSP'-AP-VF6, pAPT18 and pADST 35 which contained antisense palmitoyl-ACP thioesterase and stearoyl-ACP desaturase driven by mesocarp-specific promoter and constitutive promoter respectively. The selection marker for PSP'-AP-VF6 is hygromicin resistant gene and *bar* gene (confers resistance to phosphinotricin) for pAPT18 and pADST 35. Plant material used were immature embryos(IEs) (10 weeks after anthesis), callus culture (8 months) and embryogenic culture (1 year). Transformed IEs of *E.guineensis*, Tenera used for direct regeneration were maintained and selected on E1 media. Callus from *E.guineensis*, Tenera was initiated on C1 media. Transformed callus and embryogenic culture were maintained and selected on the same media until they become embryogenic. Regeneration of embryogenic cultures of *E.guineensis*, Tenera was done on E2 media. Rooting and hardening was done on R1 and H1 media respectively. Dura IEs used for direct regeneration were maintained and selected from CIP media. To differentiate between exogenous and endogenous δ-9-stearoyl desaturase gene, primer pairs were designed that flanks the region containing an intron within the genomic region. Stearoyl desaturase used in the vector was isolated using cDNA clones and would be without intron (Shah et al., 2000b). However, an intron (510 bp) was found within the gene between these two primers when these primers were used to detect the transgene using genomic DNA (Figure 1) (Shah et al, 2000d).

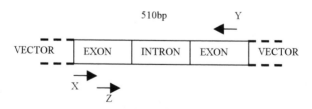

Figure 1. Strategy to analyse transgene. Primer pairs X+ Y and Y+Z were used to amplify transgene in *E. Guineensis*, Tenera and Dura respectively.

4. Results and discussion

Transformation of *E. guineensis* using the three vectors were successfully carried out. PCR and sequencing experiments was able to detect the transgene in 10% of total plantlets analysed from *E. guineensis*, Dura and 15.9% from *E. guineensis*, Tenera. The total number of plantlets tested from *E. guineensis*, Dura and Tenera were 30 and 44 respectively (Fig. 2).

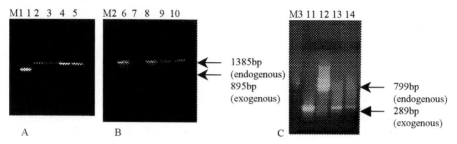

Figure 2. Agarose gel electrophoresis showing the exogenous gene by PCR. .A) &B) PCR amplification of transgene for *E. guineensis*, Tenera C) PCR amplification of transgene for *E. guineensis*, Dura. M1, M2 and M3 represent 100bps markers; lane 1 & 11 are positive control; lane 2,3 represent negative control (non transformed plants); lane 4,5,8,10 &12 represent transformed plantlets without transgene; lane 6,7,9,13 & 14 represent transformed plantlets with transgene.

Results also showed that transformation and selection strategies would influence the regeneration time (Fig 4). Overall, bombardment and selection stress had prolonged the regeneration time for all treatments. This may be because the cultures may need more time to recover from the stress condition Based on the time taken from bombardment of different target tissues to regeneration of plants to polybags stage, regeneration time was fastest from IE (8 months) as target tissues (Fig 4) followed by embryogenic cultures (13 months) and callus cultures (17 months). However, when the immature embryos are put through callusing stage with selection at the callus stage, the regeneration time is the longest (32 months) (Fig 4D). Additionally, this step could be avoided, as it has been shown that prolong culturing time will leads to abnormality (Rohani et al., 2001). Though IEs take the shortest time for regeneration, chimerism is the main problem because they are differentiated tissues. From this preliminary work, it can be concluded that the best target tissues used for oil palm transformation will be embryogenic cultures based on the total time to regenerate plants and viability rate after selection. Selected embroids that were selected at the embryogenic stage were able to produce shoots at a shorter time as compared to selected embroids that had been selected at the callus stage (Fig 4G & 4H).

Treatment	Target tissue bombarded	Stages of regeneration
A.	IE	→ Plantlet
B.	IE	→ Plantlet
C.	IE	→ Callus → Embryogenic → Plantlet
D.	IE	→ Callus → Embryogenic → Plantlet
E.	IE	→ Callus → Embryogenic → Plantlet
F.	Callus	→ Embryogenic → Plantlet
G.	Callus	→ Embryogenic → Plantlet
H.	Callus	→ Embryogenic → Plantlet
I.	Embryogenic	→ Plantlet
J.	Embryogenic	→ Plantlet

Figure 3. Transformation and selection strategy used for *E. guineensis*, Tenera transformation. Selection stress was introduced at the stage marked in box.

Additionally selection at embryogenic stage leads to higher survivor rate as compared to other stages. However, it was observed that there was no significant difference in the regeneration time when the transformed cultures were maintained in rooting and hardening media for cultures that were put through callusing stage with or without selection.

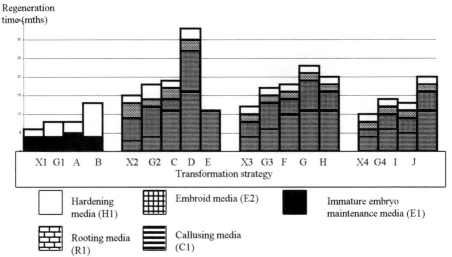

Fig 4: Graph showing the comparison of regeneration time (months) between variou transformation treatments. X1, X2, X3 and X4 indicate control (non transformed cultures) for eac transformation procedures in the group. G1, G2, G3 and G4 indicate –ve control (transformed wit gold) for each transformation procedures in the group. A→J indicated transformation and selectio strategy (refer to Figure 3).

5. References

Cartea, M.E., Migdal, M., Galle, A.M., Pelletier, G., and Guerche, P. (1998) Comparison of sense and antisense methodologies for modifying the fatty acid composition of *Arabidopsis thaliana* oilseed. Plant Science. 136: 181-194.

Cha, T.S. and Shah, F.H. (2000) β-ketoacyl-ACP synthase II (KAS II) cDNA clone from oil palm *Elaeis guineensis* var tenera. GenBank Accession No. AF143095.

Dehesh, K., Tai, H., Edwards, P., Byrne, J. and Jaworski, J.G. (2001) Overexpression of 3-Ketoacyl-Acyl-Carrier Protein Synthase III in plants reduce the rate of lipid synthesis. Plant Physiology. 125, 1103-1114.

Rangasamy,D and Ratledge,C. (2000) Genetic enhancement of fatty acid synthesis by targeting rat liver ATP : Citrate lyase into plastid of tobacco. Plant Physiology. 122, 1231-1238.

Rohani, O., Sharifah, S.A., Mohd, R.Y., Ong, M., Tarmizi, A.H. and Zamzuri, I. (2001) Tissue culture of oil palm in Basiron, Y., Jalani, B.S., Chan, K.W. (eds), Advances in Oil Palm Research. Malaysian Palm Oil Board, pp 238-283.

Shah, F.H. and Asemota, O. (2000a) The nucleotide sequence of oleoyl thioesterase from oil palm *Elaeis guineensis*. GenBank Accession No. AF220453.

Shah, F.H. and Cha, T.S. (1999) *Elaeis guineensis* var tenera palmitoyl-ACP thioesterase (PTE) mRNA, partial cds. GenBank Accession No. AF147879.

Shah, F.H., Rashid,O. and Cha, T.S. (2000b) Temporal regulation of two isoforms of cDNA clones encoding delta-9-stearoyl-ACP desaturase from oil palm (*Elaeis guineensis*). Plant Science. 152, 27-33.

Shah, F.H. and Shahabuddin, H. (2000c) The nucleotide sequence of a gene encoding *E. guineensis* (tenera). GenBank Accession No. AF143502.

Shah, F.H. and Shahabuddin, H. and Cha, T.S. (2000d) Successful integration of stearoyl desaturase gene in oil palm. Int. Conf. Plt. Mol. Biol. Quebec City, Canada.

Synder, G.W., Ingersoll, J.C. and Smigocki, A.C. (1999) Introduction of pathogen defense genes and cytokinin biosynthesis gene into sugarbeet (*Beta vulgaris* L.) by *Agrobacterium* or particle bombardment. Plant Cell Reports. 18,829-834.

ENDOGENIC AND EXOGENIC INTERACTION OF MICROALGAE LIPIDS WITH SURFACTANTS IN WATER ENVIRONMENT

T.V.PARSHYKOVA and M.M.MUSIENKO
Kiev National University, Vladimirskaya st., 64
01017, Kiev, Ukraine, E-mail: parshik@ukrpack.net

Introduction

It is well known (Gladyshev, 1999; Jordan and Gunningham, 1999) that lipids and surfactants are localized in superficial water layer and form the medium for cells as well phytoneuston as phytoplankton. It has been the lipids play a leading role in the formation of, membrane structure and stability of photosystems function (Kochubey, 2001). The goal of our experiments was investigation the influence of surfactants on the hydrophobic biopolymers of cells in connection with peculiarities of its survival under different conditions. At parallel it was considered the other adaptation provisions (motion, degree of polysaccharide sliming of cells), which are contributory factors for formation of microalgae stability to effect of stress-factors.

Materials and Methods

As a objects have been used microalgae: from *Cyanophyta* (*Cyanobacteria*) - *Microcystis aeruginosa* Kutz. emend Elenk., *M. muscicola* (Menegh.) Elenk., *M. pulverea* (Wood) Forti em. Elenk.; from *Chlorophyta* – mobile and immobile algae *Dunaliella salina* Teod., *Chlorella vulgaris* Beji., *Chlamydomonas reinhardtii* Dang. and phytoplankton of Dnieper river (Ukraine). The algae were grown on the selective cultural medium at 20-22°C and lighting 6000-7000 lux under artificial lamps. For experiments have been used the cultures on the logarithmic growth phase. From physiology-biochemical criteria of algae answer to surfactant effect we investigated the correlation between saturated and unsaturated fatty acids (coefficient of saturation – K_{sat}). No account has been taken in calculation the part of fatty acids (FA), such as isoacids and acids with odd number of atoms, which possessed to bacteria-satellite. It were compared the indices of sum content (% to sum) for saturated FA (sFA) – ($C_{sat} = C_{12:0} + C_{14:0} + C_{18:0} + C_{20:0} + C_{22:0}$) and unsaturated FA (uFA) – ($C_{unsat} = C_{14:1} + C_{16:1} + C_{18:1} + C_{18:2} + C_{18:3} + C_{20:4}$) as well in the biomass as cultural liquid. Methyl ethers of FA were analyzed on the chromatograph Chrom-4. The speed and energy of motion were measured by laser-doppler spectroscopy method. Using surfactants: cationic active (CS) – catamine; anionic active (AS) – NaSDS; nonionogenic (NS) – hydropol in concentration from 0.1 to 10 mg/l. Contact time for algae and surfactant was 1-24 hours.

N. Murata et al. (eds.), Advanced Research on Plant Lipids, 423–426.

It was shown in this manuscript the more typical changes which were registered in the native algae cells with different modifications of cells cover.

Results and Discussion

The analyze of species content for microalgae, which survived in water after contact with surfactant it was shown that stable species were characterized by widely habitats of distribution in nature. There are 21 species-cosmopolites in reservoirs, lakes, ponds and rivers of Ukraine, Belarus and Poland (Miheeva, 1999; Hutorowicz, 2001). Among them representatives of genus *Microcystis, Anabaena* from *Cyanobacteria*; genus *Coelastrum, Scenedesmus, Chlamydomonas, Pediastrum, Tetraedron* from *Chlorophyta*; genus *Asterionella, Cyclotella, Melosira, Synedra, Navicula, Nitzchia* from *Bacillariophyta*.

Microscopic analyze of degree of sliming algae cells testifies that exogenic mucilage-forming polysaccharides are of considerable importance in formation of algae stability in the presence of surfactants. Especially in *Cyanobacteria* they occupies up to 70-75% of cells mass. Exogenic mucilage capsules have most of *Bacillariophyta* and some genus of *Chlorophyta*. The significant adsorption surface and ability to ion-exchange reactions are characteristic of polysaccharides.

It was established (Parshykova, 2001), that mobility of algae vegetative cells may be considered as common biological adaptation mechanism for its survival in contaminated environment. As illustrated Figure 1 at first minutes of contact the cells of *Chlamydomonas* with surfactant (concentration from 0,1 to 3 mg/l) significantly increasing the speed of cells motion and its energy potential. It makes possible the cells survival under different conditions of environmental contamination.

Among the biochemical adaptation reactions of algae cells on the stress we estimated the correlation between sFA and uFA (K_{sat}). For example (Figure 2) for representatives of genus *Microcystis* the K_{sat} in biomass with CS (10 mg/l) was the smallest among the experimental variants (accordingly 0.90 for experiment and 1.79 for control). Sum of sFA was decreased at the sacrifice of myristinic $C_{14:0}$ (in 1.5 times), palmitic $C_{16:0}$ (in 1.5 times) and $C_{22:0}$ (in 1.3 times). Among the uFA measurably increased the content of palmitooleinic $C_{16:1}$ (in 1.7 times), arachidonic $C_{20:4}$ (in 2 times), linolic $C_{18:2}$ and linolenic $C_{18:3}$ (in 1.1 times accordingly). In cultural liquid of variant II we registered the increasing of sFA and decreasing of uFA in comparison with control. It produced the increasing of K_{sat} (1.67 in experiment against 0.30 in control). Among the sFA increased the part of palmitic $C_{16:0}$ (in 1.7 times) and stearic $C_{18:0}$ (in 2.3 times). It was found in the cultural medium for variant II the FA (such as $C_{19:0}$, $C_{22:0}$), which was not registered in the control. Among the uFA significantly decreased the content of $C_{16:3}$ (in 7 times), oleinic $C_{18:1}$ (in 1,9 times), linolic $C_{18:2}$ (in 1,5 times), linolenic $C_{18:3}$ (in 3.3 times) and arachidonic $C_{20:4}$ (in 2.4 times).

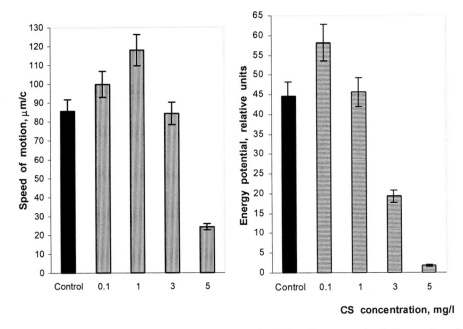

Figure 1. Changes of speed of motion and energy potential for *Chlamydomonas reinhardtii* Dang. cells under CS effect (contact time – 1 hour).

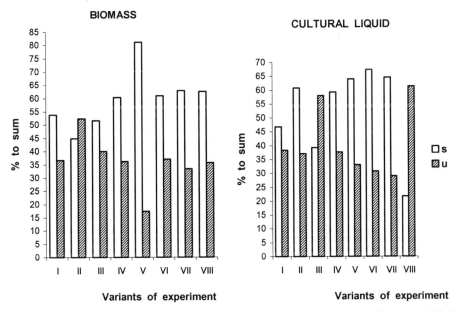

Figure 2. Correlation between saturated and unsaturated FA for *Microcystis aeruginosa* Kutz. emend. Elenk. under surfactants effect: I – AS, 10 mg/l; II – CS, 10 mg/l; III – NS, 10 mg/l; IV – AS + CS, 5 mg/l each; V – CS + NS, 5 mg/l each; VI – AS + NS, 5 mg/l each; VII – AS + CS + NS, 3.3 mg/l each; VIII – control without surfactants.

On the whole the K_{sat} in biomass changed in next sequence: CS, 10 mg/l (K_{sat} – 0.90) < NS, 10 mg/l (1.39) < AS, 10 mg/l (1.42)< AS+NS, 5 mg/l each (1.69) < AS+CS, 5 mg/l each (1.76) < control (1.79) < AS+CS+NS, 3.3 mg/l each (1.95) < CS+NS, 5 mg/l each (4.74). It was established for cultural liquid the next sequence of K_{sat}: control (0.30) < NS, 10 mg/l (0.72) < AS, 10 mg/l (1.14) < AS+CS, 5 mg/l each (1.62) < CS, 10 mg/l (1.67) < CS+NS, 5 mg/l each (1.78) < AS+NS, 5 mg/l each (2.26) < AS+CS+NS, 3.3 mg/l (2.45). It was noted in biomass and cultural liquid the stable presence from sFA only palmitic, from uFA – palmitooleinic, oleinic, linolic and linolenic. Caprinic, hendecanoic, $C_{20:0}$, $C_{22:0}$, and from uFA $C_{20:1}$, $C_{20:2}$ were registered periodically. From sFA – laurinic, tridecanic, myristinic, heptadecane, $C_{19:0}$ and uFA – arachidonic, $C_{14:1}$, $C_{14:3}$, $C_{16:3}$ were present stably in biomass, but not always in cultural liquid. The stearic acid was present stably in cultural liquid but not always in biomass. Marked changes in qualitative content and correlation between sFA and uFA testifies about significant influence of surfactants on hydrophobic algae biopolymers as a constituent part of membranes of photosynthetic cell. CS produce the most significant effect on state of lipids. It confirms the increasing of uFA content in algae biomass. More detectable changes of content were registered for oleinic, linolic, palmitooleinic and arachidonic FA. These distinctions in qualitative content of FA testifies that surfactants interaction with lipid components of native cells may be one of reasons of its negative effect on pigment apparatus and photosynthetic activity of algae under natural conditions. Thus, microalgae species, which have no these protective adaptations (such as mobility, sliming, shell outshoots and so on), are characterized by lower K_{sat} (for example, for *Chlorella* cells – from 0.47 до 0.85; for cultural medium – from 0,13 to 0,52) at sacrifice of sharp increasing (in 2 and more times) of uFA (linolic, linolenic) (Parshikova, 1998). It is evident that increasing of uFA content under stress conditions may be considered as one of adaptation biochemical reactions for microalgae cells.

References

Gladyshev M.I. (1999) Foundations of ecological biophysics for water system. Nauka, Novosibirsk (in Russian)

Hutorowicz A. (2001) Phytoplankton of the humic lake Smolak. Sorus, Poznan.

Jordan R. and Gunningham A. (1999) Surfactant-enhanced bioremediation. In Ph.Baveye, J.Block and V.Goncharuk (eds.), Bioavailability of organic xenobiotics in the environment. Kluwer Academic Publishers, Dordrecht, pp.463-496.

Kochubey S.M. (2001) Organization of photosynthetic apparatus of high plants. Alterpress, Kiev (in Russian)

Micheeva T.M. (1999) Algoflora of Belarus. Taxonomic catalog. Belarus University Press, Minsk (in Russian)

Parshikova T.V. (1998) Changes in content,correlation of fatty acids and state of CPL complex of native algae cells in the presence of surfactants. In J.Sanchez et al. (eds.), Advances in Plant Lipid Research. Universidad de Sevilla, Sevilla, pp.564-567.

Parshykova T.V. (2000) Interrelationship of microalgae mobility and its cells resistance to chemical factors effect. Ukrainian Botan. J. 6, 658-663 (in Ukrainian)

PRODUCTION OF HYDROXY FATTY ACIDS IN *ARABIDOPSIS THALIANA.*

M.A. SMITH, G. CHOWRIRA, AND L. KUNST.
Dept. of Botany, University of British Columbia.
6270 University Blvd, Vancouver, BC V6T 1Z4. Canada

Introduction

Castor bean (*Ricinus communis*) produces a seed oil that contains nearly 90% ricinoleic acid (12-hydroxyoctadeca-9-*cis*-enoic acid: 18:1-OH). This fatty acid is a valuable industrial raw material and is used in a wide variety of processes and products. In castor, 18:1-OH is synthesized by the Δ12-hydroxylation of oleate esterified to the *sn*-2 position of phosphatidylcholine (Bafor et al., 1991). The reaction is catalysed by an enzyme that is closely related to a FAD2 (ER-Δ12) desaturase (van de Loo et al., 1995). We have been using *Arabidopsis* lines, transformed with a gene construct encoding the castor hydroxylase, as a model system to study the synthesis and accumulation of hydroxy fatty acids in the seed.

Materials and Methods

The *Arabidopsis* lines described here were transformed with a DNA cassette containing a cDNA encoding the castor oleate hydroxylase under the control of the seed specific *Lesquerella* hydroxylase promoter (Broun et al., 1998). For GC analysis, FAMES were prepared by methylation of whole seeds, or lipid classes, in 1M HCl in methanol at 80°C for 90 or 60 minutes, respectively. Lipase digestion of TAGs was conducted using *Rhizopus arrizus* lipase (to remove fatty acids from the *sn-1* and *sn-3* positions of TAG) as described previously (Bafor et al., 1990).

Results

Production of ricinoleic acid in Arabidopsis seeds
The expression of the castor hydroxylase gene in wild type *Arabidopsis*, under the control of a seed specific promoter, resulted in the production of ricinoleic (18:1-OH) densipolic (18:2-OH), lesquerolic (20:1-OH) and auricolic acids (20:2-OH) in the seed oil (Figure 1A). We have previously shown that the FAD3 (Δ15) desaturase and FAE1 condensing enzyme are involved in the synthesis of 18:2-OH, 20:1-OH and 20:2-OH (Smith et al 2000). To determine whether it was possible to produce a seed oil

N. Murata et al. (eds.), Advanced Research on Plant Lipids, 427–430.
© 2003 *Kluwer Academic Publishers. Printed in the Netherlands.*

containing ricinoleic acid as the only hydroxy fatty acid, we expressed the castor hydroxylase in a *fad3/fae1* double mutant line. Analysis of the seed lipids of the resulting primary transformants indicated that ricinoleic acid was indeed the only detectable hydroxy fatty acid present (Figure 1B). The highest percentage of ricinoleic acid that we detected in this population was 7.8%. Castor oil contains nearly 90% ricinoleic acid.

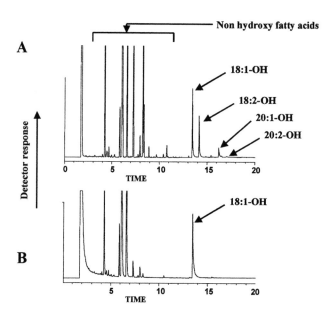

Figure 1. Gas chromatographs showing the fatty acid profile of seeds from *Arabidopsis* lines expressing the castor hydroxylase protein. **A.** Wild type **B.** *fad3/fae1* double mutant.

Lipid analysis

TLC analysis of seed neutral lipids from *Arabidopsis* lines transformed with the castor hydroxylase gene showed the presence of novel lipid species not present in seeds from untransformed plants (Figure 2). These contained almost all of the hydroxy fatty acids found in the seed (Table 1) and had similar RF values to mono-hydroxy-TAG (1-OH TAG) and di-hydroxy-TAG (2-OH TAG) from castor bean oil. We were not able to determine whether any lipid corresponding to tri-hydroxy-TAG (3-OH-TAG), the most common TAG species in castor oil, was present in these samples. A small amount of hydroxy fatty acids were found in the "polar lipid" fraction of the oil.

Table 1 shows the results of an analysis of 2 oil samples with different hydroxy fatty acid profiles: **Hydroxy-C18 oil** (18:1-OH and 18:2-OH from a transformed *fad2/fae1* double mutant line) and **Hydroxy-C18+C20 oil** (18:1-OH and 20:1-OH from a transformed *fad3* mutant line). TLC separation of hydroxy-C18 oil (11.1% hydroxy fatty acids) indicated that nearly 40% of all TAG molecules contained at least 1 hydroxy fatty acid chain (Table 1). As these results suggested that the oil should contain more than 11% hydroxy fatty acids, we determined the percentage of hydroxy fatty acids in each lipid class. The data consistently showed that the 1-OH-TAG and 2-OH-TAG fractions contained around 24% and 53% hydroxy fatty acids respectively. Similar results were obtained during the analysis of the hydroxy C18+C20 oil (Table 1). Analysis of TAG from castor oil gave hydroxy fatty acid contents of close to the predicted values of 66% and 100% for 2-OH-TAG and 3-OH TAG fractions.

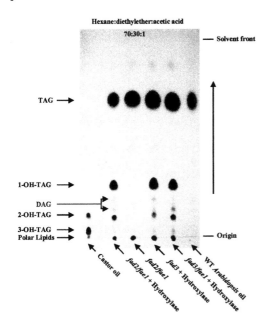

Figure 2. TLC analysis of seed neutral lipids from *Arabidopsis* lines expressing the castor hydroxylase protein.

	Hydroxy-C18 oil. (11.1% Hydroxy fatty acids contains 18:1-OH and 18:2-OH)		Hydroxy-C18+C20 oil. (7.2% hydroxy fatty acids, contains 18:1-OH and 20:1-OH)	
Lipid class	% of total lipid	% hydroxy FAs in lipid class	% of total lipid	% hydroxy FAs in lipid class
TAG	55.9	0.0	70.7	0.0
1-OH –TAG	33.0	24.5	22.8	24.4
DAG	3.2	Trace	2.0	1.6
2-OH-TAG	4.7	52.8	1.9	50.9
Polar *	3.2	1.6	2.6	1.6

Table 1. Distribution of hydroxy fatty acids in seed neutral lipids from *Arabidopsis fad2/fae1* and *fad3* mutant lines expressing the castor hydroxylase protein (hydroxy-C18 oil, and hydroxy-C18+C20 oil, respectively). Each lipid class is expressed as a percentage of total seed lipid based on fatty acid content. * "Polar" refers to lipids remaining at, or close to, the site of sample application.

To determine whether hydroxy fatty acids were present in the *sn-2* position of TAG we conducted a lipase digestion of the 1-OH-TAG and 2-OH-TAG fractions from the hydroxy-C18 oil (Figure 3). Castor oil and wild type *Arabidopsis* oil were subjected to a similar digestion to provide reference samples. A partial digestion of the two fractions resulted in the production of 2 separate species of MAG, one corresponding to MAG containing no hydroxy fatty acids and one with a similar RF value to ricinoleoyl-MAG

430

from castor bean. GC analysis of the putative hydroxy-MAG showed that this lipid only contained hydroxy fatty acid, almost all of which was ricinoleic acid. This indicates that hydroxy fatty acids do not appear to be excluded from the *sn*-2 position of TAG in *Arabidopsis*, and is in agreement with results reported earlier by Broun and Somerville (1997). Further evidence for the presence of hydroxy fatty acids at position *sn*-2 of TAG was obtained in the lipase digestion product of the 2-OH-TAG fraction where 2-OH-DAG was clearly seen. In these samples OH-MAG was the predominant form of MAG. Due to the small amount of material and the incomplete lipase digestion it is difficult to estimate the relative amount of hydroxy fatty acid in position *sn*-2 compared to *sn*-1 and *sn*-3.

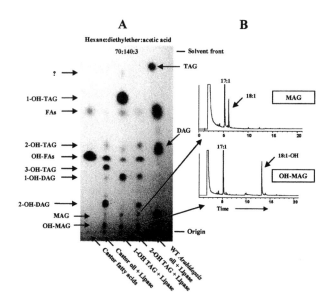

Figure 3. TLC separation of lipase digested TAG species. Seed neutral lipids from a line of *Arabidopsis fad2/fae1* double mutant plants expressing the castor hydroxylase were separated by TLC. Spots corresponding to 1-OH-TAG and 2-OH-TAG were scraped from the plate and subjected to lipase digestion. **A.** The resulting digestion products were separated by TLC and visualized with iodine. Wild type *Arabidopsis* oil and castor oil were digested with lipase for comparison. **B.** The fatty acid composition of spots identified as MAG and OH-MAG were determined by GC analysis

Conclusions

- Expression of the castor hydroxylase in an *Arabidopsis* line lacking FAD3 desaturase and FAE1 condensing enzyme activities results in the production of an oil containing ricinoleic acid as the only hydroxy fatty acid.

- In transformed *Arabidopsis* plants, hydroxy fatty acids are not excluded from position *sn*-2 of TAG.

References

Bafor, M., Jonsson, L., Stobart, A.K. and Stymne, S. (1990). Biochem. J. **272**, 31-38.

Bafor, M., Smith, M.A., Jonsson, L., Stobart, K. and Stymne, S. (1991). Biochem. J. **280**, 507-514.

Broun, P. and Somerville, C. (1997). Plant Physiol. **113**, 933-942.

Broun, P., Boddupalli, S. and Somerville, C. (1998). Plant. J. **13**, 201-210.

Smith, M., Moon, H., Kunst, L. (2000). Biochem. Soc. Trans. **28**, 947-950.

van de Loo, F.J., Broun, P., Turner, S. and Somerville, C. (1995). Proc. Natl. Acad. Sci. USA **92**, 6743-6747.

Acknowledgments.

This work was supported by The Science Council of BC and Linnaeus Plant Sciences Inc.

Author Index

432

434

Subject Index

438